Nitration

ACS SYMPOSIUM SERIES **623**

Nitration

Recent Laboratory and Industrial Developments

Lyle F. Albright, EDITOR
Purdue University

Richard V. C. Carr, EDITOR
Air Products and Chemicals, Inc.

Robert J. Schmitt, EDITOR
SRI International

Developed from a symposium sponsored
by the Division of Industrial and Engineering Chemistry, Inc.,
at the 209th National Meeting
of the American Chemical Society
Anaheim, California,
April 2–7, 1995

American Chemical Society, Washington, DC 1996

Library of Congress Cataloging-in-Publication Data

Nitration: recent laboratory and industrial developments / Lyle F. Albright, editor, Richard V. C. Carr, editor, Robert J. Schmitt, editor.

   p.   cm.—(ACS symposium series, ISSN 0097–6156; 623)

"Developed from a symposium sponsored by the Division of Industrial and Engineering Chemistry, Inc., at the 209th National Meeting of the American Chemical Society, Washington, D.C., April 2–7, 1995."

   Includes bibliographical references and indexes.

   ISBN 0–8412–3393–4

   1. Nitration—Congresses.

   I. Albright, Lyle Frederick, 1921–   . II. Carr, Richard V. C., 1947–   . III. Schmitt, Robert J., 1952–   . IV. American Chemical Society. Division of Industrial and Engineering Chemistry. V. American Chemical Society. Meeting (209th: 1995: Washington, D.C.) VI. Series.

QD63.N5N57   1996
547'.26—dc20   96–11096
                                                                              CIP

This book is printed on acid-free, recycled paper.

Copyright © 1996

American Chemical Society

All Rights Reserved. The appearance of the code at the bottom of the first page of each chapter in this volume indicates the copyright owner's consent that reprographic copies of the chapter may be made for personal or internal use or for the personal or internal use of specific clients. This consent is given on the condition, however, that the copier pay the stated per-copy fee through the Copyright Clearance Center, Inc., 222 Rosewood Drive, Danvers, MA 01923, for copying beyond that permitted by Sections 107 or 108 of the U.S. Copyright Law. This consent does not extend to copying or transmission by any means—graphic or electronic—for any other purpose, such as for general distribution, for advertising or promotional purposes, for creating a new collective work, for resale, or for information storage and retrieval systems. The copying fee for each chapter is indicated in the code at the bottom of the first page of the chapter.

The citation of trade names and/or names of manufacturers in this publication is not to be construed as an endorsement or as approval by ACS of the commercial products or services referenced herein; nor should the mere reference herein to any drawing, specification, chemical process, or other data be regarded as a license or as a conveyance of any right or permission to the holder, reader, or any other person or corporation, to manufacture, reproduce, use, or sell any patented invention or copyrighted work that may in any way be related thereto. Registered names, trademarks, etc., used in this publication, even without specific indication thereof, are not to be considered unprotected by law.

PRINTED IN THE UNITED STATES OF AMERICA

Advisory Board

ACS Symposium Series

Robert J. Alaimo
Procter & Gamble Pharmaceuticals

Mark Arnold
University of Iowa

David Baker
University of Tennessee

Arindam Bose
Pfizer Central Research

Robert F. Brady, Jr.
Naval Research Laboratory

Mary E. Castellion
ChemEdit Company

Margaret A. Cavanaugh
National Science Foundation

Arthur B. Ellis
University of Wisconsin at Madison

Gunda I. Georg
University of Kansas

Madeleine M. Joullie
University of Pennsylvania

Lawrence P. Klemann
Nabisco Foods Group

Douglas R. Lloyd
The University of Texas at Austin

Cynthia A. Maryanoff
R. W. Johnson Pharmaceutical
Research Institute

Roger A. Minear
University of Illinois
at Urbana–Champaign

Omkaram Nalamasu
AT&T Bell Laboratories

Vincent Pecoraro
University of Michigan

George W. Roberts
North Carolina State University

John R. Shapley
University of Illinois
at Urbana–Champaign

Douglas A. Smith
Concurrent Technologies Corporation

L. Somasundaram
DuPont

Michael D. Taylor
Parke-Davis Pharmaceutical Research

William C. Walker
DuPont

Peter Willett
University of Sheffield (England)

Foreword

THE ACS SYMPOSIUM SERIES was first published in 1974 to provide a mechanism for publishing symposia quickly in book form. The purpose of this series is to publish comprehensive books developed from symposia, which are usually "snapshots in time" of the current research being done on a topic, plus some review material on the topic. For this reason, it is necessary that the papers be published as quickly as possible.

Before a symposium-based book is put under contract, the proposed table of contents is reviewed for appropriateness to the topic and for comprehensiveness of the collection. Some papers are excluded at this point, and others are added to round out the scope of the volume. In addition, a draft of each paper is peer-reviewed prior to final acceptance or rejection. This anonymous review process is supervised by the organizer(s) of the symposium, who become the editor(s) of the book. The authors then revise their papers according to the recommendations of both the reviewers and the editors, prepare camera-ready copy, and submit the final papers to the editors, who check that all necessary revisions have been made.

As a rule, only original research papers and original review papers are included in the volumes. Verbatim reproductions of previously published papers are not accepted.

ACS BOOKS DEPARTMENT

Contents

Preface xi

1. Nitration: An Overview of Recent Developments and Processes 1
   Lyle F. Albright, Richard V. C. Carr, and Robert J. Schmitt

2. Role of Protosolvation in Nitrations with Superacidic Systems: The Protonitronium Dication (NO_2H^{2+}) Identified 10
   G. K. Surya Prakash, Golam Rasul, Arwed Burrichter, and George A. Olah

3. Nitrocyclohexadienones in Aromatic Nitration 19
   Robert G. Coombes, Panicos Hadjigeorgiou, Doron G. J. Jensen, and David L. Morris

4. Aromatic Nitration in Liquid Nitrogen Tetroxide Promoted by Metal Acetylacetonates 31
   Ripudaman Malhotra and David S. Ross

5. Synthesis of 2,5,7,9-Tetranitro-2,5,7,9-tetraazabicyclo[4.3.0]nonanone 43
   H. R. Graindorge, P. A. Lescop, F. Terrier, and M. J. Pouet

6. Photochemical Chlorocarbonylation: Simple Synthesis of Polynitroadamantanes and Polynitrocubanes 51
   A. Bashir-Hashemi, Jianchang Li, Paritosh R. Dave, and Nathan Gelber

7. Isolation of Electrochemically Generated Dinitrogen Pentoxide in a Pure Form and Its Use in Aromatic Nitrations 58
   M. J. Rodgers and P. F. Swinton

8. Pilot-Plant-Scale Continuous Manufacturing of Solid Dinitrogen Pentoxide 68
   T. E. Devendorf and J. R. Stacy

9. Separation of Dinitrogen Pentoxide from Its Solutions in Nitric Acid 78
   Robert D. Chapman and Glen D. Smith

10. Nitrated Hydroxy-Terminated Polybutadiene: Synthesis and Properties 97
    M. E. Colclough and N. C. Paul

11. Novel Syntheses of Energetic Materials Using Dinitrogen Pentoxide 104
    R. W. Millar, M. E. Colclough, H. Desai, P. Golding, P. J. Honey, N. C. Paul, A. J. Sanderson, and M. J. Stewart

12. A New Route to Nitramines in Nonacidic Media 122
    R. W. Millar

13. Reinvestigation of the Ponzio Reaction for the Preparation of *gem*-Dinitro Compounds 134
    P. J. Honey, R. W. Millar, and Robert G. Coombes

14. New Nitration and Nitrolysis Procedures in the Synthesis of Energetic Materials 151
    Philip F. Pagoria, Alexander R. Mitchell, Robert D. Schmidt, Clifford L. Coon, and Edward S. Jessop

15. Flow Reactor Nitrations Using Dinitrogen Pentoxide 165
    N. C. Paul

16. Modeling To Gain Insight into Thermal Decomposition of Dinitrotoluene 174
    J. L. Case, Richard V. C. Carr, and M. S. Simpson

17. Removal and Destruction of Tetranitromethane from Nitric Acid 187
    T. L. Guggenheim, C. M. Evans, R. R. Odle, S. M. Fukuyama, and G. L. Warner

18. Modeling Nitronium Ion Concentrations in HNO_3–H_2SO_4–H_2O Mixtures 201
    Lyle F. Albright, M. K. Sood, and Roger E. Eckert

19. Commercial Dinitrotoluene Production Process 214
    Allen B. Quakenbush and B. Timothy Pennington

20. **Recent Advances in the Technology of Mononitrobenzene Manufacture** 223
    A. A. Guenkel and T. W. Maloney

21. **Industrial Nitration of Toluene to Dinitrotoluene: Requirements of a Modern Facility for the Production of Dinitrotoluene** 234
    H. Hermann, J. Gebauer, and P. Konieczny

22. **Practical Considerations in Concentration and Recovery of Nitration-Spent Acids** 250
    C. M. Evans

INDEXES

Author Index 271

Affiliation Index 271

Subject Index 272

Preface

ALTHOUGH NITRATION IS CONSIDERED to be a "mature" process, important changes and advances have occurred in the past 20 years. A comparison of this book with the 1975 ACS Symposium Series book entitled *Laboratory and Industrial Nitrations*, edited by Lyle F. Albright and Carl Hanson, is just one of the many ways to document these advances.

A summary of these changes and an overview of nitration are presented in Chapter 1 of this volume. The more theoretical aspects of nitration and the production of polynitro compounds are discussed in Chapters 2 through 6. The next several chapters discuss N_2O_5 technology, including N_2O_5 production and use as a nitrating agent. Several nitrated compounds that are difficult to produce by conventional nitrations can be produced rather easily by using N_2O_5. Industrial aspects of nitration are discussed in the final chapters of the book. Important process improvements are reported.

Several authors and attendees from the 1975 ACS symposium on nitration were authors or attendees at the 1995 symposium. This book should be of value to all individuals interested in nitration: chemists, engineers, administrators, experienced scientists, students, industrial employees, academicians, and government workers.

We thank all who made possible the 1995 symposium and this publication: the ACS Division of Industrial and Engineering Chemistry, Inc., for sponsorship; equally important the authors without whom neither symposium nor book would have existed; the reviewers who helped improve the quality of the chapters; and the staff of ACS Books who provided important help and advice. We also thank our secretaries, Phyllis Beck at Purdue University and Shelia Beltz at Air Products and Chemicals, Inc., who worked behind the scenes, having been burdened with more than normal duties as a result of this book.

LYLE F. ALBRIGHT
Purdue University
West Lafayette, IN 47907

ROBERT J. SCHMITT
SRI International
Menlo Park, CA 94025

RICHARD V. C. CARR
Air Products and Chemicals, Inc.
Allentown, PA 18195–1501

December 12, 1995

Chapter 1

Nitration: An Overview of Recent Developments and Processes

Lyle F. Albright[1], Richard V. C. Carr[2], and Robert J. Schmitt[3]

[1]School of Chemical Engineering, Purdue University, West Lafayette, IN 47907
[2]Air Products and Chemicals, Inc., 7201 Hamilton Boulevard, Allentown, PA 18195-1501
[3]SRI International, 333 Ravenswood Avenue, Menlo Park, CA 94025

> Although many chemists and chemical engineers consider nitration processes as mature, if not ancient, considerable new information has been obtained in the last 10-15 years that clarifies the chemistry and indicates methods to obtain previously unattainable or hard to attain nitration products. New and improved nitration processes have also been developed that minimize production costs and promise reduced safety hazards. Even more information pertaining to the nitrations can be expected in the near future.

Nitration reactions have been investigated for many years in the laboratory, and numerous nitration products are produced commercially. Some organic compounds nitrated in large quantities include aromatics (such as toluene, benzene, phenol, and chlorobenzene), alcohols, glycols, glycerine, aromatic amines, and paraffins. Several nitrated products are important high explosives including trinitrotoluene (TNT), picric acid, nitroglycerine, nitrocellulose, and RDX. Dinitrotoluene (DNT) is another nitrated product of major importance. It is converted to toluene diisocynates that are then used to produce polyurethane foams, elastomers, fibers, and varnishes. Aniline is produced from nitrobenzene formed by nitration of benzene. Several nitroparaffins are produced commercially by the vapor-phase nitration of propane. Although the term nitration is often restricted to reactions involving organic compounds, the production of ammonium nitrate from ammonia and nitric acid is an example of the nitration of an inorganic compound.

Chemistry of Nitration

Aromatics, alcohols, glycols, glycerine, and amines are often nitrated by an ionic mechanism using acid mixtures (containing nitric acid and a strong acid such as sulfuric acid). Hughes, Ingold, et al. (1) indicated that nitronium ions (NO_2^+'s) present in the acid mixture attack these organic molecules to form an unstable complex. When a proton is ejected from this complex, the nitrated hydrocarbon is produced. A C-N bond is formed when an aromatic is nitrated, but a O-N bond forms when an alcohol (or hydrox group) is nitrated. In the case of amine nitrations, a N-N bond is formed. The strong acid in the acid mixture acts as a catalyst when nitric acid is used and promotes the formation of NO_2^+.

Olah, Malhotra, and Narang (2) have recently reviewed nitration chemistry with a strong emphasis on nitronium ion salts. These salts, used as a replacement for mixed acids, are often preferred choices for certain nitrations performed at anhydrous conditions. The two best known salts, NO_2BF_4 and NO_2PF_6, are effective nitrating agents for deactivated substrates and for obtaining highly selective nitrations. Anhydrous N_2O_5 (3) is now available, and it too has a wide range of uses, for especially the synthesis of nitrocompounds that are unstable in mixed acid media and for highly selective type nitrations. Anhydrous N_2O_5 is often prepared by electrochemical oxidation of N_2O_4 (or NO_2) dissolved in nitric acid (3). Separation of N_2O_5 from the nitric acid can be performed; Chapters 7, 8, and 9 of this book report recent developments. N_2O_5 is also readily prepared by oxidizing N_2O_4 dissolved in an aprotic solvent (3,4,5). Suzuki et al. (6,7) have been able to nitrate selectively certain aromatics using $N_2O_4/O_3/O_2$ mixtures. $(CF_3CO)_2)O/HNO_3$ can be used for selective nitrations of some deactivated substrates, but it is not recommended since its use has resulted in several explosions.

When toluene, phenol, or chlorobenzene are nitrated, three isomers of the mononitrated product can be produced in each case. Considerable effort has been expended to develop procedures to produce the more desired isomers. For example, 4-nitrotoluene or 2-nitrotoluene are commercially much more desired than 3-nitrotoluene. A low temperature process of Hill et al. (8) reduces yields of 3-nitrotoluene significantly. Recently Kwok and Jayasuriya (9) employed H-2SM-5 zeolite as catalyst and used n-propyl nitrate to nitrate toluene; 4-nitrotoluene was produced to a very high degree with essentially no production of 3-nitrotoluene. The nitration reactions occur in the pores of the zeolite catalyst. The toluene is apparently positioned in the pores in such a way that nitration occurs mainly at the para position. Unfortunately this catalyst quickly became deactive as the nitration of toluene progressed. Wright (10) found that the distribution of mono-nitrotoluene isomers can be varied to a considerable extent when toluene is nitrated with nitric acid in the presence of select ion-exchange resins. An ion-exchange resin with correct pore structure might become a highly desired catalyst.

The nitrosonium ion (NO^+) may be the ion that sometimes attacks easily nitratable aromatics (11). The attached nitroso group is then oxidized to form a NO_2 group. Much effort has also been given to better identifying by-products including some which can lead to explosions or other safety problems.

Free-radical reactions are often employed to nitrate paraffins (12). Nitrations using nitric acid or N_2O_4 are frequently performed in the gas-phase although on some occasions liquid-phase processes are used. For gas-phase nitrations using nitric acid, temperatures employed are frequently 380-420°C. Such temperatures cause a small fraction of the nitric acid to decompose producing the following free radicals: ·NO_2 and ·OH. Both radicals react with paraffins to produce alkyl radicals. These latter radicals react with nitric acid to produce nitroparaffins and ·OH, and a chain reaction then occurs. ·NO₂ can also react with an alkyl radical to form a nitroparaffin or an alkyl nitrite. These nitrites are relatively unstable, and when they decompose, lower molecular weight alkyl radicals and/or oxygenated hydrocarbons are often produced. When nitric acid is used as the nitrating agent for propane, only 35-40% of the nitric acid reacts to form nitroparaffins. In this case, 1-nitropropane, 2-nitropropane, nitroethane, and nitromethane are all produced. Oxygenated products include aldehydes, alcohols, and CO. NO and other oxides of nitrogen are also produced.

When N_2O_4 is used as a nitrating agent, temperatures of 200-250°C are generally sufficient. N_2O_4 decomposes readily producing ·NO₂, which reacts at these low temperatures to produce alkyl radicals. The alkyl radicals and ·NO₂ then combine to produce nitroparaffins.

For free-radical nitrations of paraffins, dinitroparaffins are not produced when the reaction conditions are in the 380-420°C range. 2,2-dinitropropane has, however, been produced at lower temperature using a liquid-phase nitration process. At higher temperatures, intermediate radicals apparently decompose before dinitroparaffins are produced.

Although no known attempts have been made to develop a mathematical model for the free radical nitration of paraffins, sufficient information is probably available to develop such a model for the vapor-phase nitration of cyclohexane with N_2O_4 (13). For this nitration, concentrations of nitrocyclohexane, cyclohexylnitrite, cyclohexylnitrate, cyclohexanol, and cyclohexanone have been reported. Similar models likely could be developed later for the nitration of other paraffins. Data are, however, needed on the side reactions that occur after the nitroparaffins are formed. Part of these nitroparaffins degrade or are oxidized. Some heavier nitroparaffins may react forming lower molecular weight nitroparaffins, e.g., a nitropropane may be converted to nitromethane.

Victor Meyer processes, or modifications of them, are also employed to produce nitroparaffins. Such processes are of limited industrial importance.

Physical Steps During Ionic Nitrations

During many ionic-type nitrations, two immiscible liquid phases are often present. Using the nitration of benzene with mixed acids as an example, transfer steps are obviously important in order to contact the reactants. NO_2^+ produced in the acid phase must be contacted with benzene initially present in the hydrocarbon phase. The main reactions are generally thought to occur at or at least near the interface between the two phases (12). A large interfacial area would obviously promote rapid reactions. Relatively little quantitative information has been found in the literature on the values of the interfacial area/volume of the dispersions in the reactor or on which is the preferred continuous phase and which the dispersed phase.

About 30-50 years ago, batch nitrations of benzene or toluene frequently required 30-60 minutes to complete. In some modern continuous-flow reactors, the nitration is completed within several seconds or even within one second. Yet the nitric acid in some of these units is almost completely reacted. The intensity of agitation is apparently much greater in the modern units which results in much smaller droplets of the dispersed phase. In older units, decanters were often, if not always, employed to separate the two liquid phases, but centrifuges are generally employed in more modern units. Centrifuges promote rapid separation, result in fewer side reactions, and reduce the inventories of nitrated products in the reactor portion of the process. There is still a need to learn more about the type of emulsions or dispersions formed.

New and Potential Industrial Processes

Most industrial nitrations of aromatics, alcohols, and amines currently use mixed acids (mixture of nitric and sulfuric acids) as nitrating agents. These mixed acids are relatively cheap and generally highly effective, as compared to other nitrating agents. A major drawback associated with mixed acids is the recovery and regeneration of the sulfuric acid, which becomes diluted with water formed as a by-product. Such regeneration requires considerable heat (or energy). Techniques have been developed for some nitrations to utilize the exothermic heat of nitration to provide all or at least most of the heat required for acid regeneration. Such a technique is industrially used when benzene is nitrated to produce nitrobenzene (14). A similar technique has also been patented for production of dinitrotoluenes (15). For most other nitrocompounds, safety and quality control problems are yet to be solved with this technique. Plug-flow reactors that provide high shear in the reaction stream have been a key factor in the commercialization of the new technology (16).

For the production of specialty products, certain nitrating agents offer the following advantages: high yields, selective nitration characteristics, and improved safety. N_2O_5 has been investigated to a considerable extent especially in the last 10 years. Although it is a relatively expensive compound, certain specialty-type nitrocompounds can be produced readily and safely. As a specific example, Bottaro and Schmitt (17-19) have used N_2O_5 in the synthesis of ammonium dinitramide, $NH_4N(NO_2)_2$. Several new high explosives have recently been synthesized for the first time, and more likely will be announced in the next several years.

Much effort has been made to develop viable alternatives to nitrations using mixed acids. Solid catalysts theoretically offer two major advantages: increased safety and better control of the isomers produced. Both vapor and liquid phase nitrations result when zeolites (20) and related solid acid catalysts (21) are employed. Both NO_2 (22) and nitric acid (23) have been used as nitrating agents. For monosubstituted benzenes, high levels of para selectivity are often realized with the proper solid catalysts (23) even though meta selecting may be predominant when mixed acids are used. In addition, cupric nitrate or other metal nitrates deposited on clay, such as montmorrillonite clay, has given promising results (24).

Finally a newly developed non-acid technology which may have application for specialty nitrations is the ozone mediated nitration with NO_2 (Kyodai-nitration), a process developed by Suzuki (25). This process is often facilitated by the addition of a Lewis acid catalyst (26), and aromatic ketones were found to give ortho selectivity using this technology (27).

Hazards of Nitration

Safety is always be a key concern in and around a nitration plant or when handling nitrated hydrocarbons. Fires or explosions of air/hydrocarbon gaseous mixtures are always a possibility since the feedstocks and nitrated products are highly flammable. Some nitrated products are explosives or monopropellants. Highly hazardous chemicals or acids are also present in many plants. Some feedstocks such as benzene and certain nitrocompounds are for example carcinogens.

Nitration plants or products have experienced a significant number of serious accidents in the last 40 years. One author (LFA) of this chapter was a member of an investigation team for each of several accidents listed below. Most accidents listed resulted in fatalities.

1) In 1969, a storage tank partially filled with liquid p-nitro-meta-cresol exploded (28). The temperature of the tank had, for an unknown reason, risen above the design value of 135°C. Subsequent tests indicated that instability of the melt becomes a problem as the temperature increases. Some air may have entered the gas space above the melt (through a vent line) and could have contributed to the explosion. This explosion was a surprise to some who did not think that a monoaromatic would explode.

2) In the late 1970's, a batch reactor exploded; a mononitroaromatic was being nitrated in this reactor to produce a dinitroaromatic. The reactor had been loaded with all of the reactants (both mononitroaromatic and mixed acids). Agitation was provided to mix the phases, and the temperature was adjusted to the desired value to start the run. As the run progressed, the temperature increased above the desired level. Obviously insufficient heat transfer was occurring. Shortly thereafter, the reactor exploded.

3) In 1982, a batch reactor used to nitrate an alcohol exploded; the top of the nitrator was blown a considerable distance. In this unit, all of the alcohol was added to the reactor, and the mixed acids (nitric and sulfuric acids) were then added at a relatively constant rate. The objective in such a batch unit is to react the nitric acid essentially as rapidly as it is added. When that occurs, only a limited amount of unreacted nitric acid is ever present in the reactor, and hence the rate of heat generation is always relatively low (and controllable). The following operating changes generally help meet this desired objective. First, the rate of acid addition is decreased or maintained at a relatively low level. Second, the temperature of the reactants is increased or maintained at a level sufficiently high to promote the reaction of the acid as it is added to the reactor. Tests in small laboratory reactors prior to the explosion indicated that "runaway" reactions sometimes occur. When that occurs, the temperature of the reactants rise rapidly in an uncontrolled manner. The rates of heat generation in the reactor are then greater than the rates of heat transfer. In the run that resulted in an explosion, the temperature of the reaction mixture was, however, deliberately reduced presumably in the hope that a runaway reaction would be prevented. For several minutes prior to the explosion, brown fumes

(obviously NO_2) were emitted in ever increasing amounts. Nevertheless, the feed of mixed acids to the reactor was not stopped. The presence of NO_2 clearly indicates that at least part of the nitric acid is decomposing and is acting as an oxidant.

4) In the late 1950's, two tank cars containing liquid nitromethane exploded in separate accidents, both in railroad switching yards. In these cars, the gas mixture above the liquid nitromethane contained both air and nitromethane. Subsequent tests indicated that adiabatic compression (and the resulting temperature increase) of the gas in the top of the tank car initiated the explosion of the liquid nitromethane which is a known monopropellant. Adiabatic compression occurred when a car containing nitromethane was hit and jolted by another car during switching. In both explosions, a relatively large crater was formed in the ground. Such an explosion would have been prevented if the liquid nitromethane had been blanketed with an inert gas. Liquid nitromethane is currently shipped in drums. Adiabatic compression in drums results in much smaller temperature increases that are not a problem.

5) In 1991, a major explosion occurred in a large nitroparaffin plant (29). The initial explosion apparently caused further explosions and fires resulting in severe damage and fatalities. A pipeline filled with liquid nitromethane presumably eventually exploded.

6) A British company that nitrates toluene experienced serious accidents in both 1989 and 1992. In 1989, molten DNT's at about 125-135°C exploded (30). This melt in a pipe contained dissolved nitric acid that had been extracted from mixed acids. Schiefferle, Hanson, and Albright (31) report that significant amounts of nitric acid can be extracted by various nitroaromatics. Little sulfuric acid or water are, however, extracted. The dissolved nitric acid is an excellent oxidizing agent. The cause of this explosion may be similar to that of a 1972 explosion in the plant of another company. The DNT's in this earlier explosion likely were at a much higher temperature, perhaps 210°C.

In the 1992 explosion, a batch still that had been used to separate mononitrotoluenes(MNT's), DNT's, and nitrocresols was being cleaned (32). About 1820 liters of residues or sludge had collected over a period of time in the still. This sludge exploded, and the damage was widespread resulting in several fatalities.

7) In 1991, 40 tons of trinitrotoluene (TNT) exploded in a Chinese plant (33). Records indicated that the concentration of nitric acid in the mixed acids being fed to one reactor was too high. A series of explosions resulted, causing major damage and many fatalities.

8) In December 1994, an ammonium nitrate plant experienced a severe explosion (34). Many years ago, a ship filled with ammonium nitrate exploded in the Texas City harbor causing major damage to the Monsanto plant in Texas City. Fatalities were high in the earlier explosion and four fatalities occurred in the 1994 explosion.

Modern nitration units have been designed to minimize the potential dangers and to limit or isolate any explosions. High rates of nitration have resulted in smaller reactors (containing less nitrated product). Major improvements have been made in many units to better control the operation of the nitration reactors, to reduce undesired reactions including oxidation, to obtain quicker and safer recovery of products, and to provide improved storage of products. The Design Institute for Emergency Relief Systems of the American Institute of Chemical Engineers has assimilated data for the improved design and operation of equipment such as used in nitration plants.

Environmental Concerns of Industrial Nitration Plants

The recent enactment and promulgation of EPA effluent guidelines pose serious challenges to both existing and new manufacturers of nitro compounds. The EPA list of undesired pollutants includes nitrobenzene, 2,4-dinitrotoluene, 2,6-dinitrotoluene, 4-nitrophenol, 2,4-dinitrophenol, and 4,6-dinitro-o-cresol. Regulations restrict concentrations of these pollutants to ppb concentrations in aqueous effluents.

Many nitrators have had to resort to expensive end-of-pipe treatments such as carbon adsorption or tandem peroxide treatment/carbon adsorption (35), wet air oxidation (36), photocatalytic degradation (37), and catalytic liquid phase oxidation (38) in order to achieve compliance. Pollutant source reduction may provide a better alternative to expensive end-of-pipe treatments. A good example of this approach is found in the Olin dinitrotoluene process which generates so little waste water that it is economically feasible to distill the waste water from the product (39).

Recent air quality regulations also have had a major impact on nitration facilities. The Clean Air Act amendment of 1990 lays out strict guidelines for $(NO)_x$ emissions. Chemical manufacturers have spent considerable time, effort, and money to reduce $(NO)_x$ emissions. Also the National Emissions Standards for Hazardous Air Pollutants (NESHAPS) for SOCMI chemicals regulates many nitrocompounds such as dinitrophenol, nitrobenzene, and dinitrotoluene.

Literature Cited

1. Hughes, E.D.; Ingold, C.K.; Reed, R.J., J. Chem. Soc. 1950, 2400-2473.
2. Olah, G.S.; Malhotra, R.; Narang, S.C., *Nitration Methods and Mechanism*, VCH Publishers, New York, NY, 1989.
3. Fisher, J.W., The Chemistry of Dinitrogen Pentoxide" in *Nitro Compounds: Recent Advances in Synthesis and Chemistry*, ed. H. Feuer and A.T. Arnold, Advances in Chemistry Series, pp 267-365, VCH Publishers, Inc., New York, 1990.
4. Bloom, A.J.; Fleishmann, M.; Mellor, J.M.; J. Chem. Soc. Perkins I, 1994, 1367-1369.
5. Bloom, A.J.; Fleishmann, M.; Mellor, J.M., Electrochimica Acta 1987, 32, 785-790.
6. Suzuki, H.; Murashima, T.; Shimizu, K.; Tsukamoto, K., J. Chem. Soc., Chem. Commun. 1991, 1049.

7. Suzuki, H.; Yonezawa, S.; Mori, T.; Maeda, K., J. Chem. Soc. Perkin Trans. I 1986, 79-82.
8. Hill, M.E.; Coon, C.L.; Blucher, W.G.; McDonald, G.J.; Marynowski, C.W.; Tolberg, W.; Peters, H.M.; Simon, R.L; Ross, D.L. in *Industrial and Laboratory Nitrations*, L.F. Albright and C. Hanson, Eds., ACS Symposium Series 22, American Chemical Soc., Washington, D.C., 1976, pp 257-271.
9. Kwok; T.J.; Jayasuriya, J. Org. Chem. 1994, 59, 4939.
10. Wright, O.L., Use of Ion-Exchange Resins, U.S. Patent 2,948,759 (Aug. 9, 1960).
11. Bunton, C,.A.; Hughes, E.D.; Ingold, C.K., J. Chem. Soc. 1950, 2628-2694.
12. Albright, L.F. in *Kirk-Othmer Encyclopedia of Chemical Technology*, Vol. 15, 3rd Ed., 1981, 850; Vol.17, 4th Ed., 1996, 68.
13. Lee, R.; Albright, L.F.; Ind. Eng. Chem., Proc. Des. Dev. 1965, 4, 411.
14. Alexanderson, V.; Trecek, J.B.; Vanderwart, C.M.; U.S. Patent 4,021,498, May 3, 1977.
15. Schieb, T.; Wiechers, G.; Sundermann, R.; Zarnuck, U., Canadian Pat. 2,102,587, (May 14, 1994).
16. Guenkel, A.A.; Rae, J.M.; Hauptmann, E.G., U.S. Patent 5,313,009, May 17, 1994.
17. Bottaro, J.C.; Schmitt, R.J.; Penwell, P.E.; Ross, D.S., U.S. Patent 5,198,204; 1993; International Pat. Applic. No. WO91/19669, (Dec. 26, 1991).
18. Bottaro, J.C.; Schmitt, R.J.; Penwell, P.E.; Ross, D.S., U.S. Patent 5,254,321: 1993; International Pat. Applic. No. WO91/19669 (Dec.26, 1991).
19. Schmitt, R.J.; Bottaro, J.C.; Penwell, P.E.; Bomberger, D.C., U.S. Patent 5,316,749, Aug. 19, 1993; International Pat. Applic. No. WO93/16002.
20. Malysheva, L.V.; Paukshtis, E.A.; Ionc. K.G., *Catal. Rev.-Sci. Eng.*, 1995, 37, 179.
21. Akolekar, D.B.; Lemay, G.; Sayari, A.; Kaliaguine, S., *Res. Chem. Intermed*, 1995, 21, 7.
22. a) Bertea, L.E.; Kouwenhoven, H.W.; Prins, R., *Stud. Surf. Catal.*, 1994, 84, 1973. b) Bertea, L.E.; Kouwenhoven, H.W.; Prins, R., *Stud. Surf. Catal.* 1993, 78, 607.
23. Reith, R.A.; Hoff, G.R., U.S. Patent 4,754,083, June 28, 1988.
24. Cornelis, A.; Laszlo, P., *Aldrichimica Acta*, 1988, 21, 97.
25. a) Suzuki, H.; Mori, T., *Synlett, 1995, 383*. b) Suzuki, H.; Mukrashima, T.; Shimizu, K.; Tsukamoto, K., *J. Chem. Soc. Chem. Commun.*, 1991, 1049.
26. Suzuki, H.; Murashima, T.; Tatsumi, A.; and Kozai, I., *Chem. Lett*, 1993, 1421.
27. Suzuki, H.; Murashima, T.; Kozai, I.; Mori, T., *J. Chem. Soc. Perkin Trans. I*, 1993.
28. Dartnell, R.C.; Ventrone, T.A., Chemical Engineering Progress 1971, 67, #6, 58.
29. Chem. and Engineering News, 1991, May 13, p 6.
30. Loss Prevention Bulletin, Explosion in Dinitrotoluene Pipeline, 1989, p 13.
31. Schiefferle, D.F.; Hanson, C.; Albright, L.F., in *Industrial and Laboratory Nitrations*, L.F. Albright and C. Hanson, Eds., ACS Symposium Series 22, American Chemical Soc., Washington, D.C., 1976, pp 176-189.

32. Hickson and Welch, Ltd., The Fire at Hickson and Welch; Company Report, 1992.
33. Zhang, G.S.; Tang, M.J., J. Hazardous Materials 1993, 34, 225-233.
34. Chem. and Engineering News, 1995, June 15, p. 9.
35. Peter, S.B.; Adams, K.B.; Casas, B.; Sawicki, J.E., U.S. Patent 5,456,539, October 18, 1994.
36. Mishra, V.A.; Mahajani, V.V.; Joshi, J.B, *Ind. Eng. Chem. Res.*, 1995, 34, 2.
37. Augugliaro, V.; Palmisano, L.; Schiavello, M.; Sclafani, A., *Appl. Cat.*, 1991, 69, 323.
38. a) Pintar, A.; Levec, *Chem. Eng. Sci.*, 1992, 47, 2395. b) Pintar, A.; Levec, J., *Ind. Eng. Chem. Res.*, 1994, 33, 3070.
39. Quakenbush, A.B.; Pennington, B.T., *Polyurethanes World Congress*, 1993, 484.

RECEIVED December 12, 1995

Chapter 2

Role of Protosolvation in Nitrations with Superacidic Systems
The Protonitronium Dication (NO_2H^{2+}) Identified

G. K. Surya Prakash, Golam Rasul, Arwed Burrichter, and George A. Olah

Loker Hydrocarbon Research Institute and Department of Chemistry, University of Southern California, Los Angeles, CA 90089-1661

> The reactivity of nitronium ion, the species responsible for electrophilic nitration, is greatly enhanced in strong Bronsted acid medium. Now such reactivity is attributed to the protosolvation of nitronium ion leading to protonitronium dication, NO_2H^{2+}. This paper describes our attempts to identify highly reactive protonitronium dication with the aid of ^{17}O NMR spectroscopy and theoretical methods.

Hantzsch's pioneering studies (2) in the 1930's laid the foundation for Ingold and his associates to establish nitronium ion (NO_2^+) as the reactive electrophile in the acid-catalyzed nitration with nitric acid and its derivatives (3). Olah et al. in 1956 (4) reported the simple preparation of the remarkably stable nitronium tetrafluoroborate, $NO_2^+BF_4^-$, and its use as a convenient and highly efficient nitrating agent. Nitronium salts with a large number of anions (such as PF_6^-, $CF_3SO_3^-$, etc.) since gained wide use (5) as convenient nitrating agents.

Highly deactivated aromatics such as *m*-dinitrobenzene are not nitrated by nitronium salts in aprotic solvents (nitromethane, sulfolane, methylene chloride etc.). However, in superacidic FSO_3H solution, nitronium ion nitrates *m*-dinitrobenzene to 1,3,5-trinitrobenzene (6). Nitrations of aromatics were also carried out (7) with nitric acid/trifluoromethanesulfonic acid system. Many deactivated aromatics such as nitro- and chlorobenzenes were successfully nitrated in high yields. We have also developed nitric acid/trifluoromethanesulfonic acid/phosphorous pentoxide system as a highly electrophilic nitrating system (7). More recently, nitration of various deactivated arenes (including methanesulfonyl-, nitro-and polyhalobenzenes) has been carried out in good yields with nitric-triflatoboric acid, $HNO_3:HB(OSO_2CF_3)_4$ (8). The new nitrating system gives high regioselectivity and yields under generally mild reaction conditions. The reagent system is relatively non-oxidizing and compatible with many functional groups of arenes. These systems are so powerful, even triphenylmethyl cation can be nitrated at the *m*-position (9).

Similarly, whereas nitronium ion shows no reactivity toward methane in aprotic media, it reacts, *albeit*, in low yield, in FSO_3H to give nitromethane (10). This indicates that the nitronium ion is activated by the superacidic medium compared to its reactivity in aprotic solvents. To explain the highly increased reactivity of the nitronium ion in strongly acidic medium, it was suggested (10,11) that the nitronium ion could undergo protonation to protonitronium dication (NO_2H^{2+}), **1**, an extremely reactive superelectrophile. The linear nitronium $O=N^+=O$ is only a polarizable electrophile in its ground state. It has no empty atomic orbital on nitrogen or low-lying LUMO on nitrogen (similar to the ammonium ion), and its electrophilic ability is only due to its polarizability when attracted by π-donor aromatic nucleophiles. In contrast to reactive π-donor aromatics, deactivated aromatics or σ-donor alkanes are weak electron donors and cannot bring about such a polarization. A "reactive" nitronium ion should be bent with a developing p-orbital on nitrogen. The driving force for the formation of the bent nitronium ion must be the ability of the oxygen nonbonded electron pairs to coordinate with the strong Bronsted acid in the nitrating system. Similar reactivity is also indicated using $NO_2Cl:3AlCl_3$ reagent system, wherein Lewis acid activation of the electrophile occurs (12).

$$NO_2Cl + 3\ AlCl_3 \longrightarrow \underset{AlCl_3}{O}=\overset{+}{N}=\underset{AlCl_4^-}{O}\diagup^{AlCl_3}$$

Previously no evidence has been obtained for the protonitronium dication, NO_2H^{2+}. Simonetta's lower level calculations (13) on the protonitronium dication indicated that the dication may not correspond to a minimum. However, our recent HF/6-31G* and MP2/6-31G** levels calculations (14) show that NO_2H^{2+} corresponds to a minimum. Although NO_2H^{2+} was found to be substantially more energetic than the nitronium ion, the deprotonation barrier was calculated to be 17 kcal/mol at the HF/6-31G*//HF/6-31G* level. More recently Schwarz et al. (15) were indeed able to generate the long-sought-after NO_2H^{2+} in the gas phase by dissociative electron impact ionization of HNO_3.

^{15}N NMR chemical shifts of NO_2H^{2+} and NO_2^+ were calculated with use of IGLO (16) (individual gauge for localized orbitals). Due to presence of nonbonded electrons on nitrogen, ^{15}N NMR chemical shifts are more dependent on solvents and temperature than ^{13}C NMR chemical shifts; consequently, calculated ^{15}N NMR chemical shifts were expected to differ from experimental values. Schindler's (16) IGLO calculation on ^{15}N NMR chemical shifts of nitrogen containing compounds, however, showed good agreement between theory and experiment. Our calculated ^{15}N NMR chemical shifts of nitronium ion NO_2^+ at the II/6-31G* level is $\delta(^{15}N)$ 268.3, 17 ppm deshielded form the experimental $\delta(^{15}N)$251.0 (14). The ^{15}N NMR chemical shifts of the protonitronium dication ($\delta(^{15}N)$ 272.5) calculated at the same level is, however, deshielded from the ^{15}N NMR chemical shift of the nitronium ion by only 4 ppm. On the other hand, the calculated ^{15}N NMR chemical shift of nitric acid is $\delta(^{15}N)$ 366.8, 10 ppm shielded from experimental $\delta(^{15}N)$ 377.0 (14). Attempts to experimentally observe the protonitronium dication (NO_2H^{2+}) by ^{15}N NMR spectroscopy using ^{15}N-enriched nitronium ion (98%) in a large excess of $HF:SbF_5$ did not indicate any chemical shift change from that of the nitronium ion. The result is not unexpected as the protonitronium dication (NO_2H^{2+}) is expected to be present only in low concentration in the superacid system and may undergo fast proton exchange.

We also attempted to use faster IR and Raman spectroscopy to observe NO_2H^{2+}. The nitronium ion in excess SbF_5 (containing some HF) solvent did not show any significant change in the Raman frequencies. Furthermore, no new absorptions were observed in the IR spectrum. Again low concentration of NO_2H^{2+} would not be detectable by used methods.

^{17}O NMR spectroscopy has been applied extensively to study the oxonium and carboxonium ions (17a). Attempts to study the nitronium ion NO_2^+ by ^{17}O NMR spectroscopy have been reported in the literature (17b, c). However, due to signal broadening, the ^{17}O-NMR chemical shift of nitronium ion NO_2^+ could not be assigned correctly. ^{17}O NMR spectroscopy should be useful to study these ions because it is known that ^{17}O NMR chemical shifts are very sensitive to changes in the nature of oxygen bonding.

In this paper we wish to report ^{17}O NMR chemical shift studies and application of density functional theory (DFT) (18), *ab initio* and GIAO-MP2 study on nitronium and its protonated analog. Density Functional Theory (DFT) recently become popular and reliable (18) to predict the molecular properties. Rauhut and Pulay showed (19) that the B3LYP hybrid functional together with 6-31G* basis set gives very good molecular geometries and vibrational frequencies. In addition we have used G2 method (20) which has been shown to give accurate molecular energies We have also used GIAO-MP2 method (21), which includes dynamic electron correlation in chemical shift calculations, to predict the ^{17}O NMR chemical shifts of the ions.

Results and Discussion

DFT and *ab initio* calculations were carried out by using the GAUSSIAN 94 (22) package of programs. Restricted Hartree-Fock calculations were performed throughout. Optimized geometries were obtained with the B3LYP/6-31G* and MP2/6-31G* levels. G2 method (20), which is offered by GAUSSIAN 94, is a composite method based on MP2/6-31G* optimized geometry, was used for acurate energy calculations. Total energies and relative energies are listed in Table-I. GIAO-MP2 calculation for the ^{17}O NMR chemical shifts, using tzp/dz and qz2p/qz2p basis sets (21) were performed with the ACES II program (23). The data is shown in Table II. Both experimental and calculated ^{17}O NMR chemical shifts are referenced to H_2O. Vibrational frequencies were calculated at the B3LYP/6-31G*//B3LYP/6-31G* and MP2/6-31G*//MP2/6-31G* levels (data indicated in Table-III).

^{17}O NMR Chemical Shifts. The ^{17}O enriched nitronium ion were prepared by treating nitronium tetrafluoroborate (NO_2^+ BF_4^-) with 20% ^{17}O enriched H_2O ($H_2^{17}O$) and dissolving the products in FSO_3H (Scheme 1). The ^{17}O spectra were then obtained at ambient temperature.

$$NO_2^+BF_4^- + H_2^{17}O \rightleftharpoons HN^{17}O_3 \xrightarrow{FSO_3H} N^{17}O_2^+$$

Scheme 1

The ^{17}O spectrum (Figure 1) consists of three sharp peaks at $\delta^{17}O$ 143.5, 196.6 and 461.3. The signal at $\delta^{17}O$ 196.6 were assigned for nitronium ion NO_2^+ **1** which is 217.4 ppm shielded with respect to nitric acid. The GIAO-MP2/qz2p//MP2/6-31G* calculated value for NO_2^+ is $\delta^{17}O$ 194.2 agrees very well with the experimental values of 196.6 ppm. However, GIAO-

Table I: Total energies (-Hartree) and relative energies (kcal/mol) at different levels

# Ion	B3LYP/6-31G* (ZPE)[a]	Rel. Energy[b]	G2	Rel. Energy[c]
1 NO_2^+	204.71184 (7.1)	0.0	204.48421	0.0
2 NO_2H^{2+}	204.57546 (11.3)	89.8	204.34842	85.2
3 NO_2H^{2+} (TS)	204.54999 (7.4)	101.9	204.32334	101.0
4 NO_2H^{2+} (TS)	204.49509 (3.4)	136.3	204.28706	123.7
5 NO^+	129.52979 (3.4)	0.0	129.39888	0.0
6 HNO^{2+}	129.32980 (7.5)	129.6	129.20114	124.1
7 HNO^{2+} (TS)	129.32636 (5.0)	129.3	129.20034	124.6

[a] zero point vibrational energies are at B3LYP/6-31G*//B3LYP/6-31G* level scaled by a factor of 0.96; [b] relative energies besed on B3LYP/6-31G*//B3LYP/6-31G* + ZPE (B3LYP/6-31G*//B3LYP/6-31G*); [c] based on G2 energies.

Table II: ^{17}O NMR chemical shifts

# Ion		GIAO-MP2/tzp// MP2/6-31G*	GIAO-MP2/qz2p// MP2/6-31G*	GIAO-MP2/tzp// B3LYP/6-31G*	GIAO-MP2/qz2p// B3LYP/6-31G*	Expt.
H_2O		0.0	0.0	0.0	0.0	0.0
1 NO_2^+		192.4	194.2	186.7	185.4	196.6
2 NO_2H^{2+}	O=N	176.2	195.2	238.4	244.9	
	O-H	172.2	170.3	157.3	154.2	
	average	174.2	182.8	197.9	199.6	
5 NO^+		478.8	483.7	463.0	466.1	461.5
6 HNO^{2+}		421.9	428.3	427.6	431.1	
HNO_2	O=N	739.1	751.8	720.0	727.3	
	O-H	440.5	451.4	433.5	468.3	
	average	589.8	601.6	576.8	597.8	
HNO_3	O-H	345.7		340.2	341.2	
	$O-N_t$	452.5		437.1	446.9	
	$O-N_c$	466.3		456.1	458.0	
	average	421.5		411.1	415.4	414.0

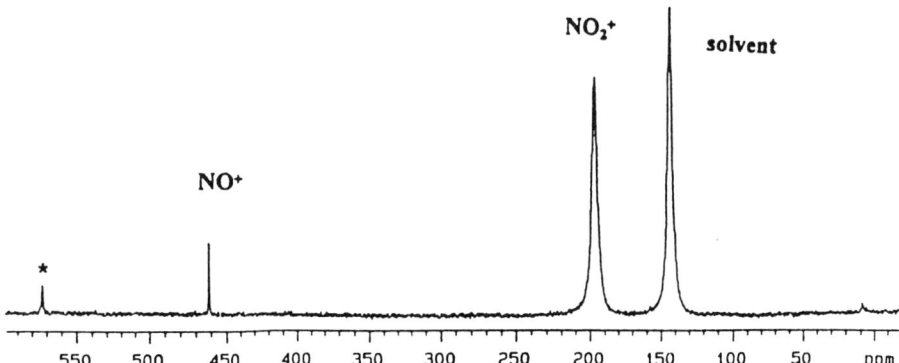

Figure 1. ^{17}O NNMR spectrum of NO_2^+ ion in FSO_3H at ambient temperature. *peak due to acetone d_6 capillary.

MP2/qz2p//B3LYP/6-31G* calculated value of 185.4 ppm is about 10 ppm shielded than experimental value. We also found a good agreement between experimental ($\delta^{17}O$ 414.0) and calculated ($\delta^{17}O$ 421.5 at GIAO-MP2/tzp//MP2/6-31G* and 411.1 at GIAO-MP2/qz2p//B3LYP/6-31G*) average $\delta^{17}O$ of nitric acid. The signal at $\delta^{17}O$ 461.5, which corresponds to the nitrosonium ion, is due to the presence of considerable amount of nitrsonium ion impurity in commercially available $NO_2^+ BF_4^-$. In fact this ^{17}O enrichment procedure could be used to detect the extent of NO^+ impurity in $NO_2^+ BF_4^-$ salts using ^{17}O NMR spectroscopy. The GIAO-MP2/qz2p//MP2/6-31G* calculated chemical shifts of NO^+ was found to be $\delta^{17}O$ 483.7 also matches with the experimentally observed chemical shift of 461.5 ppm. However, in this case GIAO-MP2/qz2p//B3LYP/6-31G* calculated value of 466.1 ppm matches better with experimental shift than that of GIAO-MP2/qz2p//MP2/6-31G* value. The remaining peak in the spectrum at $\delta^{17}O$ 143.5 is due to the solvent FSO_3H.

Upon increasing the acidity of the solution of FSO_3H by adding the SbF_5 (three fold excess), the lines broadened in the spectrum (Figure 2). The nitronium ion signal (line width is 930 Hz) moved upfield by about 5 ppm. Although viscosity of the the acid solution may contribute some to the observed line broadening it is not a significant factor at room temperature. Thus the majority of the observed broadening could be due to possible proton exchange process occuring between the NO_2^+ and the acid (even the ^{17}O peak of the acid is broadened) through small equilibrium concentration of protonitronium dication (Scheme 2). This observation could be a definite evidence for the existance of NO_2H^{2+} **2** ion in superacid solutions. The calculated ^{17}O NMR chemical shifts of NO_2H^{2+} **2** is somewhat dependent on the level of calculations used. GIAO-MP2 calculations using MP2/6-31G* geometry give an average $\delta^{17}O$ of NO_2H^{2+} ($\delta^{17}O$ 174.2) which is shielded compared to that of NO_2^+ **1** ion at the same level (Table II). In contrast, by using B3LYP/6-31G* geometries GIAO-MP2 give an average $\delta^{17}O$ of 199.6 for NO_2H^{2+} which is deshielded compared to that of NO_2^+ **1**.

$$NO_2^+ + H^+ \rightleftharpoons NO_2H^{2+}$$

Scheme 2

Similar to nitronium ion, the ^{17}O enriched nitrosonium ion was also prepared by reacting nitrosonium tetrafluoroborate ($NO^+ BF_4^-$) with 20% ^{17}O

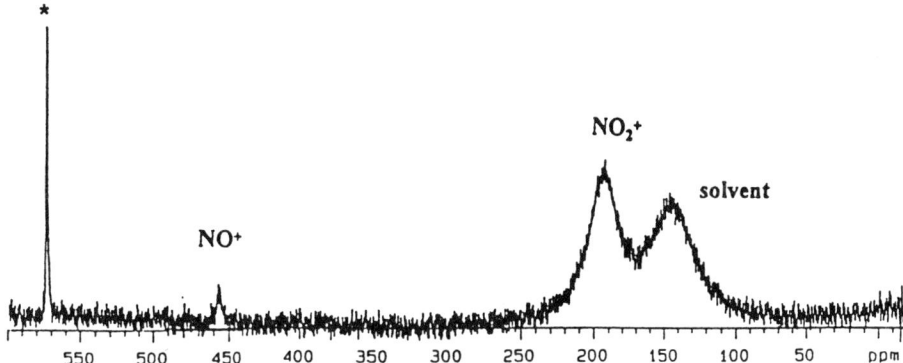

Figure 2. ^{17}O NMR spectrum of NO_2^+ ion in $FSO_3H:SbF_5$ (three fold excess) at ambient temperature. *peak due to acetone d_6 capillary.

enriched H_2O ($H_2^{17}O$) and dissolving the products in FSO_3H (Scheme 3). The spectrum contains a sharp peak at $\delta^{17}O$ 461.5 as previously identified for NO^+ **5** ion and also a minor peak at $\delta^{17}O$ 196.6 responsible for NO_2^+ **1** ion.

$$NO^+BF_4^- + H_2^{17}O \rightleftharpoons HN^{17}O_2 \xrightleftharpoons{FSO_3H} N^{17}O^+$$

Scheme 3

The peak at $\delta^{17}O$ 461.5 shifted upfield by about 5 ppm when the acidity of the FSO_3H (Scheme 3) were increased by adding the SbF_5. This observation also could be due to the presence of protonated nitrosonium dication (HNO^{2+} **6**) in low equilibrium concentration. Interestingly, in all levels of calculations the $\delta^{17}O$ of HNO^{2+} **6** is shielded compared to that of NO^+ **5** ion.

Geometries and Energies. Figure 3 displays the DFT (B3LYP/6-31G*) and MP2/6-31G* optimized geometries together with available experimental data. From the figure it is clear that the DFT geometries are comparable to those obtained from MP2. The biggest difference is in the N-O(H) bond length of transition structure **4**, where the DFT predicted value of 1.961 Å is 0.1 Å longer than the MP2 value. However, for N-O bond length in NO^+ **5** the DFT data compare more favourably with experimental values than MP2 values.

Recently G2 has become one of the most accurate methods available to calculate molecular energies. We have applied G2 method for scanning the potential energy surfaces of NO_2H^{2+} **2** and HNO^{2+} **6**. NO_2H^{2+} **2** lies 85.2 kcal/mol above NO_2^+ **1**. Dissociation barrier of NO_2H^{2+} **2** into NO_2^+ **1** and H^+ through transition structure **3** is found to be 15.8 kcal/mol. On the other hand, the dissociation of NO_2H^{2+} **2** into NO^+ **5** and OH^+ through transition structure **4** is 35.5 kcal/mol. As we mentioned in the introduction Schwarz et al[8] were able to generate the dication NO_2H^{2+} **2** in the gas phase. The meta stable (MI) spectrum of NO_2H^{2+} **2** demonstrates that the proton loss to generate NO_2^+ **1** is favoured over the loss of OH^+ to form NO^+ **5** which is consistent with our G2 calculation. Schwarz et al. (15) also calculated dissociation barrier of NO_2H^{2+} ---> $H^+ + NO_2^+$ and NO_2H^{2+} ---> $OH^+ + NO^+$ at the MRCI + D + ZPVE level using CASSCF/6-31G* geometries and found to be 16.7 and 28.4 kcal/mol, respectively, although the latter value differs by 7 kcal/mol from our G2 value of 35.5 kcal/mol.

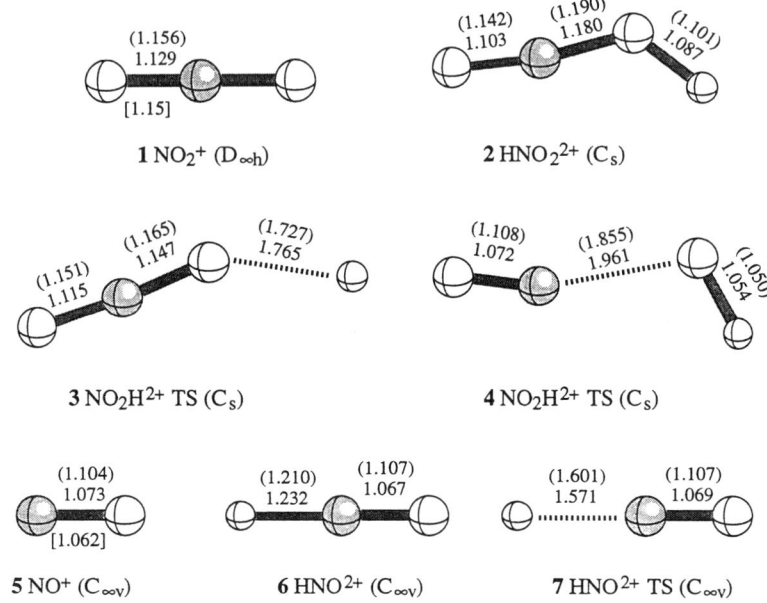

Figure 3. DFT (B3LYP/6-31G*) calculated stucture. MP2/6-31G* parameters are given in parentheses. Experimental data of **1** from ref. 24 and **5** from ref. 25 are given in bracket.

We have also calculated the dissociation barrier of HNO^{2+} at the G2 level. However, the dissociation of HNO^{2+} **6** into NO^+ **5** and H^+ through transition state **7** was found to be low (0.5 kcal/mol). This indicates that the protonitrosonium dication HNO^{2+} should be kinetically unstable. However, we must remember that these results are for isolated species and might be different when interaction with solvent occurs.

IR and Raman Frequencies. IR and Raman frequencies were calculated both at the MP2/6-31G* and DFT (B3LYP) levels (Table III). Overall the DFT frequencies were generally found to be in better agreement with the experimental data than those obtained at MP2/6-31G* level. For the nitronium ion NO_2^+ **1** the DFT results for the symmetric stretching frequency ν1, bending frequency ν2, and asymmetric stretching frequency ν3 are 1391, 603, and 2353 cm^{-1}, respectively. The DFT frequencies are in good agreement with the experimental frequencies of 1396, 571, and 2360 cm^{-1}. Calculations at MP2/6-31G* level (1207, 544, and 2381, respectively) were found to be less accurate than the DFT results, but are still in satisfactory agreement with the experimental data. Whereas MP2 predicts relatively little change in vibrational frequencies when going from the symmetrical nitronium ion (NO_2^+ **1**) to the less symmetrical protonitronium dication (NO_2H^{2+} **2**), DFT calculations predict a considerably different pattern of vibrational frequencies for the dication (Table III). It is interesting to note that MP2 completely fails to predict the N-O stretching frequency of NO^+ **5** (1966 cm^{-1}) when compared to the experimental frequency of 2377 cm^{-1}. However, DFT produces a frequency of 2379 cm^{-1}, which is in excellent agreement with the experimental data.

Conclusion

The ^{17}O NMR line broadening of nitronium ion peak in strong acid $HSO_3F:SbF_5$ medium has been attributed to proton exchange involving protonitronium dication NO_2H^{2+}. The studies are also supplemented by theoretical calculations of structures, energies and ^{17}O NMR chemical shifts at various levels.

Table III: IR and Raman frequencies (cm^{-1})

# Ion	MP2/6-31G*[a]	B3LYP/6-31G*[b]	Experiment
1[c] NO_2^+	2381	2353	2375
	1207	1391	1400[d]
	544	603	600
2 NO_2H^{2+}	2339	2329	
	1196	2315	
	531	850	
	507	565	
		533	
5[e] NO^+	1966	2379	2377
6 HNO^{2+}	1941	2462	
	1459	1369	
	659	705	

[a] *scaled by a factor of 0.93;* [b] *scaled by a factor of 0.96;*
[c] *experimental data from ref 26;* [d] *Raman active;*
[e] *experimental data from ref 27.*

Experimental

The superacids $HSO_3F:SbF_5$ were prepared from freshly distilled HSO_3F and SbF_5. All manipulations using $HSO_3F:SbF_5$ were carried out in 5 mm pyrex NMR tubes. ^{17}O enriched water (20%) was commercially available. ^{17}O enriched nitronium ($N^{17}O_2^+$) and nitrosonium ($N^{17}O^+$) ions were prepared by reacting commerically available nitronium tetrafluoroborate ($NO_2^+BF_4^-$) and nitrosonium tetrafluoroborate ($NO^+BF_4^-$), respectively, with 20% ^{17}O enriched $H_2^{17}O$ in a 5 mm pyrex NMR tube. The resulting hydrolysis products were dissolved in excess fluorosulfonic acid (FSO_3H). ^{17}O NMR spectra of the solutions were obtained at ambient temperature on Varian Associates Model VXR-300 NMR spectrometer equipped with a 5 mm variable temperature broad band probe. The ^{17}O NMR chemical shifts were referenced to external capillary $H_2^{17}O$ signal ($\delta^{17}O$ 0.0 ppm).

Acknowledgment. Support of our work by the Office of Naval Research is gratefully acknowledged.

References

1. Onium Ions Part 41. For Part 40 see reference 14. Presented at the Symposium on Nitration, Industrial and Engineering Chemistry Division, Paper No 001, 209th ACS National Meeting, Anaheim, CA, April 1995.
2. Hantzsch, A. *Ber.* **1925**, *58*, 941; *Z. Phys. Chem.* **1930**, *149*, 161.
3. Goddard, D.R.; Hughes, E.D.; Ingold, C.J., *J. Chem. Soc.,* **1950** 2559.

4. Olah, G.A.; Kuhn, S. *J. Chem. Ind.* **1956**, *98*. Olah, G.A.; Kuhn, S.J.; Mlinko, A. *J. Chem. Soc.* **1956**, 4257.
5. Olah, G.A.; Malhotra, R.; Narang, S.C., Nitration, Methods and Mechanisms, VCH Publishers, New York, 1989.
6. Olah, G.A.; Lin, H.C. *Synthesis* **1974**, 444.
7. Olah, G.A.; Reddy, V.P.; Prakash, G.K.S., *Synthesis,* **1992**, 1087.
8. Olah, G.A.; Orlinkov, A.; Oxyzoglou, A.; Prakash, G.K.S. *J. Org. Chem.* submitted.
9. Olah, G.A.; Wang, Q.; Orlinkov, A.; Ramaiah, P., *J. Org. Chem.* **1993**, *58*, 5017.
10. Olah, G.A.; Germain, A.; Lin, H.C.; Forsyth, D. *J. Am. Chem. Soc.* **1975**, *97*, 2928.
11. (a) Olah, G.A.; Prakash, G.K.S.; Lammertsma, K. *Res. Chem. Intermed.* **1989**, *12*, 141. (b) Olah, G.A., *Angew. Chem. Int. Ed. Engl.*, **1993**, *32*, 767.
12. Olah, G. A.; Orlinkov, A.; Oxyzoglou, A.; Ramaiah, P.; Prakash, G. K. S., unpublished results.
13. Cremaschin, P.; Simonetta, M. *Theor. Chim. Acta.* **1974**, *34*, 175.
14. Olah, G. A.; Rasul, G.; Aniszfeld, R.; Prakash, G. K. S.; *J. Am. Chem. Soc.*, **1992**, *114*, 5608.
15. Weiske, T.; Koch, W.; Schwarz, H.; *J. Am. Chem. Soc.*, **1993**, *115*, 6312.
16. Schindler, M.,*J. Am. Chem. Soc.* **1987**, *109*, 5950.
17. (a) Olah, G. A.; Berrier, A. L.; Prakash, G. K. S.; *J. Am. Chem. Soc.*, **1982**, *104*, 2373. (b) Andersson, L. O.; Mason, J. *J. Chem. Soc., Dalton Trans.* **1974**, 2, 202. (c) Bogachev, Y. S.; Shapet'ko, N. N.; Gorelik, M. V.; Andrievskii, A. N.; Avidon, S. V.; Kisin, A. V.; Kuznetsova, M. G. *Russ. J. Gen. Chem.* **1993**, 63, 848.
18. Ziegler, T.; *Chem. Rev.*, **1991**, *91*, 651.
19. Rauhut, G.; Pulay, P.; *J. Phys. Chem.*, **1995**, *99*, 3093.
20. Curtiss, L. A.; Raghavachari, K.; Trucks, G. W.; Pople, J. A.; *J. Phys. Chem.*, **1991**, *94*, 7221.
21 Gauss, J.; *J. Chem. Phys. Lett.*;. **1992**, *191*, 614.; Gauss, J.; *J. Chem. Phys.*;. **1993**, *99*, 3629.
22. Gaussian 94 (Revision A.1), Frisch, M. J.; Trucks, G. W.; Schlegel, H. B.; Gill, P. M. W.; Johnson, B. G.; Robb, M. A.; Cheeseman, J. R.; Keith, T. A.; Peterson, G. A.; Montgomery, J. A.; Raghavachari, K.; Al-Laham, M. A.; Zakrzewski, V. G.; Ortiz, J. V.; Foresman, J. B.; Cioslowski, J.; Stefanov, B. B.; Nanayakkara, A.; Challacombe, M.; Peng, C. Y.; Ayala, P. Y.; Chen, W.; Wong, M. W.; Andres, J. L.; Replogle, E. S.; Gomperts, R ; Martin, R. L.; Fox, D. J.; Binkley, J. S.; Defrees, D. J.; Baker, J.; Stewart, J. J. P.; Head-Gordon, M.; Gonzalez, C.; Pople, J. A., Gaussian, Inc., Pittsburgh PA, 1995.
23. Stanton, J. F.; Gauss, J.; Watts, J. D.; Lauderdale, W.; Bartlett, R. J.; *ACES II, an ab initio program system*; University of Florida: Gainesville, FL, 1991.
24. Grison, E.; Eriks, K.; DeVries, J. L.; *Acta. Cryst.*, **1950**, *3*, 290.
25. Hohle, T.; Hijlhoff, F. C.; *Rec. Trav. Chim.*, **1967**, *86*, 1153.
26. Nebgen, J. W.; McElroy, A. D.; Klodowski, H. F.; *Inorg. Chem.*, **1965**, *4*, 1796.
27. Sharp, D. W. A.; Thorley, J.; *J. Chem. Soc.*, **1963**, 3557.

RECEIVED January 10, 1996

Chapter 3

Nitrocyclohexadienones in Aromatic Nitration

Robert G. Coombes[1], Panicos Hadjigeorgiou[2], Doron G. J. Jensen[1], and David L. Morris[2]

[1]Department of Chemistry, Brunel University, Uxbridge, Middlesex UB8 3PH, United Kingdom
[2]Department of Chemistry, City University, Northampton Square, London EC1V 0HB, United Kingdom

Nitrocyclohexadienones have been identified as intermediates formed in a range of nitration reactions in recent years. The mechanisms of their subsequent reactions have been studied and discussed. Published and other progress in these areas is reviewed. Some nitrocyclohexadienones containing electron-withdrawing substituents have been suggested as mild, selective and recyclable nitrating agents for phenols, naphthols and aromatic amines which do not involve problems of substrate oxidation. Studies of certain phenols involving the observation of chemically induced dynamic nuclear polarisation effects on reaction with ^{15}N labelled 2,3,5,6-tetrabromo-4-methyl-4-nitrocyclohexa-2,5-dien-1-one have established that nitro-products are formed by a radical process. Bromination may be a concurrent and sometimes predominant process. These studies have been extended, including, for example, aniline as a substrate. With N,N-dimethylaniline, however, a main product is N-methyl-N-nitrosoaniline. Mechanisms for these processes are presented.

This paper is largely concerned with the chemistry of 4-nitrocyclohexa-2,5-dien-1-ones of which (**A**), (in Figure 1), is a typical example. Compounds of this type which are of varying stability are easily prepared by direct nitration of the phenol or sometimes its acetate (*1*) and have also been identified as intermediates in nitration reactions under other conditions (*2-5*).

Catalysed and Uncatalysed Rearrangements of Nitrocyclohexadienones

4-Methyl-4-nitrocyclohexa-2,5-dien-1-one, **A**, rearranges to 4-methyl-2-nitrophenol, **B**, in a range of solvents (*1*) and at a rate which increases with acidity (*2-5*). The behaviour (*2-4*) shown in Figure 2 for **A** is typical. The uncatalysed reaction, evident at the lowest acidities and in other solvents (*1*) is supplanted by an acid-catalysed process as the acidity is increased in aqueous sulfuric acid. The slope of the log k vs -H_o plot is around 0.7. This slope is close to that applicable to the protonation equilibria of other dienones of this type (*6*) and suggests a rate-limiting reaction of the protonated dienone that is formed reversibly and in a small amount. Figure 2 also gives two new examples of this general behaviour: the reactions of 4-ethyl- and 2,4-dimethyl- 4-nitrocyclohexa-2,5-dien-1-ones. The mechanism of the uncatalysed reaction was originally established by Barnes and Myhre (*1*) as involving homolytic

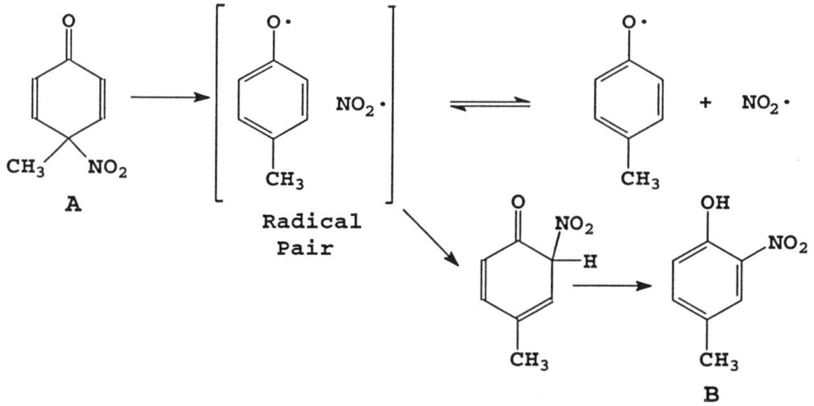

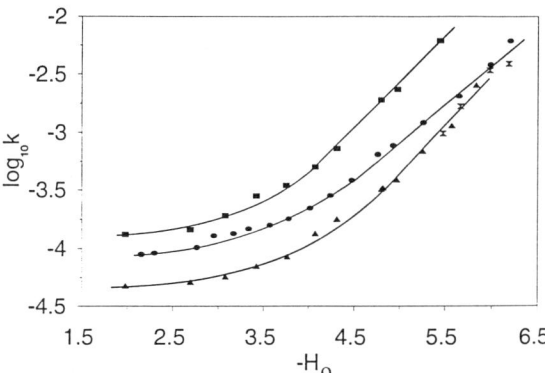

Figure 1. Mechanism for the uncatalysed decomposition of 4-methyl-4-nitrocyclohexa-2,5-dien-1-one (A).

Figure 2. Decomposition of dienones in aqueous sulphuric acid at 25°C.
● *(data from ref. 4) and* ✘ *(data from refs. 2 & 3) for 4-methyl-4-nitrocyclohexa-2,5-dien-1-one (A);* ▲ *and* ■ *for 4-ethyl- and 2,4-dimethyl- 4-nitrocyclohexa-2,5-dien-1-ones respectively.*

fission of the C-N bond and a radical pair intermediate from which escape is possible (Figure 1). This mechanism has more recently been confirmed using the technique of observation of chemically induced dynamic nuclear polarisation (CIDNP) in the ^{15}N nmr spectra of reactions involving ^{15}N labelled dienone performed in an nmr machine (7,8). This polarisation can lead to enhanced emission or absorption signals in the ^{15}N spectra of the species involved and is particularly easy to observe because of the long spin-lattice relaxation time of ^{15}N in the ^{15}NO$_2$ group (about 3 minutes). The sign of the effect in a particular species concerned in a reaction involving a radical pair intermediate, from which escape is possible, can be predicted by modified Kaptein's rules (9,10). A simplified qualitative rationalisation (11) is appropriate here and its application can be illustrated by reference to the mechanism of Figure 1.

In order to form the product the radical pair must be in the singlet state. The rate of singlet-triplet interconversion depends on the coupling of electronic and nuclear spins and, in the type of pair concerned, it is the upper spin state of the ^{15}N nucleus which facilitates interconversion. From the dienone, **A**, the pair is formed as a singlet by homolytic fission of the C-N bond. The upper spin ^{15}N nuclear state facilitates conversion into the triplet and, if escape from the cage occurs, it is the triplet species, which cannot react, which will be more likely to escape. The escaping species will therefore be enriched in the upper ^{15}N nuclear spin state and it follows that any return to the dienone or product formation within the cage will be associated with ^{15}N which is enriched in the lower ^{15}N nuclear spin state and therefore will give enhanced absorption signals for the ^{15}NO$_2$ groups. This is precisely the effect observed by Ridd, Sandall and Trevellick (7,8). There were enhanced absorption signals both in the dienone, **A**, formed by return from the pair and in the 4-methyl-2-nitrophenol product, **B**, in ^{15}N nmr spectra taken during reaction.

In a general case concerning a radical pair of the type concerned here, from which escape is possible, if the pair is formed by diffusion together of radicals with uncorrelated spins, conversion into the singlet pair necessary for reaction is facilitated by the upper ^{15}N nuclear spin state, leading to enhanced emission in the products formed within the radical pair. Examples of this type of behaviour will be discussed later.

Ridd, Sandall and Trevellick (7,8) also studied the ^{15}N nuclear polarisation during the acid-catalysed reaction and demonstrated a similar and indeed slightly more pronounced pattern of enhanced absorption consistent with the presence of a radical cation formed by homolytic fission of the C-N bond in the protonated cyclohexadienone on the reaction pathway. A similar conclusion had been reached previously on other grounds by Myhre (12) from studies of the acid-catalysed reaction of 2,4,6-trimethyl-4-nitrocyclohexa-2,5-dien-1-one [**C**, in Figure 3].

Our own studies of compound **C** are relevant here (4). The rate profile for reaction of this dienone is shown in Figure 4 establishing the occurrence of an uncatalysed and an acid-catalysed reaction, similar to the behaviour of the other dienones where both the *ortho* positions are not blocked. The product of the uncatalysed reaction studied in an organic solvent is 2,6-dimethyl-4-nitromethylphenol. The major product (about 50%) from the acid-catalysed process is 3,5-dimethyl-4-hydroxybenzaldehyde, **D**, accompanied by about 15% 2,4,6-trimethyl-3-nitrophenol. 2,6-Dimethyl-4-nitromethylphenol undergoes an acid-catalysed conversion into compound **D** in aqueous sulfuric acid (Figure 4), but the rate of this process is not sufficiently large for the former to be involved as an intermediate in the acid-catalysed reaction of **C** (13). The products of both the uncatalysed and acid-catalysed reactions are consistent with homolytic fission of the C-N bond in the protonated dienone and Figure 3 suggests a plausible mechanism for the acid-catalysed reaction.

More recent work by Ridd, Trevellick and Sandall (14) has suggested initial homolytic fission of the C-N bond in the uncatalysed and acid-catalysed regiospecific 6,2 rearrangement of related 6-nitrocyclohexa-2,4-diene-1-ones, even in the absence of observed polarisation in ^{15}N nmr studies during reaction.

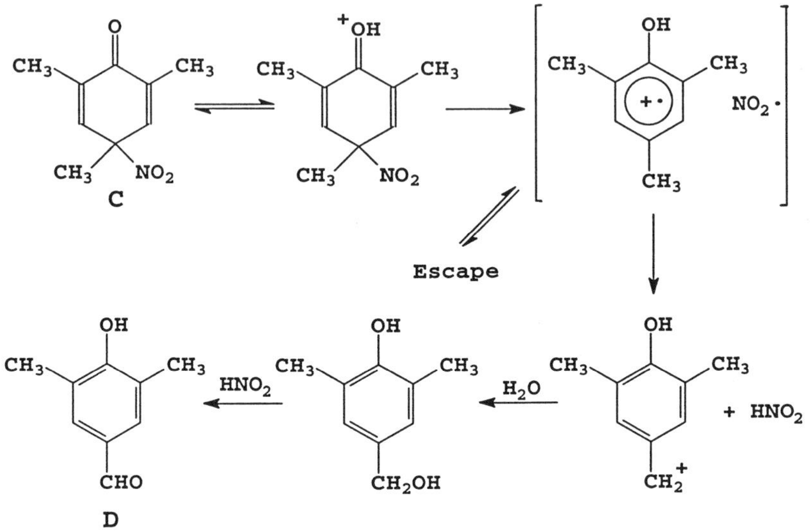

*Figure 3. Mechanism for the acid-catalysed decomposition of 2,4,6-trimethyl-4-nitrocyclohexa-2,5-dien-1-one (**C**).*

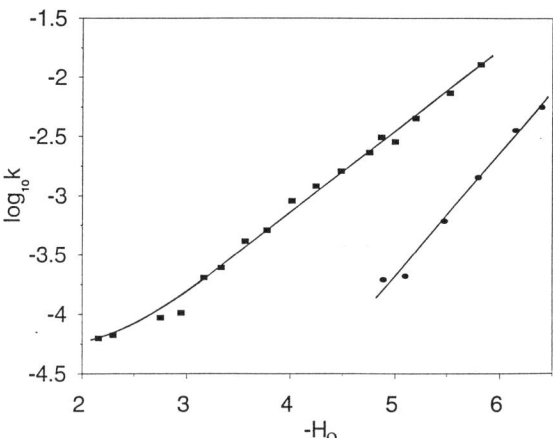

*Figure 4. Decomposition of 2,4,6-trimethyl-4-nitrocyclohexa-2,5-dien-1-one (**C**) ■ (data partly from ref. 4) and 2,6-dimethyl-4-nitromethylphenol ● (data from ref. 13) in aqueous sulphuric acid at 25°C.*

Nitrocyclohexadienones as Nitrating agents

Our interest in the chemistry of cyclohexadienones has been reawakened by the work of Lemaire and co-workers (15-17) who have suggested the use of some dienones as mild nitrating agents. Several dienones have been used including the 2,3,5,6-tetrabromo-4-methyl-4-nitrocyclohexa-2,5-dien-1-one (**E** in Figure 5). The dienones were first used to nitrate some naphthols and a phenol at room temperature in inert solvents, typically ether, avoiding problems of oxidation (15,16). For example, 1-naphthol was converted into 43% 2-nitro-1-naphthol and 57% 4-nitro-1-naphthol in a 66% overall yield. This can be compared with a reported nitration by nitric acid in acetic acid forming 8% 2-nitro-1-naphthol, 0% 4-nitro-1-naphthol and 4% 2,4-dinitro-1-naphthol with 87% oxidation products. The dienone is converted into the corresponding phenol and the suggestion has been made that this can be recycled, nitration of 2,3,5,6-tetrabromo-4-methylphenol, for example, being the method of synthesis of **E**.

Kashmiri and co-workers have also studied the reaction with a range of simple phenols and aromatic amines finding some selectivity in some systems. Phenol, for example, was reported to form 25% 2-nitrophenol and 75% 4-nitrophenol in 72% overall yield and aniline to form similar proportions of 2- and 4- nitroanilines in 73% overall yield (18,19). Lemaire and co-workers have also studied aromatic amines (17), and in particular intriguingly reported that N,N-dimethylaniline is converted by **E** in ether into N-methyl-N-nitroaniline in a 78% yield, whereas reaction with 2,3,4,5,6-pentafluoro-4-nitrocyclohexa-2,5-dien-1-one in ethanol gave 80% ring nitration. Lemaire and co-workers suggested that the dienone reagent acts by releasing a nitronium ion as the reactive species. Kashmiri and Khan suggested (19) the formation of a hydrogen-bonded complex involving bonding of the hydroxy-H or an amino-H to the dienone carbonyl group, the nitro group being transferred within this complex. The latter explanation cannot, however, cover the albeit different reactivity of N,N-dimethylaniline. These mechanistic ideas seemed at variance with the chemistry established in the previous section for nitrocyclohexadienones involving C-N bond homolysis. It is, however, possible to argue that the presence of the four electron-withdrawing bromine atoms in **E** facilitates heterolytic fission to give the nitronium ion.

Reactions of 2,3,5,6-Tetrabromo-4-methyl-4-nitrocyclohexa-2,5-dien-1-one, E, with Phenols

Nitration. In order to investigate the mechanism of action of **E**, we have studied systems of this type using the CIDNP ^{15}N nmr technique with ^{15}N labelled **E** (20). Equimolar amounts of **E** nitrate phenol rapidly at room temperature to form 60-80% yields of the nitrophenols (56% 2-, 44% 4-) and 2,3,5,6-tetrabromo-4-methylphenol. The reaction was then studied in the nmr spectrometer. The dienone, **E**, initially exhibited enhanced absorption and the 2- and 4-nitrophenols strong emission signals (>200 fold enhanced at maximum). The former signal dies away faster than the latter two and eventually the product absorption signals alone are present.

These results can be understood in terms of Figure 5. The enhanced absorption in the dienone corresponds to that we have already seen from the work of Ridd, Sandall and Trevellick (7,8) and derives from return from the radical pair where the upper ^{15}N nuclear state has facilitated escape. However, if the enhanced emission in the nitrophenols derived solely from such escape then it should die away as quickly as does the enhanced absorption. The fact that this does not happen is attributed to the presence of a second radical pair on the reaction pathway. This pair is formed by diffusion together of the phenoxy radical and $NO_2^{\bullet}$ and, as discussed above, products resulting from reaction within such a radical pair should exhibit enhanced emission in the ^{15}N signals associated with the nitro groups. The ratio of the peaks for the 2- and 4-

nitrophenols is similar to the product ratio so there is no reason to invoke any significant additional pathway. In our hands the product isomer ratio was similar to that attributed elsewhere to the reaction of the phenoxy radical with $NO_2^{\bullet}$ (*21-24*). We conclude that homolytic fission of the C-N bond is involved in the reactions of this dienone.

Concurrent Bromination. However, the situation is more complicated than the above results reveal (*25*). Although solvent effects on these reactions have been reported to be small we find that if we carry out the reaction on phenol in acetone the major product becomes 4-bromophenol (40%) and 4-bromo-2-nitrophenol (~4%) with 2,3,5-tribromo-4-methyl-6-nitrophenol (75%), and 2-nitrophenol (~5%) and 4-nitrophenol (~7%) with 2,3,5,6-tetrabromo-4-methylphenol (25%). A series of ^{15}N nmr spectra taken during a reaction again show that the dienone exhibits enhanced absorption and the nitrophenols enhanced emission. There is another peak showing enhanced absorption which is assigned to the nitro group of 2,3,5-tribromo-4-methyl-6-nitrophenol (**F** in Figure 6) and a peak which appears somewhat belatedly in enhanced emission assigned to 4-bromo-2-nitrophenol formed by nitration of 4-bromophenol formed in the reaction. The nitration products exhibit polarisation characteristic of the mechanism outlined above (Figure 5), but here the major process is bromination forming 4-bromophenol and concomitantly **F**, the latter exhibiting enhanced absorption. The bromination results are consistent with Figure 6. Recombination within the radical pair gives 2,3,5,6-tetrabromo-4-methyl-6-nitrocyclohexa-2,4-dien-1-one which acts as a brominating agent and the polarisation of ^{15}N here is carried through to **F**.

New studies of other substrates have confirmed the conclusion that the dienone, **E**, can act both as a nitrating agent and, after rearrangement, as a brominating agent, the balance between the pathways depending on the substrate and the solvent. A result for a substituted phenol is that for reaction of **E** with 4-methylphenol in acetone. Here the majority reaction is nitration with ~35% bromination. The nitration pathway measured in chloroform gives ~60% 4-methyl-2-nitrophenol, **B**, and ~40% 4-methyl-4-nitrocyclohexa-2,5-dien-1-one, **A**. With ^{15}N labelled **E**, the ^{15}N spectra (Figure 7 for results from acetone solution) indicate that the two nitro-products are formed in emission and that **E** is present, and **F** is formed, in enhanced absorption. Spectra were taken (a) 3-8 min. after mixing, (b) 8-13 min. after mixing, (c) 13-18 min. after mixing and (d) 78-83 min. after mixing. (**S**) is the peak from a reference amount of $Ph^{15}NO_2$. These ^1H-coupled spectra involved 24 pulses, 10s pulse repetition time and pulse width $20\mu s$.

A consistent pattern of behaviour and a reasonable mechanistic explanation has therefore been established for the reaction of **E** with phenols. The reaction with aromatic amines remains to be discussed.

Reaction of 2,3,5,6-Tetrabromo-4-methyl-4-nitrocyclohexa-2,5-dien-1-one, E, with Aniline and with *N,N*-Dimethylaniline

With aniline, reactions in both chloroform and acetone give a nitration to bromination ratio of around 1:1. Only about 50% of the aniline conversion products have been quantitatively identified as ~12% 2-nitroaniline, ~10% 4-nitroaniline and ~27% 4-bromoaniline from reaction in chloroform, although it is clear from ^{15}N nmr spectra that *N*-nitroaniline is another product, its amount depending on solvent. The ^{15}N spectra during reaction are similar in general to those from the phenol reactions (Figure 8 for reaction in chloroform). The dienone **E** and the 2,3,5-tribromo-4-methyl-6-nitrophenol **F** are seen in enhanced absorption and 2- and 4-nitroaniline, **G** and **H**, and *N*-nitroaniline, **I**, give enhanced emission signals. Spectra were taken (a) before addition of aniline, (b) 5-9 min. after mixing, (c) 10-13 min. after addition and (d) at the end of the reaction (solid had formed). Details of (**S**) and the spectrometer conditions are as above except that (d) involved 4402 pulses. Figure 9 outlines a

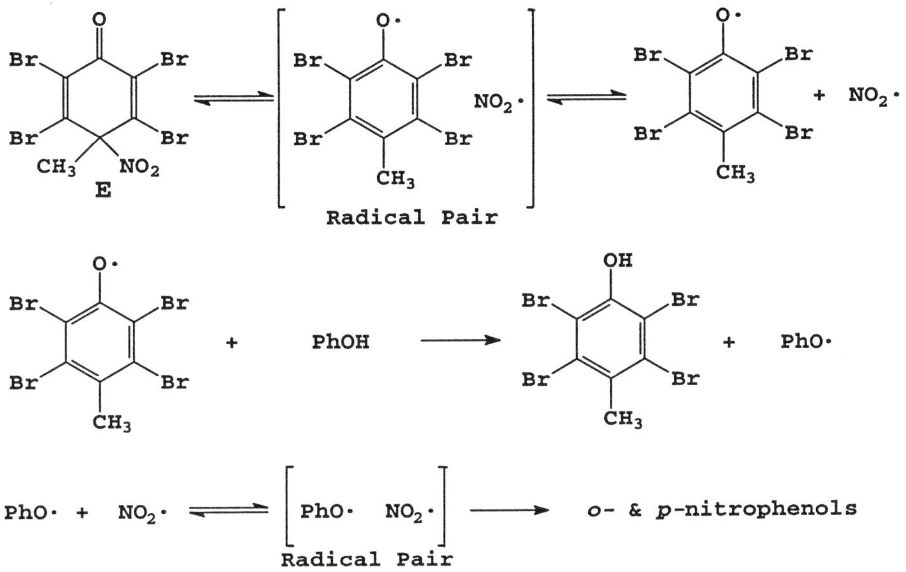

*Figure 5. Mechanism for the nitration of phenol by 2,3,5,6-tetrabromo-4-methyl-4-nitrocyclohexa-2,5-dien-1-one (**E**).*

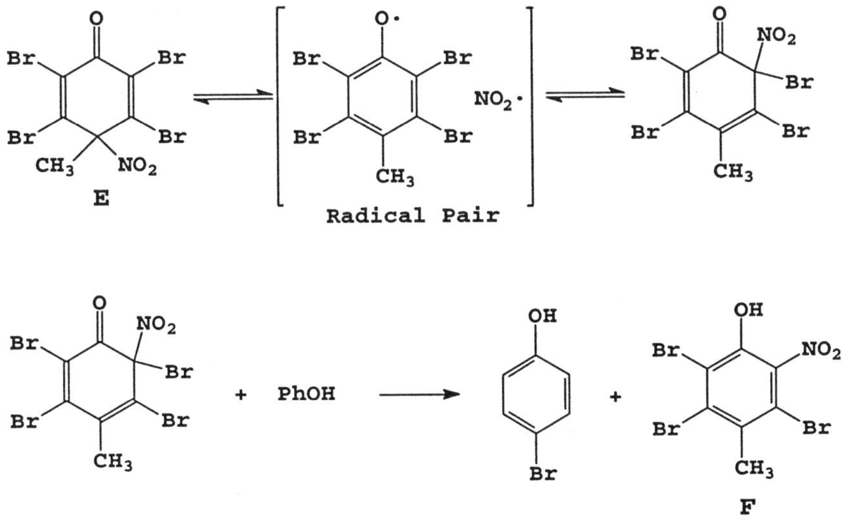

*Figure 6. Mechanism for the bromination of phenol by 2,3,5,6-tetrabromo-4-methyl-4-nitrocyclohexa-2,5-dien-1-one (**E**).*

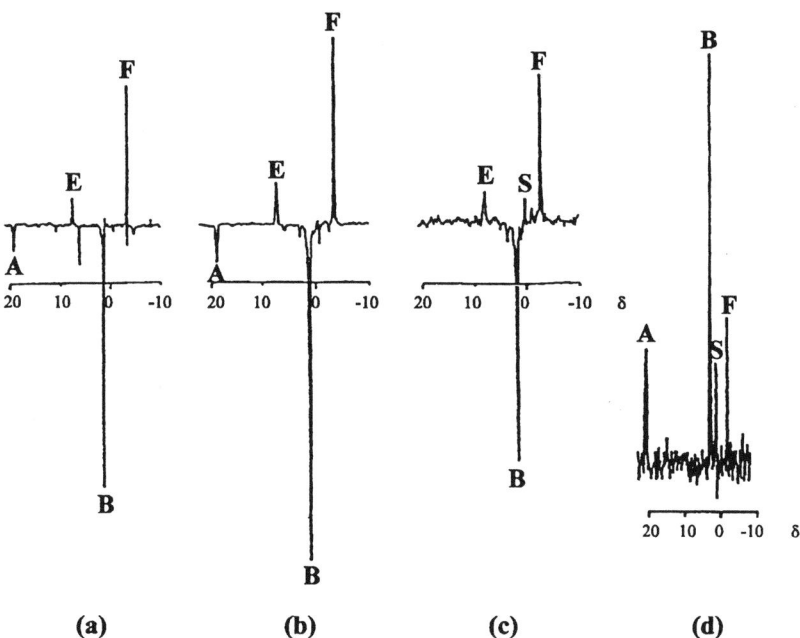

Figure 7. ^{15}N *Nmr spectra during reaction of 2,3,5,6-tetrabromo-4-methyl-4-nitrocyclohexa-2,5-dien-1-one (E) with 4-methylphenol, (a) 3-8 min. after mixing, (b) 8-13 min. after mixing, (c) 13-18 min. after mixing and (d) 78-83 min. after mixing.*

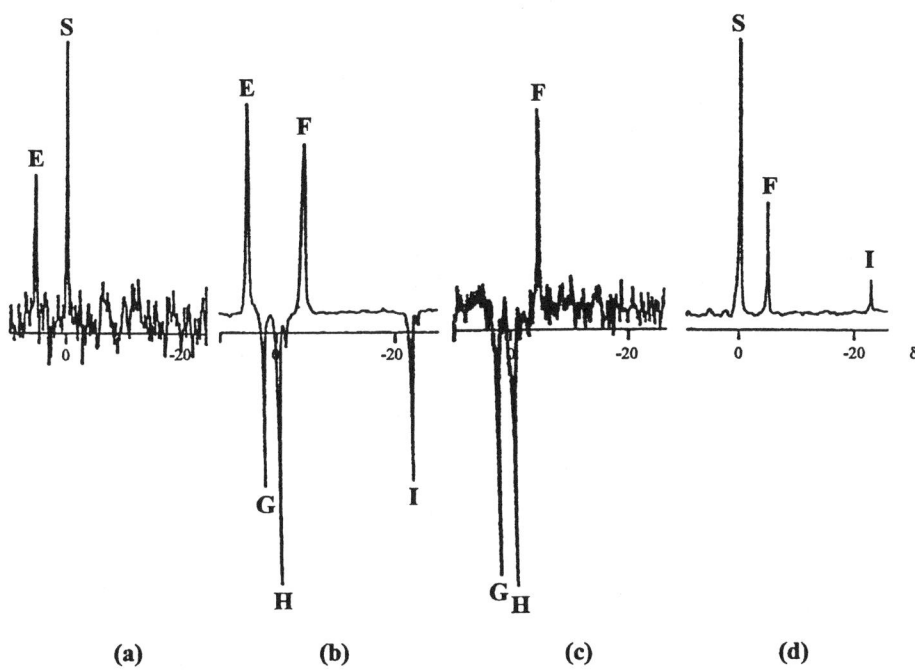

Figure 8. ^{15}N *Nmr spectra during reaction of 2,3,5,6-tetrabromo-4-methyl-4-nitrocyclohexa-2,5-dien-1-one* **(E)** *with aniline, (a) before addition of aniline, (b) 5-9 min. after mixing, (c) 10-13 min. after addition and (d) at the end of the reaction (solid had formed).*

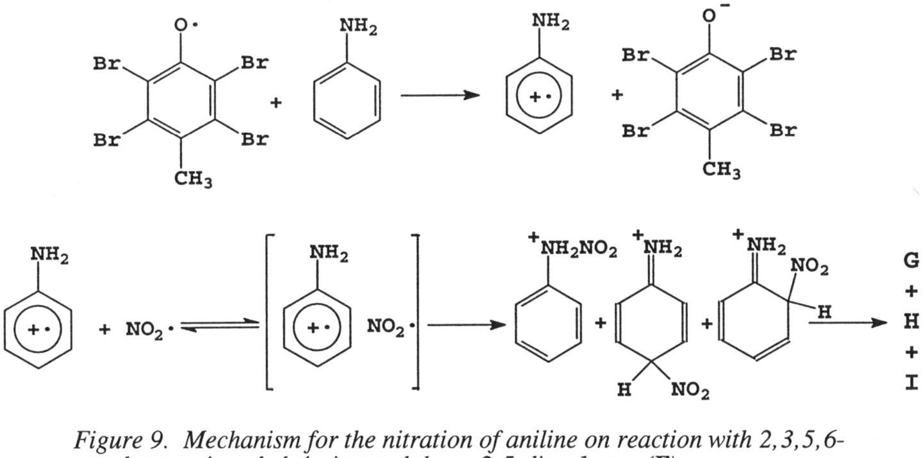

Figure 9. Mechanism for the nitration of aniline on reaction with 2,3,5,6-tetrabromo-4-methyl-4-nitrocyclohexa-2,5-dien-1-one **(E)**.

mechanism for the nitration processes, to follow the initial homolytic fission of the C-N bond of **E**, that is consistent with the polarisation results and involves the intermediacy of the aniline radical cation formed by electron transfer from aniline to the 2,3,5,6-tetrabromo-4-methylphenoxy radical. The nitrous acid catalysed nitration of anilines which involves diffusion together of amine radical cations and $NO_2^{\bullet}$ is known to give nitro-products in enhanced emission (26). Bromination is presumed to occur as outlined for the phenols (Figure 6).

In our hands the reaction of **E** with *N,N*-dimethylaniline in ether, acetone and chloroform gave mainly *N*-methyl-*N*-nitrosoaniline rather than the reported (17) *N*-methyl-*N*-nitroaniline. Bromination comprised ~25-37% of reaction. A reaction in acetone gave the ~60% of product quantitatively identified as ~40% *N*-methyl-*N*-nitrosoaniline, together with some *N,N*-dimethyl-2-nitroaniline (**J**) and 4-bromo-*N,N*-dimethylaniline. The ^{15}N nmr spectra during reaction show the expected pattern (Figure 10). The signals for **E** and 2,3,5-tribromo-4-methyl-6-nitrophenol, **F**, show enhanced absorption and the signals assigned to the nitro-products [*N,N*-dimethyl-2-nitroaniline, **J**, a tentative assignment, and the 4-nitro-isomer, **K**] are enhanced emissions. Spectra were taken (a) before addition of *N,N*-dimethylaniline, (b) 8-12 min. after mixing, (c) 19-23 min. after mixing and (d) at the end of the reaction. Details of (**S**) and the spectrometer conditions are as above, except that (d) involved 3656 pulses. No enhancement was detected in the peak assigned to the ^{15}N labelled *N*-nitroso group. This may be due to its relatively short relaxation time (~14s) and may not have mechanistic significance.

The nitration and bromination reactions are believed to occur by the pathways outlined above. A speculative but reasonable mechanism for the formation of the *N*-nitroso compound is given in Figure 11. The proposed pathway involves proton abstraction from the radical cation by the 2,3,5,6-tetrabromo-4-methylphenoxide ion giving a pathway not normally open to the former species. Other differences from the behaviour elsewhere of the radical cation (27) are attributed to the effect of solvent. The species represented as 'CH_2O' has not been identified.

The concurrent bromination again presumably proceeds by a mechanism analogous to that suggested for the phenols (Figure 6).

Although the presence of the polarisation establishes the presence of the radical pathways which can be reasonably interpreted as described for the reactions with the aromatic amines, concurrent pathways [radical or otherwise - see (28)] cannot be ruled out by this evidence. There is, however, no evidence to suggest that such other pathways are of importance.

Conclusion

The extension of the studies of reactions of 2,3,5,6-tetrabromo-4-methyl-4-nitrocyclohexa-2,5-dien-1-one, **E**, to reactions with aromatic amines confirms that homoytic fission of the carbon-nitrogen bond is involved as seems quite general for the reactions of nitrocyclohexadienones reported so far. The complications revealed by studies of **E** indicate, however, that the polyhalogenonitrocyclohexadienones may be less generally useful as nitrating agents than originally appeared.

Acknowledgments

Grateful thanks are due to Professor A. G. Davies, F.R.S., Dr. B. P. Roberts, Dr. J. P. B. Sandall and particularly to Professor J. H. Ridd for discussions and advice. Much of the latter part of the work was carried out by R.G.C. during a sabbatical period at University College London.

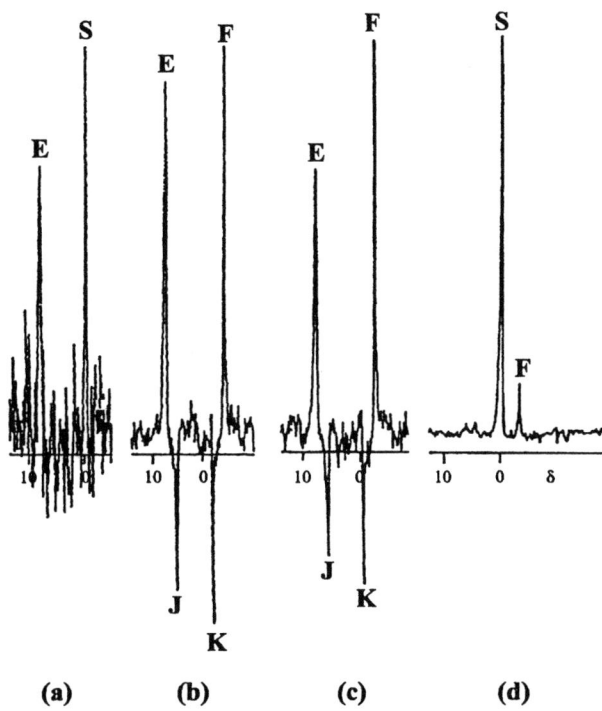

Figure 10. ^{15}N *Nmr spectra during reaction of 2,3,5,6-tetrabromo-4-methyl-4-nitrocyclohexa-2,5-dien-1-one* **(E)** *with N,N-dimethylaniline, (a) before addition of N,N-dimethylaniline, (b) 8-12 min. after mixing, (c) 19-23 min. after mixing and (d) at the end of the reaction.*

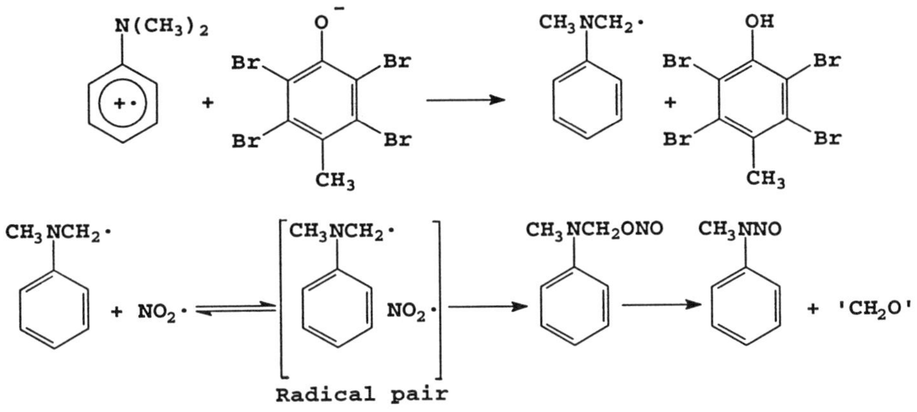

Figure 11. *Speculative pathway for the formation of N-methyl-N-nitrosoaniline from the reaction of N,N-dimethylaniline and 2,3,5,6-tetrabromo-4-methyl-4-nitrocyclohexa-2,5-dien-1-one* **(E)**.

Literature Cited

(1) Barnes, C. E.; Myhre, P. C. *J. Am. Chem. Soc.* **1978**, *100*, 973.
(2) Coombes, R. G.; Golding, J. G. *Tetrahedron Lett.* **1978**, 3583.
(3) Coombes, R. G.; Golding, J. G.; Hadjigeorgiou, P. *J. Chem. Soc. Perkin Trans. 2* **1979**, 1451.
(4) Hadjigeorgiou, P. *Ph.D. Thesis* **1979**, City University, London.
(5) Bloomfield, C.; Manglik, A. K.; Moodie, R. B.; Schofield, K.; Tobin, G. D. *J. Chem. Soc. Perkin Trans. 2* **1983**, 75.
(6) Vitullo, V. P. *J. Org. Chem.* **1970**, *35*, 3976.
(7) Ridd, J. H.; Sandall, J. P. B.; Trevellick, S. *J. Chem. Soc., Chem. Commun.* **1988**, 1195.
(8) Ridd, J. H.; Trevellick, S.; Sandall, J. P. B. *J. Chem. Soc. Perkin Trans. 2* **1992**, 1535.
(9) Kaptein, R. *J. Chem. Soc., Chem. Commun.* **1971**, 732.
(10) Porter, N. A.; Dubay, G. R.; Green, J. G. *J. Am. Chem. Soc.* **1978**, *100*, 920.
(11) Clemens, A. H.; Ridd, J. H.; Sandall, J. P. B. *J. Chem. Soc., Perkin Trans. 2* **1984**, 1659.
(12) Myhre, P. C. *Report* **1985**, ARO-17572.1-CH-H; cf. *Gov. Rep. Announce Index (U.S.)* **1985**, 85(24), Abstr. No. 555-257.
(13) Morris, D. L. *M. Phil. Thesis* **1995**, City University, London.
(14) Ridd, J H.; Trevellick, S.; Sandall, J. P. B. *J. Chem. Soc. Perkin Trans. 2* **1993**, 1073.
(15) Roussel, J.; Lemaire, A.; Guy, A.; Guetté, J. P. *Tetrahedron Lett.* **1986**, *27*, 27.
(16) Lemaire, M.; Guy, A.; Roussel, J.; Guetté, J. P. *Tetrahedron* **1987**, *43*, 835.
(17) Lemaire, A.; Guy, A.; Boutin, P.; Guetté, J. P. *Synthesis* **1989**, 761.
(18) Kashmiri, M. A.; Munawar, M. A.; Yasmin, R; Khan, M. S. *J. Natl. Sci. Math.* **1988**, *28*, 289.
(19) Kashmiri, M. A.; Khan, M. S. *Sci. Int. (Lahore)* **1989**, *1*, 177.
(20) Coombes, R. G.; Ridd, J. H. *J. Chem. Soc., Chem. Commun.* **1992**, 174.
(21) Al-Obaidi, U.; Moodie, R. B.; *J. Chem. Soc. Perkin Trans. 2* **1985**, 467.
(22) Coombes, R. G.; Diggle, A. W.; Kempsell, S. P. *Tetrahedron Lett.* **1994**, *35*, 6373.
(23) Thompson, M. J.; Zeegers, P. J. *Tetrahedron Lett.* **1988**, *29*, 2471.
(24) Thompson, M. J.; Zeegers, P. J. *Tetrahedron* **1989**, *45*, 191.
(25) Coombes, R. G. *J. Chem. Soc. Perkin Trans. 2* **1992**, 1007.
(26) Ridd, J. H.; Sandall, J. P. B. *J. Chem. Soc., Chem. Commun.* **1981**, 402.
(27) Giffney, J. C.; Ridd, J. H. *J. Chem. Soc. Perkin Trans. 2* **1979**, 618.
(28) Chatterjee, J.; Coombes, R. G.; Barnes, J. R.; Fildes, M. J. *J. Chem. Soc. Perkin Trans. 2* **1995**, 1031.

RECEIVED February 7, 1996

Chapter 4

Aromatic Nitration in Liquid Nitrogen Tetroxide Promoted by Metal Acetylacetonates

Ripudaman Malhotra and David S. Ross

SRI International, 333 Ravenswood Avenue, Menlo Park, CA 94025

Aromatics such as benzene and toluene are not nitrated by liquid N_2O_4 at 0°C to any significant extent. However, simultaneous passage of NO and O_2 through a solution of benzene in liquid N_2O_4 leads to nitration of benzene giving mono-, di-, and even trinitrobenzenes. Prompted by the possibility that one-electron oxidation of arenes followed by their reaction with NO_2 might lead to these nitrations, we examined the reaction of benzene and toluene in liquid N_2O_4 with various transition metal oxidants. Because most simple salts such as acetates and nitrates are too ionic, they could not be dissolved in liquid N_2O_4. However, inner complexes, in which the ionic and coordinating valencies are simultaneously satisfied, are soluble in this medium, and we were able to conduct a study with a range of acetylacetonate complexes. In concert with our hypothesis, nitration was readily effected by the oxidizing acetylacetonates of Fe(III), Ce(IV), Co(III), Mn (III) and Cu(II). However, nitrations were also effected by the nonoxidizing acetylacetonates of Fe(II) and Li.

By itself, N_2O_4 is sufficiently reactive to effect nitration of activated aromatic substrates such as phenol and anisole (1) as well as of polycyclic arenes such as pyrene and perylene (2). On the other hand, simple arenes such as benzene and toluene are not nitrated by liquid N_2O_4 to any significant extent. However, N_2O_4 can be activated by several means to effect the nitration of benzene, toluene, and other unactivated arenes. Suzuki and coworkers have reported on the use of ozone and N_2O_4 to nitrate benzene and toluene (3). We have previously shown that simultaneous passage of NO and O_2 through a solution of benzene in liquid N_2O_4 leads to nitration of benzene giving mono-, di-, and even trinitrobenzenes (4) In this paper we describe the results of nitrations promoted by metal acetylacetonates in liquid N_2O_4.

Our study with metal acetylacetonates was a direct offshoot of our investigations with the $NO/O_2/N_2O_4$ system and therefore it would be useful to review some of those results first. Figure 1 is a schematic of the apparatus used in that study. The reactor is designed to admit controlled flows of NO and O_2, as well as to provide for sampling during the course of a run. Typically 20-40 mL of the nitrogen tetroxide is introduced into the cooled 100-mL reactor. The tetroxide recovered from the storage tank is generally dark green due to the presence of lower oxides of nitrogen. It is stirred under an atmosphere of O_2 until the liquid turns a straw-yellow color. Figure 2 illustrates what happens when benzene (or toluene) is stirred in liquid N_2O_4 at 0°C. In a control run, when only N_2 was bubbled through the solution, no conversion to nitrobenzene was observed over a period of several hours. Passage of O_2 through the solution results in a small degree of conversion. However, when NO and O_2 were simultaneously passed, substantial quantities of nitrobenzene (or nitrotoluene) along with dinitrobenzenes (or dinitrotoluenes) were formed. Curiously enough, nitrobenzene itself does not form dinitrobenzene in this system. Thus, the dinitrobenzene forms either directly from benzene or from an intermediate on the path to nitrobenzene. Moreover, the isomer ratio of the dinitrobenzenes was very different from that observed in conventional nitration systems. The extent of *O*- and *P*-dinitration is much larger in the $NO/O_2/N_2O_4$ system: the o:m:p ratio was 13:56:31 in marked contrast to 6:92:2 in mixed acids.

A speculative mechanism for this reaction is depicted in the scheme below. The initial reaction of NO and O_2 produces an unsymmetrical NO_3 species, which in the presence of excess NO_2 radicals leads to the symmetrical NO_3. NO_3 is recognized to be a very strong one-electron oxidant and could react with the arene to give the radical cation, which in turn could react with NO_2 to give the nitrocyclohexadienyl cation, the Wheland intermediate. Alternatively, as shown in the scheme, N_2O_4 could add to the radical cation and ultimately lead to polynitrated products.

Results and Discussion

Prompted by the possibility that one-electron oxidation of arenes followed by their reaction with NO_2 might lead to the observed nitrations, we examined the reaction of benzene and toluene in liquid N_2O_4 with various transition metal oxidants. Because most simple salts such as acetates and nitrates are very ionic, they could not be dissolved in liquid N_2O_4, which has a very low dielectric constant of 2.42. However, inner complexes, in which the ionic and coordinating valencies are simultaneously satisfied, are soluble in this medium, and we were able to conduct a study with a range of acetylacetonate complexes.

Fast and Slow Modes of Nitration. We studied the effect of acetylacetonates of Co(III) and Fe(III) on the nitration of benzene and toluene in liquid N_2O_4. These runs were conducted by dissolving about 10 mmol of the acetylacetonate

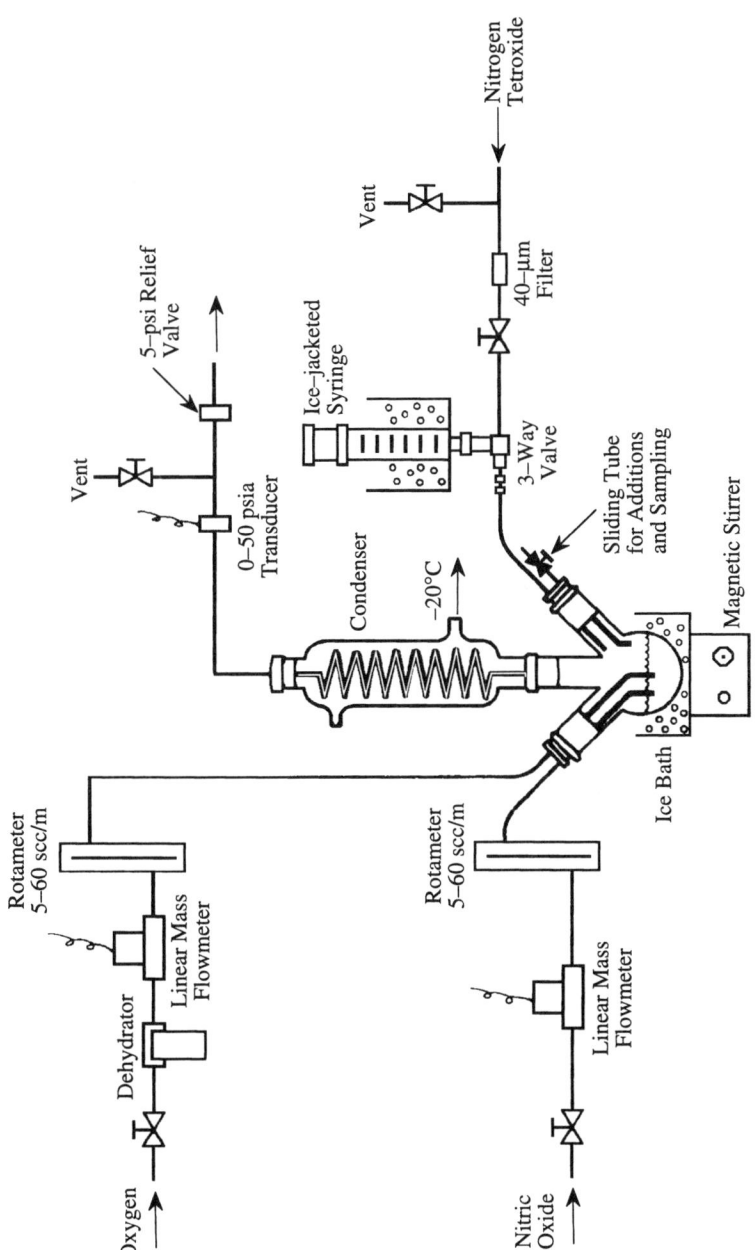

Figure 1. Apparatus for nitration of aromatic hydrocarbons in liquid N_2O_4 with NO/O_2.

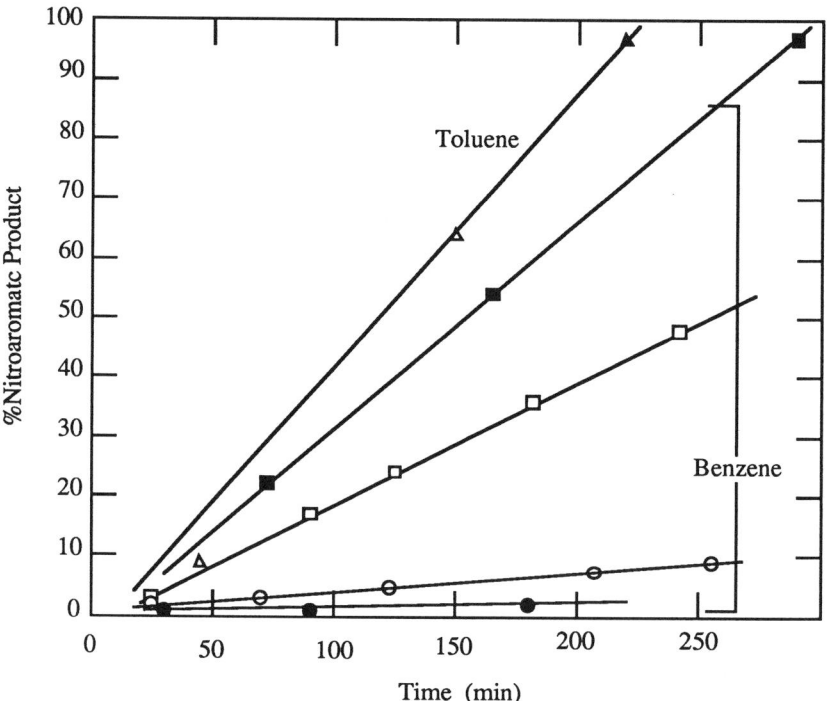

Figure 2. Nitration of benzene and toluene in liquid N_2O_4 at 0°C.
Fifty mmol aromatic substrate was used in each run.
- ● benzene, N_2 (0.22 mmol/min)
- ○ benzene, O_2 (0.32 mmol/min)
- □ benzene, O_2 (0.30 mmol/min), NO (0.17 mmol/min)
- ■ benzene, O_2 (0.49 mmol/min), NO (0.32 mmol/min)
- △ toluene, O_2 (0.62 mmol/min), NO (0.40 mmol/min)

$$NO + O_2 \longrightarrow \longrightarrow NO_3$$
$$NO_3 + ArH \longrightarrow NO_3^- + ArH^{+\bullet}$$

Reaction of arene radical cation with NO_2 and N_2O_4 then provides an array of products:

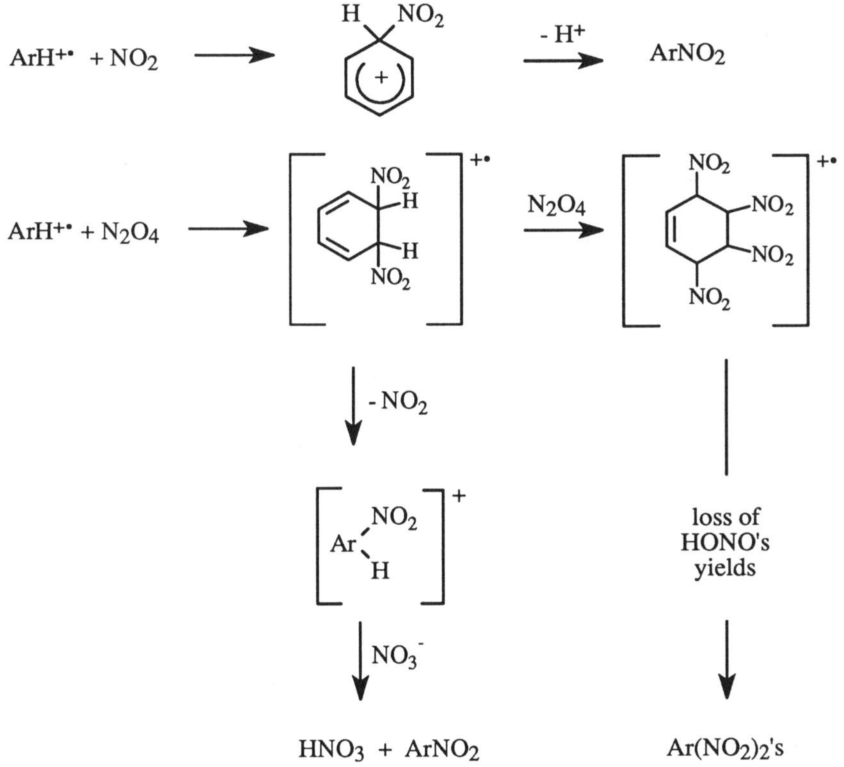

Scheme 1. One-electron oxidation mechanism for nitrations in liquid N_2O_4

(acac) in 20 mL of N_2O_4.* Following the addition of about 50 mmol of the arene, production of nitroarenes was observed to proceed over several hours. Subsequently we observed that if the arene is first dissolved in the liquid N_2O_4, addition of the metal acac results in a rapid nitration, and the reaction is essentially complete in about 15 min. We refer to the former and the latter procedures as the "slow" and "fast" modes. In the "slow" mode, we did not observe any dinitration. In the "fast" mode, some dinitration (~5%) was observed, which was completely suppressed in the presence of a cosolvent such as nitromethane. A significant result was the absence of any phenolic byproduct (< 0.4%, the detection limit).

The results of metal acac promoted nitrations are summarized in Table 1. The progress of reaction in the "fast" mode for several cases is shown in Figure 3. Rapid nitration ensued upon addition of the metal acac to the solution of benzene in liquid N_2O_4 at 0°C and then ceased in about 15 min. although a large excess of benzene was still present. In concert with our hypothesis, nitration was readily effected by the *oxidizing* acetylacetonates of Fe(III), Ce(IV), Co(III), Mn (III) and Cu(II). However, nitrations were also effected by the *nonoxidizing* acetylacetonates of Fe(II) and Li(I), and this result raises serious questions about our premise that the nitration proceeds through oxidation of the arene by the metal acac. The reaction appears to be stoichiometric and not catalytic, although the yield of nitration per mole of acac consumed varies with the metal. The stoichiometry is around 1 for Cu (II), Ce (IV), Li(I), and Fe(II); between 1 and 2 for Co (III) and Fe(III); and less than 1 for Mn(III).

Table 1. Nitration of Benzene with Various Metal Acetylacetonates in Liquid Nitrogen Tetroxide at 0°C

M(acac)/mmole	Nitrobenzene (mmole)	%DNB	DNB Isomer Distribution[b] (o:m:p)
Fe(acac)$_3$/1.85	3.2	<0.05	—
Co(acac)$_3$/2.58	3.5	3.1	0.7:81.7:17.6
Mn(acac)$_3$/2.0[a]	1.3	>2.1	0.8:87.0:12.4
Cu(acac)$_2$3.8[a]	1.2	>1.4	0.9:85.3:13.8
Ce(acac)$_4$/1.54	1.6	1.4	8.5:80.4:11.1
Fe(acac)$_2$/2.43	2.7	<0.05	—
Liacac/2.44	2.4	<0.05	—

[a]Part of the oxidant ignited on contact with N_2O_4 vapors.
[b]DNB isomer ratio under mixed acid conditions = 6:92:2; DNB isomer with $NO/O_2/N_2O_4$ = 2:74:24.

---

* Caution: In the absence of any cosolvent such as nitromethane the order of addition is important. We found that powdered metal acac's dissolve safely when dropped into a stirring quantity of liquid nitrogen tetroxide at 0°C. However, if N_2O_4 is added to several milligrams of the solid metal acac, a brilliant flame erupts!

As mentioned above, we had expected that the reaction would proceed via oxidation of the arene by the metal oxidant, and once the oxidant was consumed the reaction would cease. We had further planned to study the possibility of reoxidizing the metal *in situ* by passing oxygen through the reaction mixture. Even though the reaction does not appear to proceed as surmised, we nevertheless tested the system's response to oxygen. Figure 3 also shows the effect of admitting oxygen into the reactor after the reaction with the metal acac

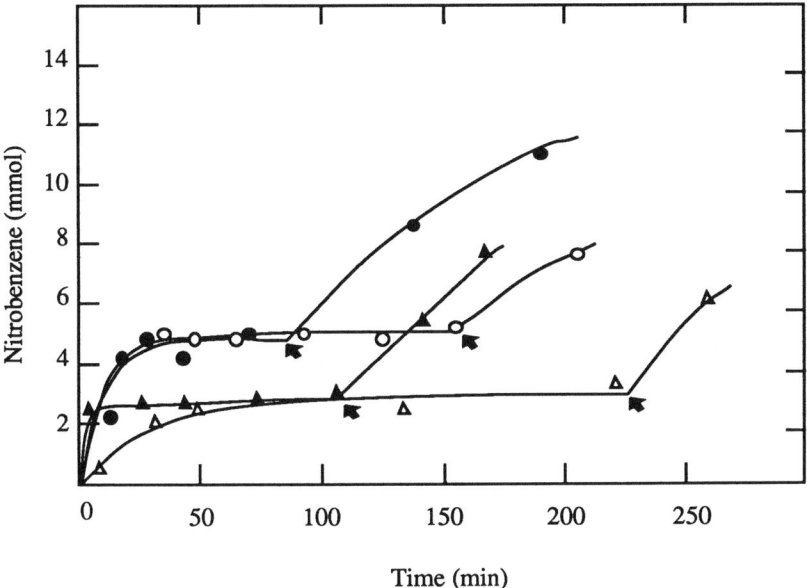

Figure 3. Nitration of benzene at 0° C in liquid N_2O_4 with metal acetylactonates.

- ● $Fe(acac)_3$ (2.36 mmol)
- ○ $Fe(acac)_3$ (2.33 mmol)
- ▲ $Fe(acac)_2$ (2.43 mmol)
- △ Li.acac (2.44 mmol)
- ↖ O_2 added (0.3 to 0.4 mmol/min)

is over: nitration of the substrate resumes at a fairly rapid rate. The amount of nitroarene produced per mole of oxygen varied between 0.9 and 1.5. Substantial phenol formation (10%) accompanied this renewed nitration with oxygen. Recall that passage of oxygen alone through a solution of benzene in N_2O_4 does not result in any significant nitration (Figure 2). Thus, the observed nitration in this case must be due to some as yet unidentified product of the reaction of metal acac in this system.

Nitrations with Nitrato-Complexes. We were extremely puzzled by the above findings and wondered if the metal acac was being converted to some other species which was the *de facto* nitrating agent. One candidate is the nitrosonium salt of the metal nitrato complex such as $NO^+[Fe(NO_3)_4]^-$. Addison and coworkers had previously reported on the formation of nitrobenzene from the reaction of benzene (and toluene) with such complexes of Fe(III) and U(VI) but not with those of Cu(II) or Zn(II) (5). To test whether the metal acac was transformed into a nitrato complex in liquid N_2O_4, we dissolved Fe(acac)$_3$ in N_2O_4 and then dried the mixture. Examination of the recovered solid by infrared spectroscopy showed some of the bands characteristic of the nitrato complexes, but the expected peak due to NO^+ was very weak.

Because we could not conclusively prove to ourselves that the metal acac was being transformed to the nitrato complex, we prepared an authentic sample of $NO^+[Fe(NO_3)_4]^-$ by following the procedure of Addison, and then studied nitration of benzene and toluene with it in N_2O_4. The results are shown in Figure 4 along with the data from a run with Fe(acac)$_3$ for comparison. Two features are noteworthy. First, the reaction is somewhat slower than the reaction with the acac itself. Second, unlike the acac-promoted nitration which was stoichiometric, the nitrations promoted by the nitrato complex appear to be catalytic. The arrows in the figure show the molar amounts of the reagent used in the reaction, and the slowing down of the reaction at long time reflects the depletion of the arene substrate. Only mono nitration products were observed with benzene, but with toluene a small amount (0.5-2%) of dinitrotoluenes (DNTs) was also observed. The mononitrotoluenes (MNTs) were produced with an isomer distribution showing preferential nitration at the *ortho* and *para* positions: $o:m:p$ = 53:3:44. Among the DNTs the 2,4- and the 2,6- isomers amounted to greater than 90% of the product in a ratio of 2,4-:2,6- = 3.7.

The literature reports on the nitrations with the nitrato complexes described the reactions in the arene itself and not in N_2O_4. Those studies showed the reaction to be stoichiometric and extremely sensitive to moisture. We suspect that the excess N_2O_4 allows for regeneration of the nitrato complex and also renders the system more tolerant to moisture. In accord with the literature report, we too found that the nitrato complex of Cu(II) did not effect nitration of benzene even in N_2O_4.

To summarize, the nitrations with nitrato complexes are catalytic and those promoted by the metal acac's are not. Furthermore, nitration is promoted by Cu(acac)$_2$ but not by the $NO^+[Cu(NO_3)_3]^-$. Together, these two facts strongly suggest that the nitrato complexes are not the reactive intermediates in the case of the metal acac-promoted nitrations. Thus, although the catalytic nitrations by the nitrato complexes in liquid N_2O_4 are quite interesting in their own right, they do not lead us to the identity of the agent(s) responsible for metal acac-promoted nitrations. However, this negative answer prompted us to examine the effect of the organic ligands. The β–dicarbonyl ligand could react with N_2O_4 giving a reactive species, such as an enol nitrate, which was responsible for the nitrations with the metal acac's.

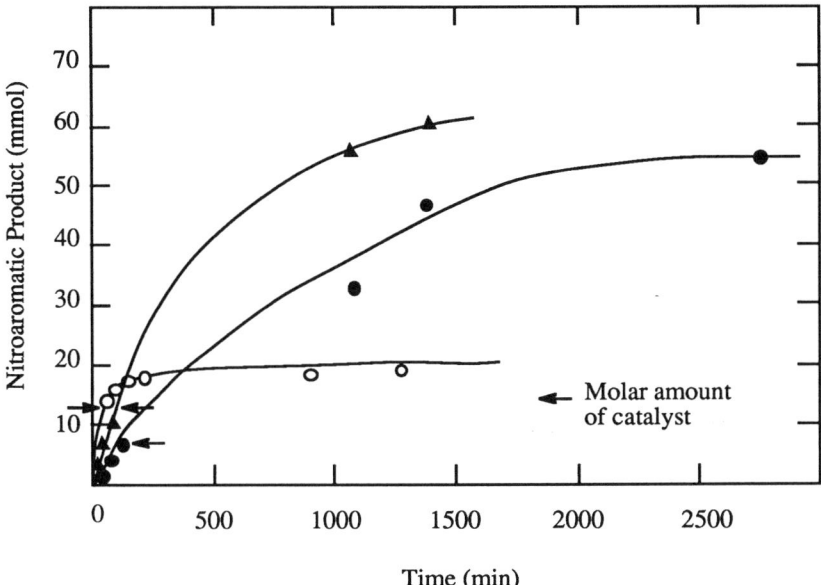

Figure 4. Nitration of benzene and toluene with NO[Fe(NO$_3$)$_4$] in N$_2$O$_4$ nitromethane solutions.

- ○ Benzene, 61.8 mmol; Fe-adduct, 8.5 mmol; N$_2$O$_4$, 30.0 mL; nitromethane, 20.0 mL; Temp.: 0-465 min 0°C, > 465 min 25°C.
- ● Toluene, 60.0 mmol; Fe-adduct, 14.9 mmol; N$_2$O$_4$, 5.0 mL; nitromethane, 20.0 mL; Temp.: 0-120 min 0°C, > 120 min 25°C.
- ▲ Toluene, 60.0 mmol; Fe(acac)$_3$, 14.9 mmol; N$_2$O$_4$, 30.0 mL; nitromethane, 20.0 mL; Temp., 25°C.

Nitrations Promoted by β-Dicarbonyls. We examined the ability of a variety of β-dicarbonyl compounds to effect nitrations in liquid N$_2$O$_4$. Included in this study were compounds such as 2,4-pentanedione, ethyl acetoacetate, and dimedone. We also tested compounds like acetic anhydride and succinic anhydride, which are not β–dicarbonyls in the common parlance, but which do have carbonyls β to each other. As controls, we studied the nitrating ability of comparable amounts of nitric acid and acetic acid in liquid N$_2$O$_4$. The yield data are summarized in Table 2, and the kinetic data for some of the runs are shown in Figure 5. Of the various compounds investigated, three stand out as being extremely potent: 2,4-pentanedione, ethyl acetoacetate, and acetic anhydride. These three compounds promote nitrations at rates at least two orders of magnitude greater than by 100% HNO$_3$ in this medium. Curiously enough, dimedone and succinic anhydride, which are structurally related to 2,4-pentanedione and acetic anhydride respectively, were not very effective. The

Table 2. Nitration of toluene in nitrogen tetroxide at 0°C promoted by various reagents

Reagent name	Structure	Amount (mmol)	N_2O_4 (mL)	Time (min)	MNT Yield (mmol)	MNT Isomer ratio
Background[a]			20	101	0.54	53:4:43
Background			5	109	0.06	62:5:34
Acetic anhydride[b]	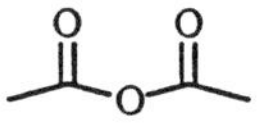	7.34	10	66	7.3	59:2:39
2,4-Pentanedione		14.3	5	104	7.0	61:4:35
Ethyl acetoacetate	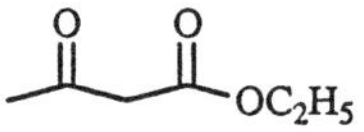	12.0	5	102	6.1	61:3:36
100% Nitric acid	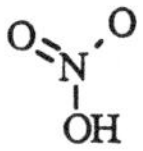	15.0	5	75	0.06	62:3:35
Acetic acid[c]	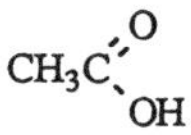	14.4	5	149	0.06	63:3:34
Succinic anhydride[d]	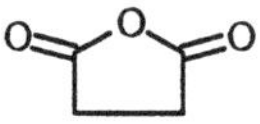	14.3	5	108	0.24	69:7:24
Diethyl oxalate	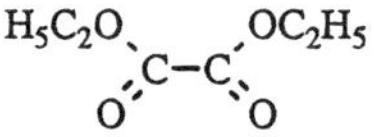	14.3	5	108	0.24	60:5:35
Dimedone	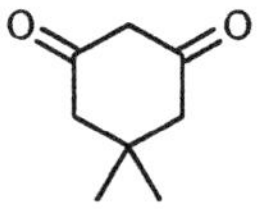	14.3	5	145	0.18	62:5:33
Tetranitromethane	$C(NO_2)_4$	13.9	5	89	0.24	18:8:74

[a] Nitromethane –
[b] Nitromethane yield - 10
[c] Conducted at 25°C
[d] Did not dissolve completely

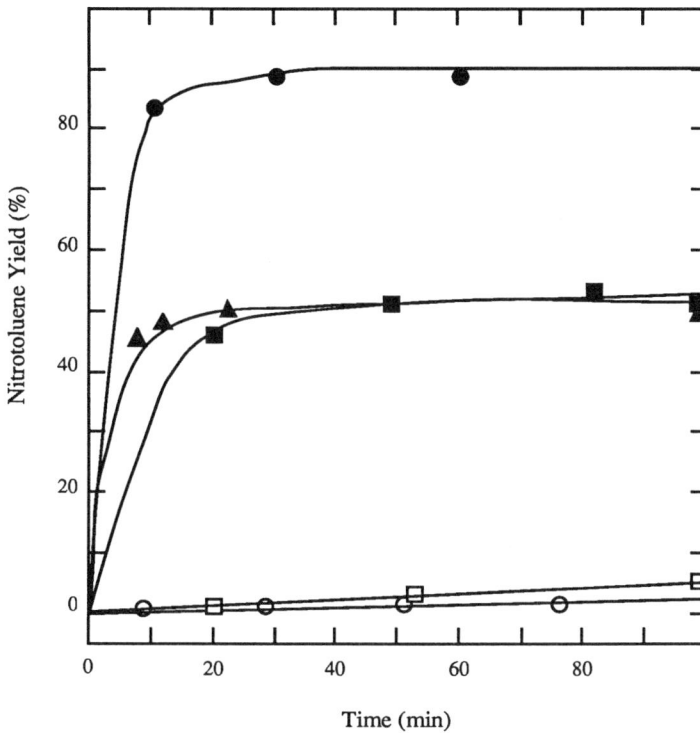

Figure 5. Nitration of toluene in N_2O_4 with various β-dicarbonyl compounds.
- ● Acetic anhydride, 25°C
- ▲ 2,4-Pentadione, 0°C
- ■ Ethyl acetoacetate, 0°C
- □ Dimedone, 0°C
- ○ Nitric acid, 0°C

difference may be due to the fact that each of the effective agents can adopt a conformation in which the carbonyl groups are parallel to each other, while dimedone and succinic anhydride are cyclic and cannot adopt such an orientation. Also worth noting is the ineffectiveness of acetic acid, particularly in view of the extreme potency of acetic anhydride. When added to liquid N_2O_4, 2,4-pentanedione, ethyl acetoacetate, and dimedone underwent a rapid reaction and they could not be recovered from the solution. We expected the formation of enol nitrates, or acyl nitrates, but were unable to characterize the decomposition products by GC or GC/MS techniques.

Summary

We have shown that a variety of metal acetoacetonates promote nitrations of arenes such as benzene and toluene in liquid N_2O_4. The nitration appears to be stoichiometric, albeit not 1:1. The effectiveness of metal acac's does not depend on the oxidation state of the metal, and the nitration is not likely to be effected via one-electron oxidation of the arene. The rates of nitration, and to some extent the product distribution, depend upon the order of addition of the reagent. We also showed that the nitrato complexes of several metals are effective in promoting nitrations in liquid N_2O_4, and that these agents act as catalysts for the nitration reaction. The mechanism by which metal acac's promote nitration remains a mystery, and our efforts at uncovering it were rewarded only with even more surprising findings that simple organics such as 2,4-pentanedione and acetic anhydride bring about nitrations in liquid N_2O_4 at rates many orders of magnitude faster than that by nitric acid!

Acknowledgment

We gratefully acknowledge financial support by Air Products and Chemicals Inc. and able assistance in the laboratory by Mr. Robert Johnson.

Literature Cited

1. Underwood, G. R.; Silverman, R. S.; Vanderwalde, A. *J. Chem. Soc., Perkin II* **1973**, 1177.
2. (a) Radner, F. *Acta. Chem. Scand.* Series B, **1983**, *37*, 65.
   (b) Eberson, L.; Radner, F. *Acta. Chem. Scand.* Series B, **1985**, *39*, 343.
3. Suzuki, H.; Murashima, T.; Shimizu, K., and Tsukumoto, K. *J. C. S. Chem. Comm.* **1991**, 1049.
4. Blucher, W. and Ross, D. S. 174th National ACS, Chicago, Ill., "Studies in Aromatic and Amine Nitration," Final Report, U.S. Army Research Office, Contract No. DAAG29-76-C0040, **1979**.
5. (a) Addison, C. C. *Chem. Rev.* **1980**, *80*, 21.
   (b) Addison, C. C., Boorman, P. M., and Logan, N. *J. Chem. Soc.* **1965**, 4978.

RECEIVED October 24, 1995

Chapter 5

Synthesis of 2,5,7,9-Tetranitro-2,5,7,9-tetraazabicyclo[4.3.0]nonanone

H. R. Graindorge[1], P. A. Lescop[1], F. Terrier[2], and M. J. Pouet[2]

[1]Centre de Recherches du Bouchet, Société Nationale de Poudres et Explosifs, B.P. 2, F−91710 Vert le Petit, France
[2]Ecole Nationale Supérieure de Chimie de Paris, 75231 Paris, Cedex 05, France

The non nitrated title ring system was synthesized for the first time by acid-promoted condensation of urea with 1,4-diformyl-2,3-dihydroxy-piperazine. Nitration of the polycycle occured first at the piperazine nitrogens and further nitration led to tri- and tetranitroderivatives. Different nitrating conditions leading to di-, tri-, and tetranitro derivatives will be discussed. Pure tetranitro-tetraaza-bicyclononanone can be obtained directly in one nitration step.

Nitramines of azaheterocycles are dense highly energetic molecules used in propellants, explosives and pyrotechnics *(1-5)*. The most famous derivatives are the six-membered ring RDX, with three N-NO$_2$ functionalities and its eight-membered homologue, HMX, with four N-NO$_2$ functionalities. For the past four decades, RDX and HMX have been the standards against which any other new energetic compound is compared. Since the discovery of HMX there has been an ongoing effort to synthesize new energetic materials superior to HMX, with increased energy and density, to meet the growing needs and wider applications of energetic materials in rocket propulsion, gun propellants, high explosives and pyrotechnics. A large research effort was focused on analogs of HMX, incorporating and developing the chemistry of the elementary fragment, -(-N(NO$_2$)-CH$_2$-)-, expecting that new combinaisons would lead to more energetic and denser compounds.

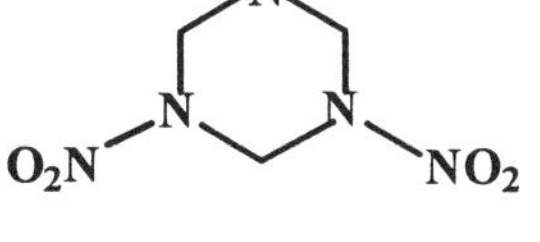

RDX

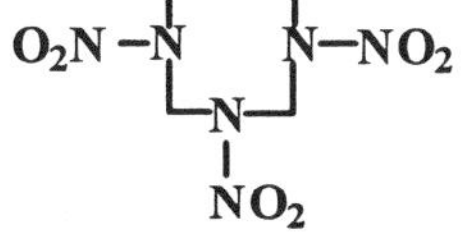

HMX

0097−6156/96/0623−0043$15.00/0
© 1996 American Chemical Society

The first example was reported in 1973 with the synthesis of tetranitroglycoluryl or SORGUYL, a bicyclo [3.3.0] octadione derivative *(6)*. More recently synthetic routes to the hemi-reduced **5** *(7)* and the fully reduced BICYCLO-HMX *(2)* homologues of SORGUYL were described along with the tricyclic bis-urea **6** *(3)*. Finally, the title ring system **4**, was cited *(2,9)* and its X-ray crystal structure was determined *(8)*, but no synthetic route was reported.

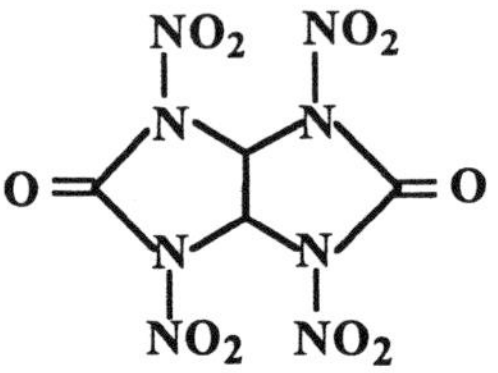

SORGUYL 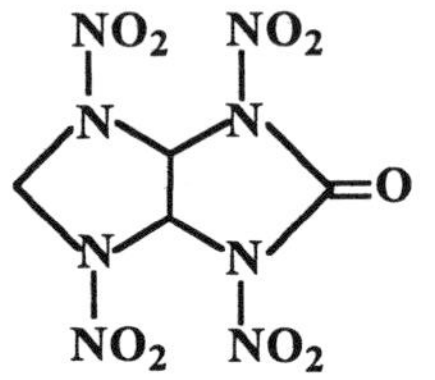   **5** 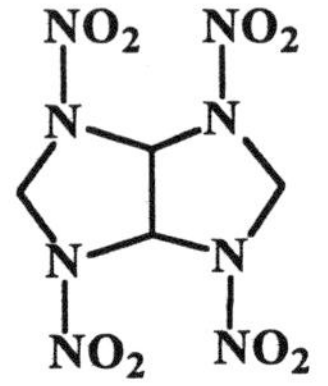   **BICYCLO-HMX**

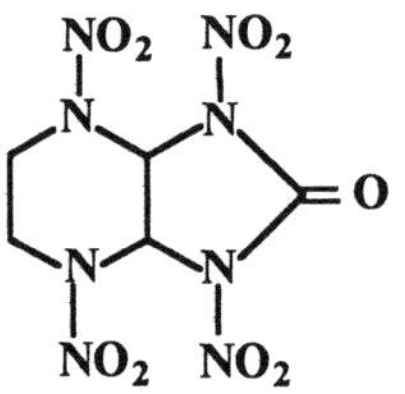

4 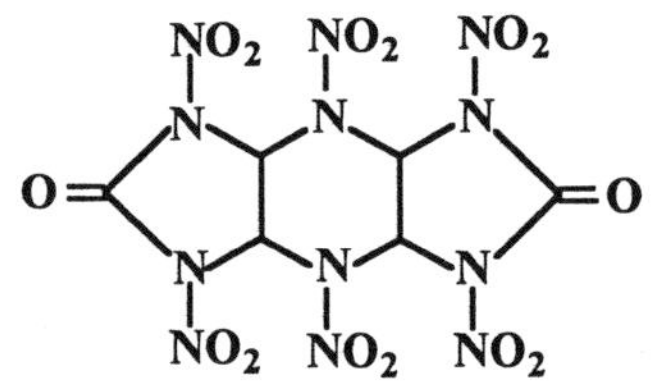   **6**

By contrast with HMX and RDX which are derived by nitrolysis under different conditions from the same reagent : hexamethylene tetramine *(10)*, the pre-cited polycyclic structures are obtained in one or several steps, and nitrated directly by nitration of a NH group or undirectly by nitrolysis of a NR group.

Synthesis of 2,5,7,9 -Tetraazabicyclo [4.3.0] Nonanone

The condensation of glyoxal with ureas is a well-established route to tetraazabicyclo [3.3.0] octanediones *(11)*. The reaction conditions used in the condensation of urea with 1,3-diformyl-4,5-dihydroxyimidazolidine *(7)* proved to be successful for the condensation of urea with 1,4-diformyl-2,3-dihydroxypiperazine which has been synthesized earlier in two steps from ethylenediamine, methyl-formate and glyoxal *(12)*. Thus **1** was prepared according to the synthetic route shown in Figure 1.

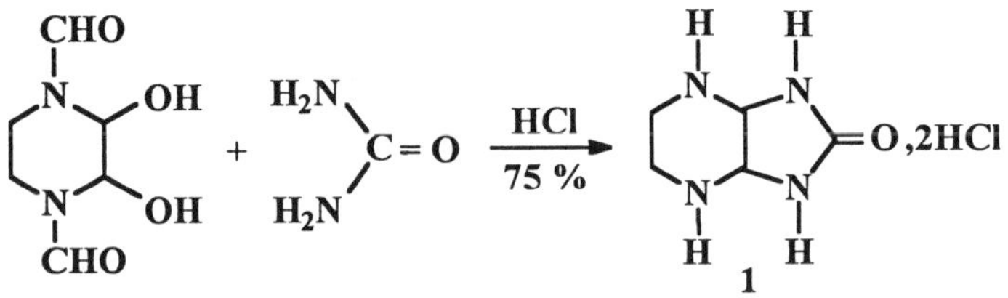

Figure 1 : Synthesis of **1**

Compound 1 was identified as a dihydrochloride salt on basis of elemental analysis, IR and NMR spectral data. The IR spectrum of 1 (in KBr pellet) shows characteristic bands at 3230 cm^{-1} (ν NH) and one carbonyl absorption at 1720 cm^{-1}. The ^{1}H NMR spectrum of 1 in DMSO shows the methine protons as a singlet at δ 5.1 and methylene protons as a multiplet centered at δ 3.1. There is also a singlet at δ 8.1 for the NH which are exchangeable with D_2O.

Nitration of 2,5,7,9-Tetraazabicyclo [4.3.0] Nonanone 1

The use of different nitrating agents such as 100 % HNO_3, HNO_3/Ac_2O, HNO_3/N_2O_5 and NO_2BF_4 combined with some specific operating conditions leads to the di-,tri- and tetranitroderivatives as it is shown in Figure 2.

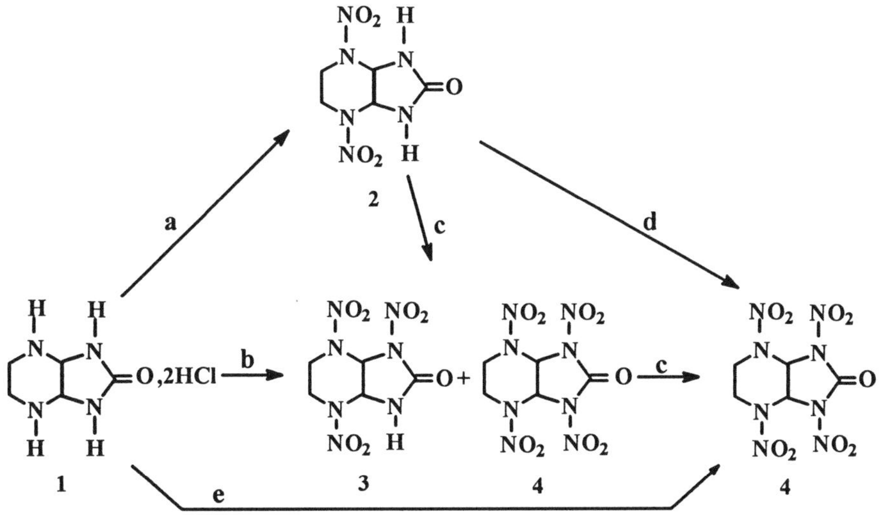

Figure 2 : Scheme of nitration of 1

Nitration of 1 with NO_2BF_4 in nitromethane *(route a)* at a temperature which didn't exceed ambient, gave nitration at the piperazine ring nitrogens only. Increased quantities of NO_2BF_4 or higher temperatures didn't modify the reaction : only compound 2 was obtained. Replacing at this stage nitromethane with acetonitrile had a dramatic effect, leading to no isolable product. The nitration of 1 with HNO_3/Ac_2O at a temperature which didn't exceed ambient or with 100 % HNO_3 at 60 °C *(route b)* gave a mixture of 3 and 4 (with 4 as a major product). Pure compound 3 was isolated from the aqueous diluted waste acids from which it was crystallized. After compound 2 had been isolated, its nitration, performed with NO_2BF_4 in acetonitrile *(route c)*, also led to a mixture of 3 and 4, once more with 4 as a major product, in low yields (less than 30 %). Further nitration of the mixture 3/4 with NO_2BF_4 in acetonitrile led quantitatively to pure 4. In contrast nitration of 2 with HNO_3/N_2O_5 *(route d)* gave directly 63 % of pure 4. The nitric oleum HNO_3/N_2O_5 was prepared by Frejacques process *(13)* and contained 25-28% of N_2O_5.

Compound **4** was also prepared in one step in 78-82 % yield from the dihydrochloride **1** by nitration with HNO_3/N_2O_5 at ambient temperature or with HNO_3/Ac_2O at 60 °C *(route e)*. The use of HNO_3/N_2O_5 for nitration provided safer reaction conditions than Ac_2O/HNO_3 where the temperature of the reaction mixture rose quickly and a chilled bath had to be applied to maintain the temperature under 60 °C. The general reaction conditions and results are given in Table I.

Table I : General Operating Conditions of Nitration

Starting Material	Nitration agent	Product	Yield (%)
1	NO_2BF_4 / CH_3NO_2	2	86
1	HNO_3/Ac_2O (20 °C)	3 / 4 7/93[a]	53[b]
1	100 % HNO_3 (60 °C)	3 / 4 10/90[a]	40[b]
2	NO_2BF_4 / CH_3CN	3 / 4 4/96[a]	25[b]
2	HNO_3 / N_2O_5 (20 °C)	4	63
3 / 4	NO_2BF_4 / CH_3CN	4	100
1	HNO_3/Ac_2O (60 °C)	4	78
1	HNO_3 / N_2O_5 (20 °C)	4	82

[a] Estimated from the ^1H NMR spectrum of the crude nitration product.
[b] Calculated from crude product and taking into account the proportions of **3** and **4**.

The structures of compounds **2 - 3** and **4** are supported by the IR, ^1H and ^{13}C NMR data respectively compiled in table II, III, and IV. The ^1H NMR spectrum of trinitroderivative **3** showed two doublets for the methine protons at δ 7.90 and δ 6.11 with a J_{HH} of 8.7 Hz indicating a torsion angle H-C-C-H of approximatively 0 degree corresponding to a cis junction between the two heterocyclic rings. A weak long distance coupling, between methine protons and methylene protons was also observed (J_{HH} = 0.8 Hz). The structure of **4** was also determined by X-ray diffraction. The results were in agreement with those already published *(8)*: the conformation showed that the five-member ring is a flattened enveloppe and the six-member ring is twisted. It crystallizes in $P3_1$ space group (trigonal), with a crystal density of 1.975 (reported 1.969). Compound **4** is quite stable to hydrolysis. No hydrolysis was observed in water at 30 °C after 48 hours, 45 %were lost at 40 °C after 48 hours and **4** disappeared completely at 70 °C after 2 hours. By cooling, compound **4** was partially recorvered in mixture with N, N' -dinitro-ethylene diamine issued from hydrolysis of **4**.

Table II : IR Spectral Data [a] for Compounds 2, 3 and 4

Compound	ν (cm^{-1})		
(KBr pellet)	N-H	C = O	N-NO$_2$
2	3270	1710	1550
3	3360	1810	1565
4	-	1840	1630 and 1550

[a] Recorded on a Perkin-Elmer Model 1310 spectrometer.

Table III : ^{1}H NMR Spectral Data for Compounds 2, 3 and 4

Compound	ppm[a] (multiplicity ; J_{HH}, Hz ; integration)		
(Solvent)	CH	CH$_2$	NH
2[b] (Me$_2$SO-d$_6$)	6.45 (s ; - ; 2)	3.70 (m ; - ; 2)	7.90 (s ; - ; 2)
		4.40 (m ; - ; 2)	
3[c] (acetone d$_6$)	7.90 (dd; 8.7, 0.8 ; 1)	4.76 (m ; - ; 2)	8.56 (s ; - ; 1)
	6.11 (dd ; 8.7, 0.8 ; 1)	4.58 (m ; - ; 2)	
4[c] (acetone d$_6$)	7.97 (s ; - ; 2)	4.78 (m ; - ; 2)	
		4.22 (m ; - : 2)	

[a] TMS internal standard. [b] Obtained with a Bruker AM 200 SY spectrometer.
[c] Obtained with a Bruker AC 400 spectrometer.

Table IV : ^{13}C NMR Spectral Data [b] for Compounds 2, 3 and 4

Compound	ppm[a]		
(Solvent)	CO	CH	CH$_2$
2 (Me$_2$SO-d$_6$)	159	65	41
3 (Me$_2$SO-d$_6$)	152	65 (beside NH)	46
		71 (beside NNO$_2$)	
4 (Me$_2$SO-d$_6$)	140	64	41

[a] TMS internal standard. [b] Obtained with a Bruker AM 200 SY spectrometer.

Experimental section

CAUTION ! Compounds 2-4 are sensitive explosives and should be handled with appropriate precautions.

Melting points were determined in open capillary tubes on a Büchi capillary apparatus and are uncorrected. Infrared spectra were recorded on a Perkin-Elmer Model 1310 spectrophotometer. ^{1}H NMR spectra were obtained at 200 and 400 MHz respectively on a Brucker AM200SY and Brucker AC400 spectrometers (Me_4Si internal standard). ^{13}C NMR spectra were recorded on a Brucker AM200SY spectrometer (Me_4Si international standard). Mass spectra were recorded on a Ribermag R10 10C (70ev) spectrometer.

2,5,7,9-Tetrahydro-2,5,7,9-tetraazabicyclo [4.3.0]Nonanone (1). 1,4-diformyl-2,3-dihydroxypiperazine (227g, 1.59 mol), prepared according to the literature method *(12)*, was added to a stirred solution of urea (111g, 1.85 mol) in concentrated hydrochloric acid (37 % w/w, 1100 ml). There was no exothermicity and the dissolution was complete. The mixture was stirred for 1h at room temperature and then warmed at 35 °C for 1h 30 min. A white solid appeared. After cooling the mixture to 15-20 °C, the solid was collected by filtration, washed 3 times with 1000 ml of ethanol and dried in vacuo and over P_2O_5 as a dessicant to give a white solid (256g, 75 %) as a dihydrochloride salt of **1**. mp 215 °C ; IR (KBr) 3230 (NH), 2940 (br), 1720 (C=O) cm^{-1} ; ^{1}H NMR (Me_2SO - d_6) δ 8.1 (s,2,NH) 5.1 (s,2, CH) 3.1 (m,4,CH_2) ; ^{13}C NMR (Me_2SO-d_6) δ 170 (C=O) 93.3 (CH) 39.4 and 38.6 (CH_2).

2,5-Dinitro-7,9-dihydro-2,5,7,9-tetraazabicyclo [4.3.0] Nonanone (2). To a stirred solution of NO_2BF_4 (3.75g, 28.2 mml) in dry nitromethane (40 ml) at 3-5 °C was slowly added the dihydrochloride **1** (1.5g, 7 mmol). The stirred mixture was allowed to warm slowly to room temperature. The mixture slight yellow and slightly turbid, was stirred for 3 h. and then poured onto ice. A white solid precipitated immediately. After stirring for 30 min., it was filtered, washed 3 times with water (150 ml) and dried in vacuo and over caustic potash as a dessicant to give a white solid (1.4g, 86.5 %). mp 213°C explosive dec. ; IR (KBr) 3270 (NH), 1710 (C=O), 1550 (N-NO_2) cm^{-1} ; ^{1}H NMR (MeSO-d_6) δ 7.90 (s, 2, NH) 3.70 (m, 2, CH_2) 4.40 (m, 2, CH_2) 6.45 (s, 2, CH) ; ^{13}C NMR (Me_2SO-d_6) δ 159 (C=O) 65 (CH) 41 (CH_2) ; MS *m/z* (relative intensity) 233 (MH+), 187, 169, 141, 126, 112, 97, 87, 85 (100), 75.

2,5,7,9,-Tetranitro-2,5,7,9-tetraazabicyclo [4.3.0] Nonanone (4). A - From compound 2 and HNO_3/N_2O_5. To a stirred nitric oleum mixture HNO_3/N_2O_5 75/25 w/w (20 g) prepared according to Frejacques process *(13)*, was slowly added at 3-5 °C the dinitro compound **2** (3.5g, 15 mmol). The clear mixture was then allowed to warm slowly to room temperature, stirred for 1 h. and poured onto ice. After stirring for 5 min., it was filtered, washed and dried to give a white solid (3.05 g, 63 %). mp 207 °C explosive dec. ; IR (KBr) 1840 (C=O), 1630 (N-NO_2), 1550 (N-NO_2) cm^{-1} ; ^{1}H NMR (acetone-d6) δ 7.97 (s, 2, CH), 4.78 (m, 2, CH_2) 4.22 (m, 2, CH_2) ; ^{13}C NMR (Me_2SO-d_6) δ 140 (C=O), 64 (CH) 41 (CH_2). Crystal density was determined by X-ray diffusion : 1.975 (reported *(8)* : 1.969).

B - From compound 2 and NO_2BF_4 in acetonitrile. To a stirred solution of

NO_2BF_4 (4g, 30 mmol) in dry acetonitrile (65 ml) at 3-5 °C was slowly added the dinitro compound 2 (2.3 g, 10 mmol). The stirred mixture was allowed to warm slowly to room temperature, strirred for 1 h and then poured onto ice. After stirring for 30 min., the precipitate was filtered washed, and dried to give a white solid (0.8g). ^1H NMR spectrum showed that it was a mixture of the trinitroderivative 3 and the tetranitroderivative 4. The molar proportions estimated from ^1H NMR spectrum were 4 % of 3 and 96 % of 4.

A further nitration of this crude product with NO_2BF_4 in acetonitrile led to pure tetranitroderivative 4. Thus, to a stirred solution of NO_2BF_4 (2.25 g) in dry acetonitrile (37 ml) at 3-5 °C was slowly added the mixture 3/4 obtained just before (0.8 g). The stirred mixture was allowed to warm slowly to room temperature, stirred for 1 h and then poured onto ice. After stirring for 30 min., it was filtered, washed and dried to give a white solid (0.8g) which was characterized as being the pure tetranitroderivative 4.

C - From compound 1 and HNO_3/Ac_2O. The dihydrochloride of 1 (1.5g, 7mmol) was added in portions with stirring to nitric acid (100 %, 18 ml, 0.43 mol) at 0-3°C. The solution was clear. Acetic anhydride (18 ml) was slowly added dropwise while the temperature was maintained between 1 and 5 °C. At the two third of the addition of acetic anhydride a slight yellow solid appeared. The mixture was stirred 2h. at 5 °C, the solid became white and all the mixture was poured onto ice. The solid was collected by filtration, washed and dried to give a white solid (1.2g) which was characterized as a mixture of 3/4 (7/93 molar). The mixture of 3/4 was converted in pure 4 by further nitration with NO_2BF_4 in acetonitrile as described above in method B.

D - From compound 1 and 100 % HNO_3. The dihydrochloride of 1 (1.5g, 7 mmol) was added in portions with stirring to nitric acid (100 %, 21 ml, 0.5 mol) at 3-5°C. The mixture was allowed to warm slowly to room temperature and warmed at 60 °C for 30 min. After cooling the mixture to room temperature, it was poured onto ice, the solid was filtered, washed and dried to give a solid (0.9g) which was characterized as a mixture of 3/4 (10/90 molar). Further nitration of the mixture of 3/4 with NO_2BF_4 in acetonitrile as described in method B, led to pure 4. A solid was cristallized again from the aqueous filtrate which had been allowed to stand for several hours at room temperature. It was characterized as the trinitroderivative 3. mp 194 °C, IR (KBr) 3360 (NH), 1810 (C=O), 1565 (N-NO_2) cm^{-1} ; ^1H NMR (acetone-d_6) δ 8.56 (s,1, NH) 7.90 (dd, 1, CH, J_{HH} = 8.7 and 0.8 Hz) 6.11 (dd,1, CH, J_{HH} = 8.7 and 0.8 Hz) 4.76 (m, 2, CH_2) 4.58 (m, 2, CH_2) ; ^{13}C NMR (Me_2SO-d_6) δ 152 (C=O) 65 (CH beside NH) 71 (CH beside NNO_2) 46 (CH_2) ; MS m/z (relative intensity) 277 (100), 263, 248, 231, 216, 186.

E - From compound 1 and HNO_3/N_2O_5. To a stirred mixture of nitric oleum HNO_3/N_2O_5 75/25 w/w (11.2g) prepared according to Frejacques process *(13)*, was slowly added at 3-5 °C the dihydrochloride of 1 (1.5 g, 7 mmol). The solution was allowed to warm slowly to room temperature, was stirred again for 1 h. and then poured onto ice. The solid was collected by filtration, washed and dried to give pure compound 4 (1.85 g, 82 %). The same result was obtained with HNO_3/Ac_2O as nitrating agent (18 ml 100 % HNO_3, 18 ml Ac_2O) but needed higher temperature of reaction (60 °C) which was then hard to control.

Acknowledgments

This work was performed under support of the Direction des Recherches et Etudes Techniques (DRET) and the Societe Nationale des Poudres et Explosifs (SNPE).
We are grateful to L. Ricard from the laboratory of the Ecole Polytechnique at Palaiseau who determined the X-ray structure of 2,5,7,9-tetraaza -2,5,7,9-tetranitro-bicyclo [4.3.0] nonanone.

Literature Cited

1 - Quinchon, J. *Les poudres, propergols et explosifs* ; Lavoisier, **1984-1991** ; volumes 1,2,3 and 4, Tech. et Doc.
2 - Coon, C. *Energy and Technology Review*, **1988,** 18.
3 - Vedachalam, M.; Ramakrishnan, V.T.; Boyer, J.H.; Dagley, I.J.; Nelson, K.A.; Adolph, H.G.; Gilardi, R.; George, C. and Flippen-Anderson, J.L. *J. Org. Chem.* **1991,** 56, No.10, 3413.
4 - Davenas, A. and Finck , B. *Revue Sci. et Tech. de la Défense*, **1992,** 17, 145.
5 - Miller, R.S. *International Defense & Technologie*, **Sept 1994,** Special serie, 22.
6 - Boileau, J.; Emeury, J.M. and Kehren J.P. *Ger. Offen 2 435651*, **1975.**
7 - Li, W.; Hua G. and Chen, M. *Proceedings of the Symposium on Pyrotechnics and Explosives* ; Beijing, China, **1987**
8 - Flippen-Anderson, J.L.; George, C. and Gilardi, R. *C. Acta Crystallogr.* **1990,** C46, 1122
9 - Delpeyroux, D.; Blaive, B.; Gallo, R.; Graindorge, H. and Lescop, P. *Propellants, Explos.* **1994,** 19, 70.
10 - Siele, V.I.; Warman, M.; Leccacorvi, J.; Hutchinson, R.; Motto, R.W.; Gilbert, E.E.; Benzinger, T.M.; Coburn, M.D.; Rohwer R.K. and Davey R.K. *Propellants, Explos.***1981,** 6, 67.
11 - Petersen, H. *Synthesis*, **1973,** 243
12 - Vail, S.L.; Moran, C.M.; Moore, H.B. and Kullman, R.M.H. *J. Org. Chem,* **1962,** 27, 2067 and 2071,
13 - Frejacques, C. *FR 1.060-425*, **18 Nov 1953.**

RECEIVED February 7, 1996

Chapter 6

Photochemical Chlorocarbonylation: Simple Synthesis of Polynitroadamantanes and Polynitrocubanes

A. Bashir-Hashemi[1], Jianchang Li[1], Paritosh R. Dave[1], and Nathan Gelber[2]

[1]Geo-Centers, Inc., 762 Route 15 South, Lake Hopatcong, NJ 07849
[2]U.S. Army Research, Development, and Engineering Center, Picatinny, NJ 07806-5000

A novel photochemical methodology for the selective functionalization of adamantane and cubane has been developed. In this method, versatile chloro-carbonyl groups, precursors to nitro groups, are introduced at the 1,3,5,7- positions of the adamantane and cubane skeletons leading to efficient syntheses of 1,3,5,7-tetranitroadamantane and 1,3,5,7-tetranitrocubane.

Polynitrocage molecules are central to the current efforts aimed at the development of new energetic materials to meet modern requirements for fuels, propellants, and explosives. These systems are particularly attractive because strain energy incorporated in the cage combined with the accumulation of nitro groups tend to bolster energy output, while the molecular compactness produces high density materials favorably increasing detonation velocity. Simultaneously, high crystal density materials are advantageous in volume-limited applications(*1*).

Although only a relatively few polynitropolycyclic compounds are known, reports describing the successful synthesis of members of this class of materials are appearing with increasing frequency(*1*). Experimental verification of the theoretical predictions regarding the usefulness of these high-energy, high-density compounds as fuels or explosives awaits the accumulation of sufficient amounts of the necessary test compounds.

There has been renewed interest in the chemistry of polysubstituted adamantanes since some of its derivatives, particularly nitroadamantanes, have shown promise as high density energetic materials(*2*). Several synthetic methods have been applied for the synthesis of nitroadamantanes. The direct nitration of adamantane with nitric acid leads to the formation of nitrate esters of substituted adamantanols(*3*). Photochemical reactions using N_2O_4(*4*) or N_2O_5(*5*) give mixtures of products and produce 1-nitroadamantane as a minor product. Only recently, Olah and his coworkers have obtained 1-nitroadamantane in 60% yield from the slow reaction of adamantane with nitronium tetrafluoroborate at room temperature in purified nitrile-free nitromethane(*6*). No evidence of any dinitroadamantane was obtained under excessive concentration of the nitrating reagent.

NITRATION

Adamantane → NO$_2$BF$_4$ → Nitroadamantane

In general, polynitroadamantanes have been obtained by introducing the proper functionalities on the adamantane skeleton which then can be converted to the corresponding nitro groups(*1-2*). The methodology used in the recent synthesis of 1,2,2-trinitroadamantane exemplifies the introduction of nitro groups on tertiary and secondary carbon atoms on the adamantane skeleton(*7*).

Trinitroadamantane

The starting functionalized adamantane for the above synthesis, 2-oxoadamantane carbonyl chloride is readily available by cyclization of 3,3,1-bicyclononanedicarboxylic acid(*8*). The carbonyl chloride function was converted to the corresponding amine via the Curtius rearrangement of the derived acyl azide. Oxidation of the amino group was using dimethyldioxirane yielded nitroadamantanone. The carbonyl group was converted to the geminal dinitro function by oxidative nitrolysis of the derived oxime. The order of conversion of the functional groups on adjacent carbon atoms was chosen so as to avoid the intermediacy of a 1,2-amino nitro grouping, which is known to lead to cage fragmentation in other systems. This methodology also represents the first example of the conversion of vicinal functional groups on any polycycle to vicinal nitro groups. The conversion of multiple carbonyl functionality to geminal dinitro groups has been successfully achieved in the synthesis of 2,2,4,4,6,6-hexanitroadamantane, the most highly nitrated adamantane known to date(*9*).

The synthesis of 1,3,5,7-tetranitroadamantane from adamantane was accomplished in multiple synthetic steps(*2*). Tetrahalogenation of adamantane at bridgeheads followed by Photo-Ritter reaction of tetraiodoadamantane with acetonitrile introduced four acetamide groups on the adamantane skeleton.

Subsequent hydrolysis of the tetraamide with hydrochloric acid and then oxidation with permanganate gave tetrahedrally nitrated 1,3,5,7-tetranitroadamantane(*10*). As interesting as this approach looks, the multi-step synthesis process and low yields limit the practical large-scale synthesis of this compound.

In the present work, a different approach for the synthesis of tetranitroadamantane was taken starting from 1,3,5,7-tetracarbonyladamantane. Several synthetic methods have been applied to the synthesis of adamantane tetracarboxylic acids(*11-12*). The most recent modified method for the synthesis of T_d-tetraester **3** requires multiple synthetic steps, requiring high-pressure, high-temperature bomb reactions, and therefore, is greatly limited for scaled-up production(*12*).

Very recently, the efficient photochemical chlorocarbonylation of a series of cyclic and acyclic carbonyl compounds with oxalyl chloride has been reported(*13*). Several carboxycubanes have been synthesized by employing this photochemical process(*14*). Here, the one-pot synthesis of 1,3,5,7-tetrakis-(chlorocarbonyl)adamantane **2** from commercially-available 1-adamantane-carboxylic acid **1** and oxalyl chloride, and its conversion to 1,3,5,7-tetranitro-adamantane **6** is reported.

Photochemical reaction of adamantane with oxalyl chloride produced only a small amount (<5%) of 1,3,5,7- tetra substituted adamantane along with a mixture of many other products(*15*). This is due to the high ratio of secondary hydrogens to tertiary hydrogens (12H to 4H) and low selectivity of the chloro-carbonylation between the 1- and 2-positions of adamantane. Positional selectivity (relative reactivity of bridgehead H/bridge H) for chlorocarbonylation is in the order of 3.7 for adamantane (*16*).

The prepositioning of a carboxy function at the bridgehead increases the possibility of chlorocarbonylation at bridgeheads since hydrogens closest to a chlorocarbonyl group are least vulnerable to radical abstraction(*13-14*). This fact has been attributed to polar effects and demonstrated by the product composition from the photochemical reactions of cyclopentanone and isovaleric acid with oxalyl chloride. In these cases, chlorocarbonylation occurred at the least electron-deficient carbons (β or γ). The regioselectivity is further affected by steric effects as well as statistical factors.

Irradiation of a solution of 1-adamantanecarboxylic acid **1** in oxalyl chloride in a Rayonet photochemical reactor (2537 A°) was followed by ^{1}H NMR. and esterified with methanol to give 1,3,5,7-tetracarbomethoxyadamantane **3** as a colorless solid in 20-30% yield(*17*).

When commercially available 1,3-adamantanedicarboxylic acid **4** was treated with oxalyl chloride under UV light. After 1h, the reaction mixture was concentrated and triturated with cold ether to give compound **2** in 40% yield. Tetraester, **3,** was obtained by methanolysis of the reaction mixture and isolated by triturating the crude mixture with methanol. As expected, the GC-MS of the reaction mixture showed a ratio of **3** to unidentified products of 60/40.

Tetrakis(chlorocarbonyl)adamantane **2** (*18*) was converted directly to the corresponding adamantane tetraisocyanate **5** using azidotrimethylsilane. 1,3,5,7-Tetranitroadamantane **6** was obtained in 40-50% yield from the oxidation of compound **5** with wet dimethyldioxirane (DMDO).

1,3,5,7-Tetranitroadamantane exhibits moderate power output and low shock and impact sensitivity. It has a very high thermal stability (DSC= 350 °C) and high crystal density(d= 1.63 g / cm^3).

This methodology has successfully been applied to the synthesis of 1,3,5,7-tetranitrocubane(*14,19*). Photochemical reaction of cubane carboxylic acid with oxalyl chloride gave 1,3,5,7-tetrakis (chlorocarbonyl)cubane in 60% yield. The solid product was treated with azidotrimethylsilane in chloroform, and, after heating the reaction mixture for 1h under reflux, the resulting tetraisocyanate was treated with wet dimethyldioxirane to give tetranitrocubane in 50% yield.

The photochemical chlorocarbonylation of nitrocubane gives predominantly 3, 5, 7-tris(chlorocarbonyl)nitrocubane, emphasizing the importance of polar effects in determining the regioselectivity in this class of reactions. As expected, tetranitrocubane did not react under a variety of reaction conditions.

In conclusion, a novel photochemical methodology for the selective functionalization of adamantane and cubane has been developed. In this method, versatile chlorocarbonyl groups, precursors to nitro groups, are introduced at the 1,3,5,7- positions of the adamantane and cubane skeletons leading to efficient syntheses of 1,3,5,7-tetranitroadamantane and 1,3,5,7-tetranitrocubane. In this approach, the direct placement of four acyl chlorides on adamantane and on cubane obviates the use of reagents such as thionyl chloride, necessary for converting acid functions to acyl chlorides (20). Additionally, the excess oxalyl chloride can be recycled and reused continuously. This methodology may also be advantageously extended to other cage systems.

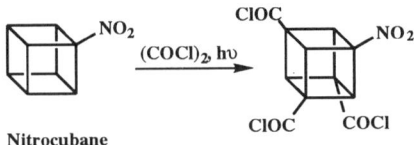

Acknowledgments: This article is respectfully dedicated to Dr. Everet Gilbert, who originated nitrocage chemistry at ARDEC. We sincerely thank Dr. N. Slagg for his support and inspiration throughout this work. Financial support provided by ARDEC to GEO-CENTERS, INC. is gratefully acknowledged.

References:
1. Marchand, A. P.; *Tetrahedron*, **1988**, *44*, 2377. Bashir-Hashemi, A.; Iyer, S.; Slagg, N.; Alster, *J. Chem. & Industry*, July 17, **1995**, 551.
2. Sollott, G. P.; Gilbert E. E. *J. Org. Chem*. **1980**, *45*, 5405; also see *Chemistry of Energetic Materials*; G.A. Olah; D.R. Squire; Ed, Academic Press, Inc., San Diego, CA, 1991.
3. Moiseev, I. K.; Klimochkin, Yu. N.; Zemtsova, M. N.; Trakhtenberg, P. L. *J.Org. Chem. USSR* **1984**, *20*, 1307.
4. Umstead, M. E., Lin, M. C. *Appl. phys*. **1986**, *B39*, 61.
5. Tabushi, I.; Kojo, S., Yoshida, Z. *Chem. Lett.* **1974**, 1431.
6. Olah, G. A.; Ramaiah, P.; Rao, C. B.; Sandford, G.; Golam, R.; Trivedi, N. J.; Olah, J. A. *J. Am. Chem. soc.* **1993**, *115*, 7246.
7. Dave, P.R.; Axenrod, T.; Qi, L.; Bracuti, A. *J. Org. Chem.* **1995**, *60*, 1895. For similar examples see: Archibald, T.; Baum, K. *J. Org. Chem.* **1988**, *53*, 4645; Dave, P. R.; Ferraro, M.; Ammon, H. L.; Choi, C. S. *J. Org. Chem.* **1990**, *55*, 4459.
8. Peters, J.A.; Remijnse, J.D.; van der Wiele, A.; van Bekkum, H. *Tetrahedron Lett.* **1971**, 3065.
9. Dave, P.R.; Bracuti, A.; Axenrod, T.; Liang, B. *Tetrahedron* **1992**, *28*, 5839.
10. For an improved synthesis of tetranitroadamantane using sodium percarbonate as oxidant see: Zajac, Jr. W. W.; Walters, T. R.; Woods, J. M. *J. Org. Chem.*, **1989**, *54*, 2468.
11. Stetter, H.; Bander, O.-E.; Neumann, W. *Chem. Ber.* **1956**, *89*, 1922. Landa, S.; Kamvcek, Z. *Collect. Czech. Chem. Commun.* **1959**, *24*, 4004.
12. Newkome, G. R.; Nayak, A.; Behera, R.K.; Moorefield, C.N.; Baker, G.R. *J. Org. Chem.* **1992**, *57*, 358.
13. Bashir-Hashemi, A.; Hardee, J.R.; Gelber, N.; Qi,L.; Axenrod, T. *J. Org. Chem.* **1994**, *59*, 2132.
14. Bashir-Hashemi, A. *Angew. Chem. Int. Ed. Engl.* **1993**, *32*, 612. Bashir-Hashemi, A.; Li, J.; Gelber, N.; Ammon, H. *J. Org. Chem.* **1995**, *60*, 698.

15. For a similar approach see.; Tabushi, I.; Hamuro, J.; Oda, R. *J. Org. Chem.* **1968**, *33*, 2108. For photoacetylation of adamantanes see; Fukunishi, K., Kohno, A., Kojo, S., *J. Org. Chem.* **1988**, *53*, 4369.
16. March, J. Ed. *Advanced Organic Chemistry*, Fourth Edition, Wiley Interscience, New York, 1992, pp 684-685. For parameters determining the regioselectivity of radical reactions see; Raymond C. Fort, Jr. Ed. *Adamantane, the Chemistry of Diamond Molecules*, Marcel Dekker, Inc. New York, NY 1976, pp 233-265.
17. Bashir-Hashemi, A.; Li, J.; Gelber, N. *Tetrahedron Letters,* **1995**, *36*, 1233.
18. For other uses of this compound in dendrimers see; Newkome, G.R., Moorefield, C.N., Baker, G.R., *Aldrichimica Acta*, **1992**, 25(2), 31, as diamandoid molecules; Ermer, O. *J. Am. Chem. Soc.* **1988**, 110, 3747, and in combinatorial chemistry; Carell, T.; Wintner, E.A.; Bashir-Hashemi, A.; Rebek, Jr. J. *Angew. Chemie, Int. Engl.* **1994**, *20*, 2059.
19. Eaton, P. E.; Xiong, Y.; Gilardi, R. *J. Am. Chem. Soc.* **1993**, *115*, 10195.
20. There have been occasional explosions reported using this method. *All cubane and adamantane derivatives prepared in this article are relatively stable at room temperature. Nevertheless, care should be taken when handling these energetic materials.*

RECEIVED September 26, 1995

Chapter 7

Isolation of Electrochemically Generated Dinitrogen Pentoxide in a Pure Form and Its Use in Aromatic Nitrations

M. J. Rodgers and P. F. Swinton[1]

Aerospace and Automotive Research Department, ICI Explosives, Ardeer Site, Stevenston, Ayrshire KA20 3LN, United Kingdom

> The use of N_2O_5 as a nitrating agent is relatively well known but has been somewhat limited by the lack of availability of the raw material in a pure form. In contrast, N_2O_5 in Nitric Acid solution is available in abundant supply via an electrochemical generation system. Combining the requirements of such a system with the solubility and phase behavior of N_2O_5, N_2O_4 and HNO_3 has allowed the invention of a method to electrochemically produce N_2O_5 and extract it in its pure form in large quantities. This new process is described along with some examples of interesting nitration reactions carried out using N_2O_5. The example of the unsymmetrical nitration of 1,3 dihydroxy benzene is described in detail.

Although N_2O_5 has been known for almost 150 years (1), little use was made of it in the early years, except perhaps as an exemplar of unimolecular reaction mechanisms (2). The first systematic nitration study using N_2O_5 was that of Haines and Adkins (3) and was largely inspired by the work of Daniels (2) and the availability of pure N_2O_5 made using his method. Ingold and his co-workers used essentially the same method to prepare N_2O_5 (4), as reported in the seminal series of papers describing nitration reaction mechanisms and kinetics (5,6). The existence of the Nitronium ion was proved unambiguously in this series of papers by isolation of various Nitronium salts (7) as well as by spectroscopic methods.

Nitration using nitronium salts has an extensive literature extending over a long period of time (8-10), and now with the advent of improved electrochemical methods of production (11-14) interest in N_2O_5 nitrations has re-emerged (11,12,15,16). Electrochemical generation of nitronium salts and their use in nitration reactions has also emerged (17-21) and still continues (22).
Over a period of time the use of batch (23-25) and continuous flow nitrations (26) to synthesise energetic materials has also been reported by the U.K. Defence Research Agency, both as part of their

---

[1]Corresponding author

own research programmes (27) and in collaboration with ICI Explosives (28,29). We wish to report here on the results of this collaboration with regards to the production of essentially pure, solid N_2O_5 and to point to novel nitration reactions using N_2O_5 in organic solvents.

Manufacture of N_2O_5.

Solid N_2O_5 may be prepared on a laboratory scale by the reaction of chlorine on solid silver nitrate (1) or similar metathesis reactions between nitryl chloride or fluoride and a metal nitrate, or by dehydration of nitric acid by phosphoric oxide, but the yields available are limited.

Solid N2O5 can also be prepared in larger quantities via the gas phase reaction between N_2O_4 and Ozone (30), but the production rate is limited by the output of the Ozone generators available and the efficiency of the trapping process. N_2O_5 in solution in organic solvents can be prepared in a similar manner using liquid phase reaction between N_2O_4 and ozone (31). While this process is more efficient, the production rate is still limited by the capacity of the ozone generator.

While solid nitric acid solvates of N_2O_5 can be obtained via the electrochemical synthesis of N_2O_5 in potentially much higher yields (32-34), the presence of the nitric acid can prove troublesome in the nitration of acid sensitive substrates. As part of a collaboration agreement with the U.K. Defence Research Agency we have now developed a process in which essentially pure solid N_2O_5 can be obtained from the output anolyte solution of an electrochemical cell (35).

Description of Process. In this process the electrochemical cell is run in such a manner as to yield an anolyte mixture rich in both N_2O_5 and N_2O_4 while still operating with good electrical efficiency. Portions of the anolyte stream are tapped off in a continuous manner, N_2O_4 added if necessary and cooled to give phase separation.

This part of the process may be illustrated with reference to Figure 1 which shows portions of the three component phase diagram at different temperatures in the usual triangular graphical form. Here the areas above the lines near the pure HNO_3 and N_2O_4 apices are the zones of complete miscibility, and the area between them the zone in which two phases exist. The portion to the right near the pure N_2O_5 apex represents the zone in which solid material exists. The temperature dependence of the boundary lines of the miscibility zones are shown over the range of interest.

As the temperature decreases the boundary line of the upper (Anolyte) phase moves towards lower N_2O_5 content and the immiscible (two phase) zone increases, with concommitant mass transfer of N_2O_5 from the miscible to the two phase zone. Similar behavior is shown for the N_2O_4 rich lighter phase.

As shown in Figure 2, in the immiscibility zone, the mixture separates out along a tie line to give two phases, whose composition correspond to their position on the phase boundary lines.

The separated lighter phase (essentially N_2O_5 in N_2O_4 solution with a little dissolved nitric acid) is then chilled to effect crystallisation of essentially pure N_2O_5 with some surface contamination with the crystallisation mother liquor. This is illustrated by the lower crystallisation tie line shown in Figure 2.

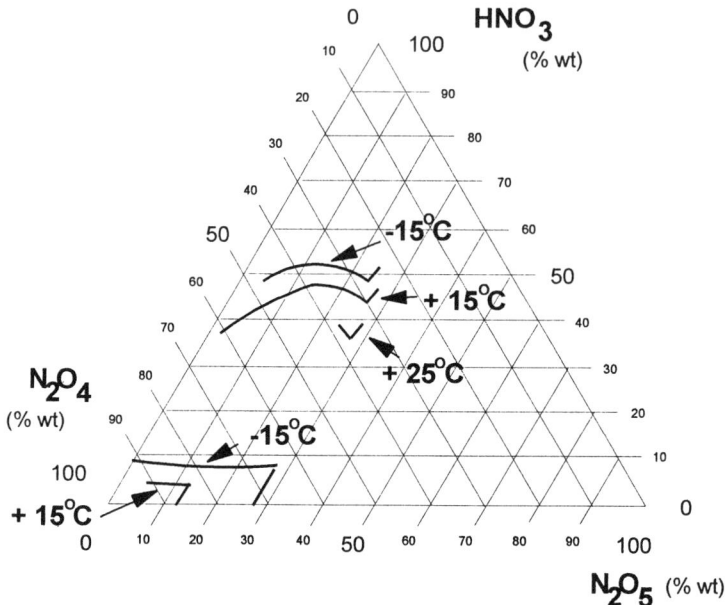

Figure 1. Partial phase composition diagram.

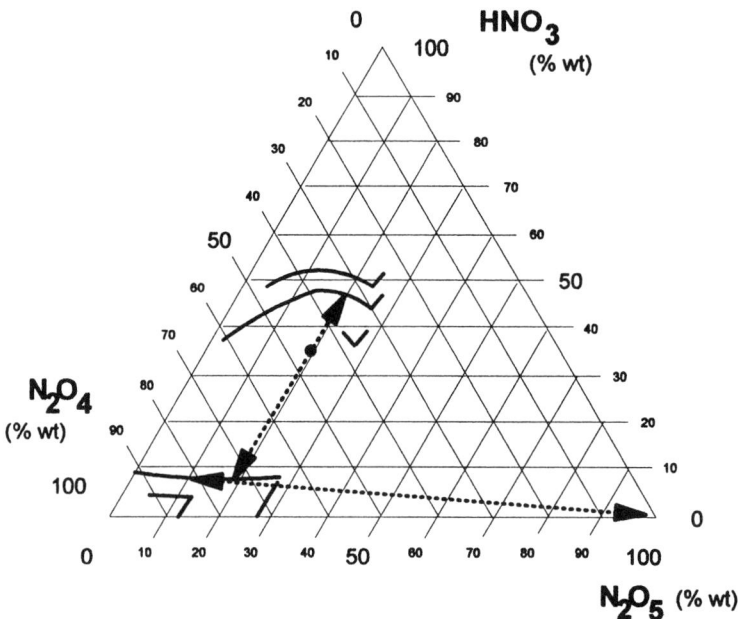

Figure 2. Liquid–liquid and liquid–solid separation.

The differential temperature behavior of the two limbs of the phase boundary curves in the region of interest is such that, taken in conjunction with the slope of the tie lines linking the upper (HNO_3 rich) and lower (N_2O_4 rich) miscible phase boundary lines, good transfer of N_2O_5 to the N_2O_4 solvent phase can be achieved. Similarly, the crystallisation tie line behavior is such as to allow efficient extraction of N_2O_5 from this phase on chilling to effect crystallisation.

The resulting solid N_2O_5 can then be essentially freed from troublesome liquid phase contamination, dissolved in organic solvent and used in the nitration of even HNO_3 or N_2O_4 sensitive substrates.

Because the process is a continuous one, with self contained closed loops at each process stage, the only liquid "*effluent*" is 67 % nitric acid from the catholyte recovery unit. Recovered 99% nitric acid and N_2O_4 are recycled back into the electrolytic cell anolyte loop. The only process inputs are 99% nitric acid and "*top up*" N_2O_4, the overall process loops being otherwise run in balance. In essence, the whole process could be scaled to give any required N_2O_5 output by adjusting the number of cells in the electrolyser, but in practice, with the ozone generators currently developed for water treatment facilities, only outputs in excess of 50 tonne per annum would justify this new electrochemical process, solvent plants being more cost effective up to this limit (*36*).

Nitration Using N_2O_5.

1,3 - Dihydroxy Benzene (Resorcinol). The heavy metal salts (mainly lead salts) of the mono-, di- and tri-nitro resorcinols have long been used in the explosives industry as ignition compositions or pyrotechnic delay compositions. Conventional nitration using sulphuric acid as the solvent gives progressive substitution to give 2-nitro, 2,4-dinitro and then the 2,4,6- trinitro derivatives, corresponding to the normal substitution rules for nitration of phenols. In order to prepare 4,6 dinitro resorcinol nitration is normally carried out at low temperature (below -20 C) using 96% nitric acid as solvent. The final yield of the 4,6 dinitro isomer is normally less than 50% and extensive purification is necessary. In none of the above mentioned reactions is 4-nitro resorcinol obtained.

Haines and Adkins (*2*) reported that bromobenzene was nitrated by N_2O_5 in CCl_4 to *p*-bromo nitro benzene as "*apparently the only product*", noting without comment that "*nitric acid, of course, gives some of the ortho compound as well as the para.*" Similarly, nitration using Acetyl Nitrate in Acetic anhydride (*37*) also gives both *ortho* and *para* nitro bromo benzenes. While Ingold *et al* were aware of Haines and Adkins result for bromo benzene (*4*), and discussed the orientation effects of molecular N_2O_5 versus nitronium ion in terms of the *ortho /para* ratio in the mono nitration of chlorobenzene (*4*, p2465), no mention was made of possible steric effects in addition to solvent effects on the reaction kinetics.

Given our interest in nitro resorcinols, more especially the preparation of 4,6 dinitro resorcinol in a safe, more controlled reaction, it was decided that flow nitration of resorcinol using N_2O_5 in aprotic solvents would be of interest (*38*). Acetonitrile was used as the solvent since resorcinol has good solubility and nitrations were carried out at 0.45M resorcinol and different N_2O_5 to resorcinol mole ratios. The results of this study are reported here for the first time.

Experimental. The N_2O_5 used in this study was prepared via the ozone route and was white, and free from any N_2O_4 contamination. The solid material (*ca -80 C*) was dissolved in spectroscopic grade acetonitrile as was the resorcinol (Analytical Reagent grade). The cold (*ca 0 C*) reactant solutions were synchronously pumped through a chilled jacketed glass bead filled flow reactor into an excess of water to both decompose any unreacted N_2O_5 and to dilute any nitric acid formed. The quenched mixtures were then evaporated to dryness under vacuum with addition of a large excess of acetonitrile to aid water and nitric acid removal by low temperature azeotroping. The maximum temperature differences between inlet and outlet were 28 degrees C for the 1:1 ratio and 40 degrees C for the 4:1 N_2O_5 to resorcinol ratio, both for the longest residence time.

The reaction products obtained were reddish brown in colour in the case of the 1:1 N_2O_5:resorcinol mole ratio and pale yellow in the case of the higher ratios. The former is typical of a mono nitro resorcinol (similar to the 2-nitro resorcinol reference compound) and the latter of di and tri nitro resorcinols.

The crude reaction mixtures were then analysed by HPLC using readily available resorcinol, 2-nitro, 2,4-dinitro,4,6-dinitro and 2,4,6-trinitro resorcinol reference samples. The unattributed peak obtained with the 1:1 mole ratio samples with a retention time of approximately 12.2 minutes was ascribed to the only other possible reaction product, 4-nitro resorcinol, since no 2- nitro peak was obtained. Since a standard reference sample was not available, 4-nitro resorcinol was estimated by difference whenever it was found, and hence is subject to the greatest uncertainty. An attempt was made to estimate the response factor for the 4-nitro resorcinol peak by fitting the measured response factors for the other standard compounds in terms of the number and relative positioning (*ortho* or *meta*) of OH - NO_2 and NO_2 - NO_2 pairs and using the best fit values for these to calculate the missing value. When the calculated response factor was applied to the peak area the calculated concentration of the 4-nitro resorcinol was in good qualitative agreement with the value shown in Table I.

Results. The results obtained from the analysis of the crude reaction products are shown in Table I and are illustrated in Figure 3 as a function of reactor residence time and N_2O_5 to resorcinol mole ratio.

While the presence of the 2-nitro derivative on a few occasions is indicative that, at least in some runs, some decomposition of the N_2O_5 had occurred, overall the results are consistent with nitration first occurring in the 4- position. Whether the anomalous results are due to N_2O_5 reacting with moisture to produce nitric acid and so catalysing nitronium ion production, or thermal decomposition leading to N_2O_4 and so to nitrosation side reactions it is not possible to say. Nevertheless, the fact that mono nitration overwhelmingly occurs at the 4 -position is indicative of nitration via molecular N_2O_5 rather than nitronium or nitrosonium ion, in as much as mono nitration in the 4 position has previously only been accomplished via unusual nitration agents - *eg.* 4-methyl 1,4 -nitro tetrabromo cyclohexadienone (*39,40*). It would thus appear that in this case steric effects are significant in that the less activated 4- position is substituted first (*cf.* Haines and Adkins results for bromo benzene (*3*)).

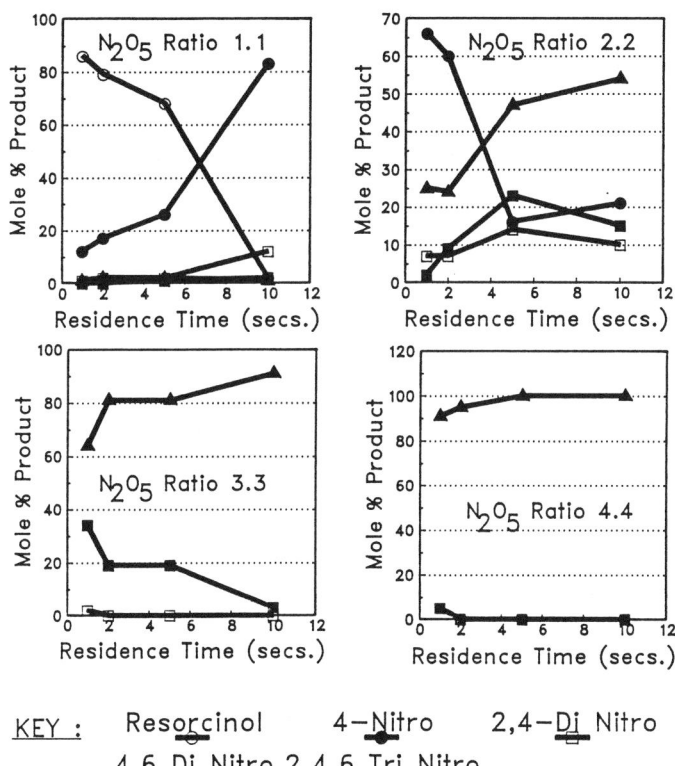

Figure 3. Isomer proportion in the N_2O_5 flow nitration of resorcinol as a function of reactor residence time and N_2O_5 to resorcinol ratio.

Thereafter, at higher N_2O_5 to resorcinol ratios the results are consistent with kinetic control. The results shown in Table I and illustrated in Figure 1, although somewhat variable, are consistent with conversion of the 4- nitro isomer to both 2,4 and 4,6 dinitro isomers as the contact time increases, with concomitant conversion of both to the 2,4,6 trinitro derivative. The difference in the proportion of the 2,4 and 4,6 isomers at an N_2O_5 to resorcinol mole ratio of 3.3 would indicate that the 2,4 dinitro derivative is formed slower and converts to the tri- nitro derivative faster than is the case for the 4,6 dinitro derivative. However the degree to which possibile side reactions arising from N_2O_5 decomposition products and the nitric acid formed during nitration contribute to these trends is not quantifiable. It is possible that the nitric acid formed during formation of 4-nitro resorcinol promotes nitronium ion formation from remaining N_2O_5 and so promotes further nitration in the 2 position, leading to the 2,4-dinitro resorcinol found. Similarly, N_2O_4 produced by decomposition of N_2O_5 may also promote nitration in the 2- position via initial nitrosation in the usual manner for phenols.

Table I. Product Distribution in N_2O_5 Nitration of Resorcinol in Acetonitrile

N_2O_5 Mole Ratio	React. Time (secs)	Mole % (Resorcinol)	Mole % (2 Nitro)	Mole % (4 Nitro)	Mole % (2,4 Di Nitro)	Mole % (4,6 Di Nitro)	Mole % (2,4,6 Tri Nitro)
4.4	10						100
	5						100
	2						100
	1					5	95
3.3	10		6			3	91
	5					19	81
	2					19	81
	1				2	34	64
2.2	10			21	10	15	54
	5			16	14	23	47
	2			60	7	9	24
	1			66	7	2	25
1.1	10	2		83	12	2	1
	5	68	1	26	2	1	2
	2	79	1	17	2		2
	1	86		12	1		1

With hindsight, mono nitration of the 4- nitro resorcinol formed at N_2O_5 to resorcinol mole ratios of about 1 would have answered this question in an unambiguous manner, but these experiments were not carried out during the lifetime of the project. Nevertheless, these results indicate the versatility of N_2O_5 in organic solvents as a nitration agent for reactive substances where steric effects could be used to direct the position of nitration.

In this particular case the results obtained indicate that it should be possible to preferentially form the 4,6 di nitro derivative with good purity and in good yield by utilising first steric control and then steric and/or kinetic control in two successive "*mono*" nitrations with a 1:1 N_2O_5 to substrate ratio. This should be easier to control than the nitrous acid free nitration of resorcinol di acetate (*41*) currently recommended for 4,6 dinitro resorcinol preparation.

Other Novel Systems. Recently the power of N_2O_5 as a novel nitrating agent has been illustrated by the first preparation of *gem*- dinitro aryl methoxy methanes from the corresponding oximes (*42*) and in the C- nitration of Pyridine and Methyl Pyridines using N_2O_5 in liquid SO_2 (*43,44*). Also, Suzuki and his co-workers have shown interesting isomer distributions in what they call ozone mediated nitration with Nitrogen Dioxide (*45-51*).

The direct aromatic nitration of pyridine has long been a synthetic chemists dream, the N- nitro derivative being normally obtained (*52*), and the use of the N_2O_5/SO_2 system thus opens up new synthetic pathways in heterocyclic chemistry. Further, the indications are that this system is even more powerful than N_2O_5 in HNO_3 (*43,44*). It is possible that the use of other unusual solvent systems may extend the utility of N_2O_5 even further.

The results of Suzuki *et al* are of relevance to general aromatic nitration chemistry, in as much as that they indicate that the mode of action of N_2O_5 in organic solvent may not be as simple as was once thought (*4*). Thus while initially speculating that nitration occurred directly via molecular N_2O_5 formed *in situ* by reaction between NO_2 and Ozone (*45,50*), it was later concluded that nitration could also be initiated by NO_3 formed from the initial stages of this reaction (*46-49*), as well as by nitronium ion from the N_2O_5 formed later (*46,50*).

In order to explain the differences in isomer ratios obtained it was therefore postulated that the reaction between NO_2 and ozone in solution could be given by the following scheme (*53,54*) :

$$NO_2 + O_3 \longrightarrow [NO_2.O_3] \longrightarrow NO_3 + O_2 \quad (rapid)$$
$$NO_3 + NO_2 \longrightarrow N_2O_5$$

Thus in the presence of an oxidisable substrate the reaction can take place via attack of the NO_3 initially formed to give a radical cation which then captures an NO_2 molecule, the overall process being a charge transfer nitration. In the absence of easily oxidisable substrates nitration may take place via the molecular N_2O_5 formed, with concomitant formation of nitric acid. As the concentration of nitric acid builds up ionisation of the N_2O_5 may then occur leading to nitronium ion nitration.

The implication of this is that N_2O_5 itself could also be considered to be in equilibrium with NO_3 and NO_2, and hence, depending on the nature of the solvent and solute, could also have three possible modes of action (49). This opens up the possibility of tailoring the N_2O_5 environment to give any desired isomer ratio.

Interestingly, in their studies of the nitration of the halo benzenes, the *ortho/para* mono nitro derivative isomer ratios could be varied between 0.3 to 1 by varying the solvent or the substrate concentration (47,48). It should be noted that bromobenzene, either neat or in dichloromethane solution always gave relatively high meta nitration (1.7 - 7.7 %), increasing with substrate concentration. The *ortho/para* ratio decreased from 1.1 to 0.68 over the same range. Comparing these findings with the results for the nitration of bromo benzene by N_2O_5 in CCl_4 solution reported by Haines and Adkins (3) would indicate that the Suzuki NO_2/O_3 system is much less selective than is N_2O_5.

Literature Cited.
1. St-Claire Deville, H *Compt. Rend.* 1849, 29,257.
2. Eyring,H.;Daniels,F. *J. Amer. Chem. Soc.* 1930, 52, 1472 ; 1486.
3. Haines,L.B.;Adkins,H. *J. Amer. Chem. Soc.* 1925, 47, 1419.
4. Gold,V.;Hughes,E.D.;Ingold,C.K. *J. Chem. Soc.* 1950, 2452.
5. Hughes,E.D;Ingold,C.K.;*et.al*. *J. Chem. Soc.* 1938, 929.
6. Hughes,E.D.;Ingold,C.K.;*et.al*. *J. Chem. Soc.* 1950, 2400; 2441; 2452; 2467; 2628; 2657; 2678
7. Goddard,D.R.;Hughes,E.D.;Ingold,C.K. *J. Chem. Soc.* 1950, 2559
8. Olah,G.A. In *Industrial and Laboratory Nitrations*; Editors, Albright,L.F.; Hanson,C., ACS Symposium Series 22,1976,1.
9. Olah,G.A. *Nitrations Conference, Mambo Park, CA.* 1983 - SRI Report A135-822
10. Olah,G.A.;Malhotra,R.;Norang,S.C.In *Nitration : Methods and Mechanisms* Editor, Feur,H.; VCH :New York. 1989.
11. Harrar,J.E.;Pearson,R.K. *J. Electrochem. Soc.* 1983, 130, 108.
12. McGuire,R.R.;Coon,C.L.;Harrar,J.E.;Pearson,R.K. *U.S. Patents Nos*.4,432,902 (1984) and 4,525,252 (1985)
13. Marshall,R.J.;Shriffrin,D.J.;Walsh,F.C.;Bagg,G.E.G. *U.K. Patent No.* 2,229,449 ; *European Patent Application No.* 88 305440.5 (1988)
14. Bagg,G.E.G. *U.K. Patent No.* 2,245,003; *International Patent Application No.* PCT/GB89/01497 (1990)
15. Siele,V.I.;Worman,M.;Leccacorvi,J.;Hutchinson,R.W.;Motto,R.;Gilbert,E.E.;Benzinger,T.M.;Coburn,M.D.;Rohwer,R.K.;Davey,R.K. *Propellants, Explosives* 1981,6,67.
16. Fischer,J.W. In *Nitro Compounds : Recent Advances in Synthesis and Chemistry*, Editor,Feur,H;Nielson,A.T.; VCH,1990,267.
17. Bloom,A.J.;Fleischmann,M.;Mellor,J.M. *Tetrahedron Letters*, 1984,25,4971.
18. Bloom,A.J.;Fleischmann,M.;Mellor,J.M. *J. Chem. Soc. Perkin Trans.1*,1986,79.
19. Bloom,A.J.;Fleischmann,M.;Mellor,J.M. *Electrochimica Acta*, 1987,32,785.
20. Bloom,A.J.;Mellor,J.M *Tetrahedron Letters*,1986,27,873.
21. Bloom,A.J.;Mellor,J.M. *J. Chem.Soc. Perkin Trans 1*,1987,2737.
22. Boughriet,A.;Wartel,M. *Electrochimica Acta*,1991,36,889.
23. Golding,P.;Millar,R.W.;Paul,N.C.;Richards,D.H. *Tetrahedron Letters*, 1988,22,2731 ; 2735.
24. Golding,P.;Millar,R.W.;Paul,N.C.;Richards,D.H. *Tetrahedron Letters*, 1991,32,4985.
25. Golding,P.;Millar,R.W.;Paul,N.C.;Richards,D.H. *Tetrahedron*,1993,49,7037 ;7051 ; 7063.

26. Millar,R.W.;Colclough,M.E.;Golding,P.;Honey,P.J.;Paul,N.C.;Sanderson,A.J.;Stewart,M.J. *Phil. Trans. Roy. Soc. Lond. A*,**1992**,*339*,305.
27. Paul,N.C.;Arber,A.;Bagg,G.;Colclough,E.;Desai,H.;Millar,R.;Salter,-D. *Proc. 21st Ann. Conf. of ICT on Technology of Polymer Compounds and Energetic Materials*, **July 1990**, Fraunhoffer Institut Fur Trieb - und Explosivestoffe, Kalsruhe, Germany.
28. Bagg,G.;Leeming,W.B.H.;Swinton,P.F. *Proc. Joint Internat. Symp. on Compatibility of Plastics and Other Materials with Explosives, Propellants, Pyrotechnics and Processing of Explosives*, ADPA, San Diego, CA, **April 1991**,108.
29. Debenham,D.;Leeming,W.B.H.;Marshall,E.J. as Ref. 28, p 119.
30. Millar,R.W.;Paul,N.C.;Richards,D.H. *U.K. Patent Applications* 2,181,124 and 2,181,139 **(1987)**.
31. Bagg,G.E.G.;Arber,A.W. *International Patent Application No.* PCT/GB91/01250 **(1992)**.
32. Bagg,G.E.G.;Salter,D.A.;Sanderson,A.J. *U.S Patent No.* 5,128,001 **(1992)**
33. Bagg,G.E.G.;Salter,D.A.;Sanderson,A.J. *International Patent Application No.* PCT/GB90/01784 **(1991)**.
34. Bagg,G.E.G.;Arber,A.W. *International Patent Application No.* PCT/GB91/01249 **(1992)**.
35. Rodgers,M.J.;Swinton,P.F.;Bagg,G.E.C.;Arber,A.W. *International Patent Application No.* PCT/GB94/01050 **(1994)**.
36. Arber,A.W.;Cree,S.A.;Hammond,J.K. Personal Communication,1994.
37. Bird,M.L.;Ingold,C.K. *J. Chem. Soc.* 1938, 918.
38. Paterson,D.H.;Swinton,P.F. ICI Explosives,unpublished data.
39. Kashmiri,M.A.;Munawar,M.A.;Yasmin,R.;Khan,M.S. *J. Nat. Sci. and Math. (Pakistan)*,**1988**,*28*,289.
40. Kashmiri,M.A.;Khan,M.S. *J. Sci. Int. (Lahore)*,**1989**,*1*,177.
41. Schmitt,R.J.;Ross,D.S.;Hardee,J.R.;Wolfe,J.F. *J. Org. Chem.*, **1988**,*53*,5568.
42. Luk'yanov,O.A.;Pokhvisneva,G.V. *Bull. Soc. Sciences USSR, Div, of Chem. Science*,**1991**,*40*,1906.
43. Bakke,J.M.;Hegbom,I. *Acta Chemica Scandinavica*,**1994**,*48*,181.
44. Bakke,J.M.;Hegbom,I. *International Patent Application No.* PCT/NO93/00065 **(1993)**.
45. Suzuki,H.;Murashima,T.;Shimizu,K.;Tsukamoto,K. *J. Chem. Soc. Chem. Comm.* ,**1991**,1049.
46. Suzuki,H.;Ishibashi,T.;Murashima,T.;Tsukamoto,K. *Terahedron Letters*,**1991**,*32*,6591.
47. Suzuki,H.;Mori,T.;Maeda,K. *J. Chem. Soc. Chem. Comm.*,**1993**,1335.
48. Suzuki,H.;Mori,T. *J. Chem. Soc. Perkin Trans. 2*,**1994**,479.
49. Suzuki,H.;Murashima,T.;Mori,T. *J. Chem. Soc. Chem. Comm.* ,**1994**,1443.
50. Suzuki,H.;Murashima,T. *J. Chem. Soc. Perkin Trans. 1*,**1994**,903.
51. Suzuki,H.;Yonezawa,S.;Mori,T.;Maeda,K. *J. Chem. Soc. Perkin Trans. 1*,**1994**,1367.
52. Schofield,K. *Aromatic Nitrations*; Camb. Univ. Press,**1980**, Ch.6.5.2, p 88.
53. Wulf,O.R.;Daniels,F.;Karrer,S. *J. Amer. Chem. Soc.*,**1922**,*44*,2398.
54. Jones,E.D.;Wulf,O. *J. Chem. Phys.*,**1939**,*3*,873.

RECEIVED December 27, 1995

Chapter 8

Pilot-Plant-Scale Continuous Manufacturing of Solid Dinitrogen Pentoxide

T. E. Devendorf and J. R. Stacy

Energetic Materials Research and Technology Department, Indian Head Division, Naval Surface Warfare Center, Indian Head, MD 20640–5035

The Indian Head Division, Naval Surface Warfare Center (NSWC), has completed design, specification, installation, and start-up test/evaluation of a pilot plant scale system capable of producing solid N_2O_5 at a rate of 360 grams per hour (ca. 130 pounds per 160 hour month). N_2O_5 is produced by the gas phase oxidation of dinitrogen tetroxide (N_2O_4) with ozone (O_3) (i.e., $N_2O_4 + O_3 \rightarrow N_2O_5 + O_2$). Ozone is produced on site at an approximate rate of one pound per hour with a concentration of 8% in oxygen. The N_2O_4-ozone gas phase reaction takes place in a plug flow reactor. Four reactors in parallel are available to accommodate various production rates. The N_2O_5 collects as a solid in four parallel trains of three traps in series, each trap suspended in a dry ice/acetone solution. The solid N_2O_5 is dissolved in various solvents for use as a non-aqueous nitrating agent.

The Indian Head Division, Naval Surface Warfare Center (NSWC), was funded to develop a process for pilot plant scale continuous manufacturing of solid dinitrogen pentoxide (N_2O_5). The principle objectives of the program were to: (1) Transition the technology of N_2O_5 gas phase synthesis from the White Oak Detachment, NSWC, to the Indian Head Division, (2) Develop and demonstrate the technology to manufacture N_2O_5 in the gas phase with a 2.5 kilogram per day capacity, and (3) Manufacture sufficient N_2O_5 to support the parallel development effort of novel nitration processes at the Indian Head Division.

Synthesis of N_2O_5

N_2O_5 is a white crystalline solid which sublimes without melting. N_2O_5 is thermally unstable, and readily decomposes to oxygen and nitrogen dioxide. The decomposition of N_2O_5 is thermally dependent with a half life of approximately one hour at 0°C and one week at -20°C. At temperatures below -60°C, N_2O_5 has been stored for up to one year (Paul, N., Defense Research Agency, personal communication, 1993). N_2O_5 is the anhydride of nitric acid, and readily converts to the acid in the presence of moisture.

N_2O_5 is produced by the gas phase oxidation of dinitrogen tetroxide (N_2O_4) with dry ozone (O_3) according to literature methods (1-3).

$$N_2O_4 + O_3 \rightarrow N_2O_5 + O_2 \qquad (1)$$

The anhydrous N_2O_5 is trapped as a solid at low temperatures (ca. -78°C).

Indian Head Process. The pilot plant scale facility produces ozone with a PCI Ozone and Control Systems model G-28S ozone generator. This generator produces ozone by corona discharge at an approximate rate of one pound per hour with a concentration of 8% ozone in anhydrous oxygen.

The N_2O_4-ozone gas phase oxidation reaction takes place in a plug flow reactor. To accommodate various production rates, four reactors in parallel are available. Maximum capacity requires all reactors be operated simultaneously. The reactors are cooled externally by water.

The N_2O_5 collects as a solid with at an approximate rate of 360 grams per hour in four parallel trains of three glass traps in series. Each N_2O_5 collection trap is suspended in a dry ice and acetone solution. The N_2O_5 is dissolved in solvents such as methylene chloride for use as a non-aqueous nitrating agent. N_2O_5 is valuable as a nitrating agent when the use of nitric acid would not be possible or effective.

Bench Top Scale Process Development

The initial N_2O_5 bench top scale process development conducted at the Indian Head Division was based on the nitric acid dehydration reaction (4-5):

$$6 HNO_3 + P_2O_5 \rightarrow 3 N_2O_5 + 2 H_3PO_4 \qquad (2)$$

This method produced low yields of poor quality N_2O_5 contaminated with N_2O_4 and small quantities of HNO_3. In addition, the waste stream contained 1.5 grams of H_3PO_4 per gram of N_2O_5 produced. Both the N_2O_5's poor quality and the high ratio of waste to product made this method unacceptable for scale-up. A higher quality product was produced with reduced waste when the gas phase reaction (1-3) was pursued:

$$N_2O_4 + O_3 \rightarrow N_2O_5 + O_2 \qquad (3)$$

Original Process. Figure 1 is a schematic of the original N_2O_5 synthesis apparatus which was a large scale modification of an apparatus developed at China Lake (Fisher, J., NAWC China Lake, actual N_2O_5 manufacturing apparatus provided). The ozone generator used was a PCI Ozone and Control Systems model GL-1. This ozone generator produces ozone in a cell consisting of an inner stainless steel grounded electrode and a silver plated glass dielectric. The stainless steel electrode is cooled internally by water and the glass dielectric is cooled externally by an insulating oil. High voltage is applied to the glass dielectric. Oxygen passes between the electrodes through a high intensity corona discharge which converts a portion of the oxygen in the gas stream to ozone. The approximate rate of ozone production is one pound per day at a concentration of 5-8% in oxygen. The oxygen is provided as a compressed gas with a dew point of less than -60°C. When used with this N_2O_5 synthesis apparatus, the oxygen feed rate to the ozone generator was 10 SCFH at 15 psig.

Operation of this N_2O_5 synthesis apparatus was semi-continuous. The N_2O_4 feed reservoir was filled with liquid N_2O_4 prior to each use. When introduced, the ozone stream was split in two with the flow rate of each stream controlled by valves (2) and (3) respectively (Figure 1). Valve (3) controlled the ozone sweep through the N_2O_4 reservoir where the reaction was initiated. Partially reacted N_2O_4, along with any remaining ozone and oxygen, was swept into the primary reaction zone where the unreacted N_2O_4 contacted a second ozone stream. The flow rate of the second ozone stream was controlled by valve (2). The reaction zone was cooled using a water condenser. The reaction rate was manually controlled by varying the ratio of ozone passing through valves (2) and (3). To determine what adjustments to valves (2) and (3) were needed, the color and temperature in the primary reaction zone were monitored. Visually observing a brown tint in the reaction zone indicated incomplete N_2O_4 oxidation. The temperature in the reaction zone was maintained between 90 and 100°C.

The N_2O_5 product (90% yield based on N_2O_4) was collected in two glass traps suspended in a dry ice and acetone solution. The N_2O_5 was introduced to the traps through a nozzle extending half way into the trap. An optimized production rate of 43.1 grams of N_2O_5 per hour was achieved using this apparatus.

The limitations of this synthesis technique became apparent during scale-up of the process. The production rate was limited by the geometry of the N_2O_4 reservoir and the fixed length and diameter of the reaction zone, which did not provide sufficient retention time at higher flow rates. The reaction rate was difficult to control because there was no way to maintain a constant N_2O_4 vaporization rate. N_2O_4 vaporization depended on the ozone/oxygen sweep rate, N_2O_4 reservoir temperature, and the head space and N_2O_4 level in the reservoir. The large number of variables associated with the N_2O_4 vaporization rate along with the two valves controlling ozone addition, made this process extremely difficult to control. Consequently, this apparatus was unsuitable for scale-up.

Modified Process. To develop an N_2O_5 process which could more easily be scaled up, several modifications were made to the original apparatus. The resulting configuration is shown in Figure 2. The N_2O_4 reservoir was replaced with a gas phase reactor. Both the ozone and the N_2O_4 were introduced through two separate nozzles extending half way into the reactor. The N_2O_4 feed rate was controlled using valve (4) independent of the ozone addition rate which was controlled using valves (2) and (3). The larger reaction zone increased the residence time allowing the reaction to go to completion over a wide range of reactant flow rates. The N_2O_4 reservoir was placed in a constant temperature bath to control the vaporization rate and feed pressure of the N_2O_4.

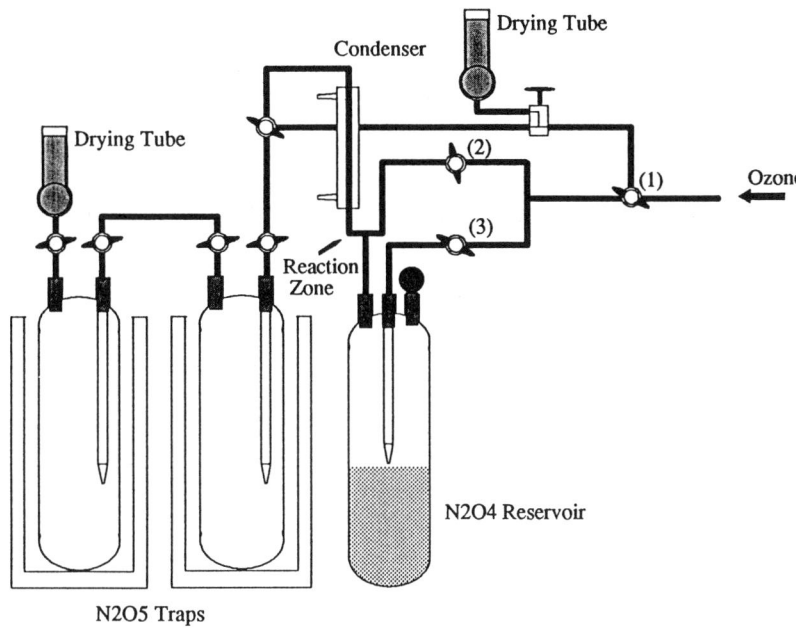

Figure 1. Original N_2O_5 Process.

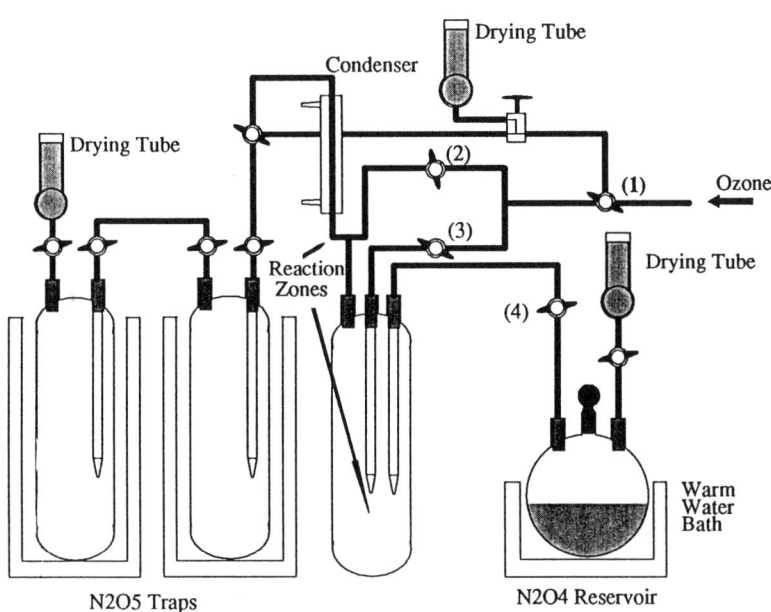

Figure 2. Modified N_2O_5 Process.

This apparatus was run under a variety of conditions. These included variations in ozone feed rate and split between the two reaction zones, variations in N_2O_4 feed rate, and different N_2O_4 bath temperatures. An optimized production rate of 70.1 grams per hour was obtained using this apparatus. This result was obtained by operating the ozone generator near maximum capacity and supplying as much N_2O_4 as possible without causing incomplete conversion to N_2O_5.

To achieve the optimized production rate, the model GL-1 ozone generator was operated with an oxygen feed rate of 20 SCFH at 8 psig. The entire ozone output was injected into the reactor through the nozzle with out any split to the second reaction zone. The N_2O_4 bath was maintained at a constant temperature of 36°C and the N_2O_4 reservoir feed pressure was 6 psig. In this manor the entire process was controlled by adjusting the N_2O_4 feed valve and observing the color and temperature of the resulting reaction products as they left the reaction zone.

The limitations of the apparatus were primarily a result of the reactor configuration. The reactor did not provide sufficient mixing of the reactants which limited the overall yield and increased the amount of ozone required to completely convert the N_2O_4 to N_2O_5. In addition, un-reacted N_2O_4 accumulated in the reactor dead spaces.

Plug Flow Reactor Process. To achieve better mixing and flow patterns, and thereby improve the N_2O_5 production rate, the original tank type reactor was replaced with a plug flow reactor. The new reactor was a modified Allihn condenser with a 500 mm jacket. Nine thermal wells were inserted into the condenser as illustrated in Figure 3, to more clearly track the extent of reaction. The reactor had two inlet paths. The first allowed the introduced gas to flow in a direct path with the only restrictions in the flow pattern caused by the bulbs in the Allihn condenser. The second inlet was through an 8 mm inside diameter concentric tube that tapered to a 4 mm inside diameter tip at the outlet. Turbulence and mixing were dramatically increased in this reactor as a result of the bulbs in the Allihn condenser and the increased gas velocity caused by the tapered tip of the second inlet. Tests showed no significant variation in N_2O_5 production rate when the ozone and N_2O_4 were introduced through either inlet. Therefore, to prevent unnecessary back pressure on the ozone generator, the N_2O_4 was introduced through the tapered inlet.

The operation of this apparatus was also semi-continuous with the N_2O_4 supplied from a two liter stainless steel pressure vessel which was filled with liquid N_2O_4 prior to each operation. The pressure vessel was maintained in a steam heated water bath with the temperature controlled at 37°C. The feed rate of N_2O_4 was controlled by adjusting a stainless steel needle valve. This valve provided better flow control than the glass stopcock used with the previous apparatus.

An optimized production rate of 86.0 grams per hour was obtained using this apparatus. This result was obtained by operating the ozone generator near its maximum capacity and feeding as much N_2O_4 as possible while ensuring complete conversion to N_2O_5. To achieve the optimized production rate, the model GL-1 ozone generator was operated with an oxygen feed rate of 20 SCFH at 13 psig. As before, the N_2O_5 was collected in two glass traps suspended in a dry ice and acetone solution. This apparatus was considered suitable for scale-up and the pilot plant scale process was developed based on this model.

Tests indicated that when no external cooling water was running, the reactor temperature reached a maximum of 73°C at the first thermal well, and rapidly cooled to ambient temperature in the later thermal wells. The majority of tests were conducted with cooling water running to maintain the temperature between 50 and 60°C at the first thermal well of the reactor. Measurements of the cooling water flow rate and the inlet and outlet temperatures provided a rough estimate of the heat of reaction. A value 246 calories per gram of N_2O_5 produced was determined in this manner which compares with a value of 294 calories per gram using handbook heats of formation.

Pilot Plant Scale Process Development

The introduction of the plug flow reactor provided a method of manufacturing N_2O_5 that was more appropriate for scale-up. However, it was not possible to determine the maximum capacity of the plug flow reactor or a scaling factor, because the model GL-1 ozone generator was operating at its maximum capacity during most of the bench top scale tests. For the pilot plant scale process four reactors identical to the one used for the bench top scale process were installed in parallel to provide the required capacity. In this manner various production rates less than the maximum capacity could also be accommodated. The reactors are cooled externally by water. Figure 4 illustrates the pilot plant scale process.

Ozone Generation. One of the earliest tasks undertaken was to purchase an ozone generator capable of providing sufficient ozone to meet the 2.5 kilogram per day N_2O_5 manufacturing requirement. Calculations showed the amount of ozone required to be approximately one pound per hour. The ozone generator selected was a PCI Ozone and Control Systems model G-28S which produces ozone by corona discharge as previously described at an approximate rate of 28 pounds per day with a concentration of 8% ozone in oxygen. The model G-28S ozone generator is operated with an oxygen feed rate of 120 SCFH at 15 psig. At the time it was purchased, this ozone generator was the largest capacity generator available that was not custom built. Several oxygen cylinders are connected to a manifold which supplies the ozone generator. This ensures continuous operation while empty oxygen cylinders are replaced by full cylinders.

An ozone monitor was purchased for use with the model G-28S ozone generator to provide continuous display of the ozone concentration in oxygen as it is fed to the N_2O_5 reactors. The sampling system consists of inlet valves for the sample and zero gas along with a solenoid valve which changes position to allow either the sample or zero gas into the sample chamber. The UV absorption of the gas in the sample chamber is measured and the ozone concentration is calculated using Beer's Law. The ratio of intensities is determined and the resulting ozone concentration displayed. Since the concentration determined by the photometer is based on the ratio of light intensities, the actual intensity of the light is not important.

N_2O_4 Supply. A continuous N_2O_4 feed system was achieved by applying moderate heat to the bulk storage cylinder using electric heat tape and a temperature controller with a set point between 30 and 35°C. This induces flow of liquid N_2O_4 from the cylinder bottom outlet (dip tube) into an expansion chamber. The expansion chamber is an empty two liter pressure cylinder maintained in a steam heated water bath with the

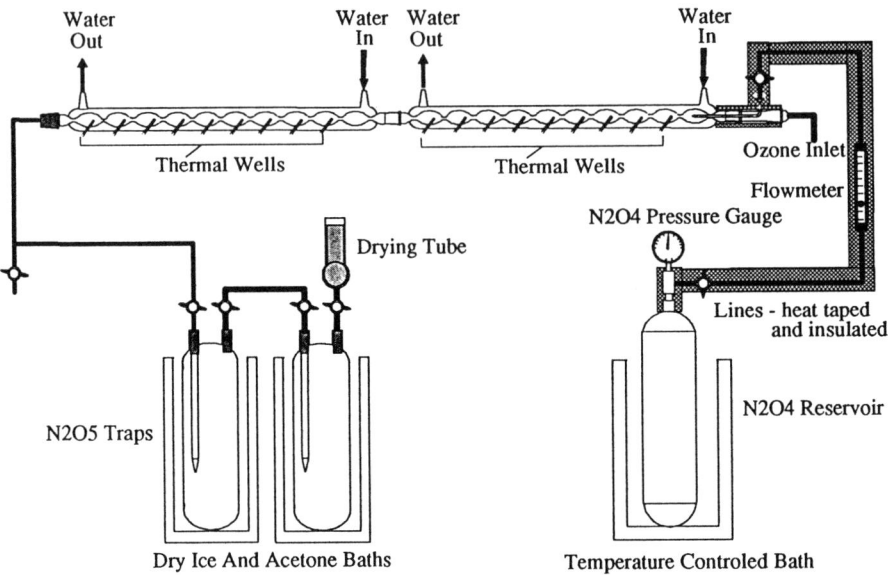

Figure 3. Plug Flow Reactor N_2O_5 Process.

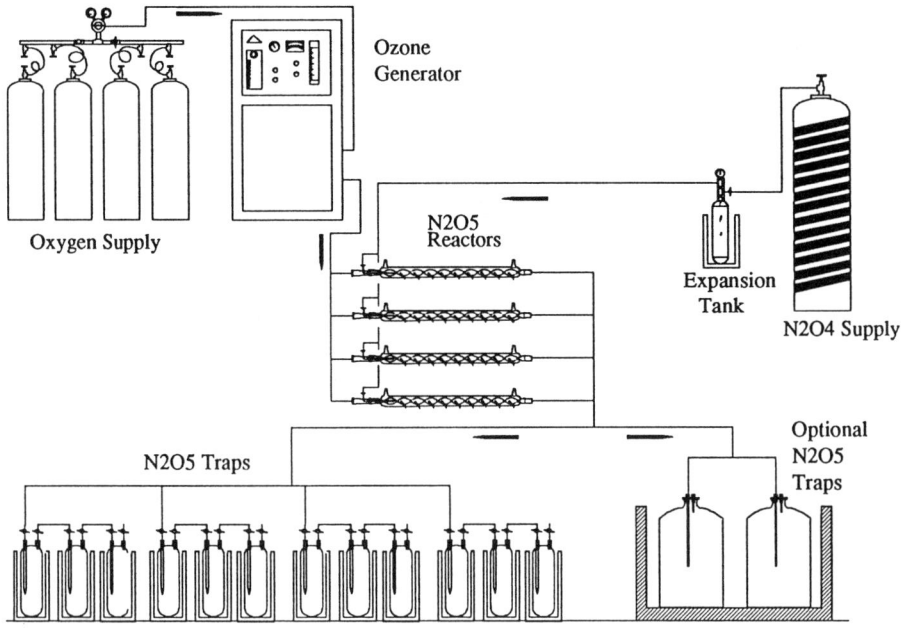

Figure 4. Pilot Plant Scale N_2O_5 Process.

temperature controlled between 35 and 40°C. The N_2O_4 is vaporized in the expansion chamber generating approximately 10 psig pressure which is used to drive the N_2O_4 into the four reactors. An advantage of this feed system was the elimination of the need to handle the dangerous liquid N_2O_4 because the small feed cylinders no longer were required.

Both the N_2O_4 feed rate and the ozone feed rate to each of the four reactors is controlled by adjusting a stainless steel needle valve and by observing the accompanying flow meter. The typical ozone in oxygen feed rate to each reactor when all four are being used is 30 SCFH with an accompanying N_2O_4 feed rate of 1.0 SCFH. An individual reactor can be shut down and the flow of ozone and N_2O_4 diverted to the other three reactors if a problem arises. Individual reactors have been operated successfully with an ozone in oxygen feed rate of as much as 60 SCFH and an accompanying N_2O_4 feed rate of 2.0 SCFH.

Solid N_2O_5 Trapping. The N_2O_5 collects at an approximate rate of 360 grams per hour as a solid in glass traps suspended in a dry ice and acetone solution. Four trains in parallel of three glass traps each in series are used to collect the N_2O_5 when operating at maximum capacity. The flow rate of N_2O_5 to each of the four trains of traps is controlled by adjusting a stainless steel needle valve and observing the accompanying flow meter. This ensures a balanced quantity of N_2O_5 collects in each of the traps. An individual train of traps can be shut down and the flow of N_2O_5 diverted to the other three trains if a problem occurs or if full traps need to be replaced by empty ones. Operating by this technique the pilot plant scale manufacture of N_2O_5 is essentially continuous and has in fact operated non-stop for 24 or more hours on several occasions.

The solid N_2O_5 manufactured is stored in a low temperature freezer at -70°C. When needed, the required amount of solid N_2O_5 is dissolved in a solvent for use as a nitrating agent.

There are several advantages to producing solid N_2O_5. First, is the flexibility of being able to choose a particular diluent solvent because the N_2O_5 is not manufactured in solution. Second, is the reduced volume needed for low temperature storage. Finally, N_2O_5 dissolved in a solvent requires higher storage temperatures to prevent the N_2O_5 from precipitating out of solution, or temperature cycling to re-dissolve the N_2O_5 prior to use. Because N_2O_5 is thermally unstable, storage at higher temperatures and/or temperature cycling will increase the rate of N_2O_5 decomposition and shorten the shelf life of the solution.

Analytical Determination of N_2O_5 Solutions. To support N_2O_5 manufacturing, an analytical method for determining the shelf life N_2O_5/methylene chloride solutions was developed (6). This method uses Fourier Transform Infrared spectroscopy (FTIR) using 0.1 mm fixed path length liquid KBr cells to determine the concentration of N_2O_5, N_2O_4, and nitric acid. The absorbence maxima for N_2O_5 is 560 cm^{-1} and for N_2O_4 is 428 cm^{-1}. Calibrations curves were established and the decomposition of N_2O_5 with time was studied at 0°C.

An estimate of the quality of N_2O_5 in solution with methylene chloride was obtained using a Milton Roy Company "Spectronic 20" UV/visible light spectrophotometer. Solutions with a known weight of N_2O_4 in methylene chloride

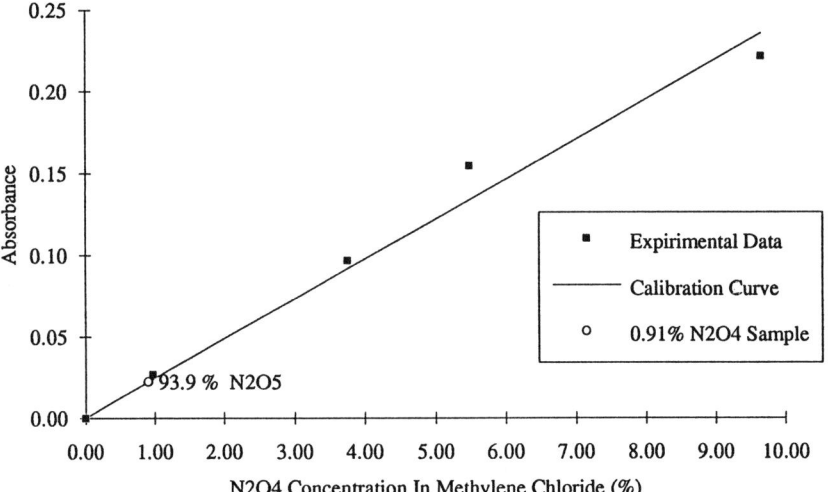

Figure 5. N_2O_4 Concentration vs. Absorbence.

A calibration curve (Figure 5) was constructed using standard regression techniques. A 15 % solution of N_2O_5 in methylene chloride was tested using this technique and the minimum purity of the product determined to be 93.9 % (92.9 mol %) N_2O_5 and 6.1 % (7.1 mol %) N_2O_4.

Conclusion

The Indian Head Division, Naval Surface Warfare Center (NSWC), currently is capable of manufacturing solid dinitrogen pentoxide (N_2O_5) at the pilot plant scale. The N_2O_5 is produced by the gas phase oxidation of dinitrogen tetroxide (N_2O_4) with ozone (O_3). During the past two years the rate of N_2O_5 manufacturing improved by an order of magnitude, from approximately 35 grams per hour to the current rate of 360 grams per hour. This represents a significant improvement in the cost of N_2O_5.

In addition to developing a process to manufacture N_2O_5 in the gas phase with a 2.5 kilogram per day capacity, the other objectives of the sponsor were met. These included transitioning the technology of N_2O_5 gas phase synthesis from the White Oak Detachment to the Indian Head Division, and manufacturing approximately 60 pounds of N_2O_5 to support the parallel development of a novel nitration processes.

The Indian Head Division is one of two domestic sources of solid N_2O_5. The advantages of solid N_2O_5 are greater flexibility in preparing various solutions in a particular solvent, reduced storage volume at low temperatures, and the ability to store the solid product for indefinite time periods at temperatures below -70°C.

Acknowledgments

The authors wish to express their appreciation to Dr. Richard Miller of the Office of Naval Research who funded this work. They would also like to thank Walter Carr, Ricky Cox, Tim Dunn, Jody Lang, and Al Stern for their assistance in developing and operating the facility.

References

1. Harris, A. D.; Trebellas, J. C.; Jonassen, H. B. *Inorganic Synthesis*; McGraw-Hill: New York, NY, 1950; Vol 9.; pp 83-88.
2. Guye, P., United States Patent 1,348,873, 1920.
3. Gruenhut, N. S.; Goldfrank, M.; Caesar, G. V.; Cushing, M. L. *Inorganic Synthesis*; McGraw-Hill: New York, NY, 1950; Vol 3.; pp 78.
4. Caesar, G. V.; Goldfrank, M. *J. Am. Chem.* **1946**, *vol 68*, pp 372-375.
5. Russ, F.; Pokorny, E. *Monatshefte Fur Chemie*. **1913**, *vol 34*, pp 1051-1060.
6. Smith, V. G. *JANNAF Propellant Development and Characterization Subcommittee Meeting*; CPIA: Columbia, MD, 1994; No. 609.

RECEIVED February 29, 1996

Chapter 9

Separation of Dinitrogen Pentoxide from Its Solutions in Nitric Acid

Robert D. Chapman[1] and Glen D. Smith

TPL, Inc., 3768 Hawkins St., NE, Albuquerque, NM 87109

A convenient process for the separation of nitrogen pentoxide from its solutions in nitric acid (as commercially prepared by electrolysis of N_2O_4–HNO_3) involves the chemical absorption of the nitric acid by sodium fluoride (producing sodium hydrogen fluoride) in inert organic solvents, such as acetonitrile, suitable for use with many nitrolyzable substrates. Contact time between the N_2O_5–HNO_3 solution and the sodium fluoride solid is critical due to a competing sorption of the N_2O_5 in the later stages of the process, speculated to be due to hydrogen bonding between N_2O_5 and $NaHF_2$. During this stage, continued contact of the solution with the partially spent acid absorbent (sodium fluoride) removes even small amounts of residual nitric acid but also lowers the recoverable yield of N_2O_5.

The objective of this work was the development of a process for an improved preparation or separation of acid-free nitrogen pentoxide. Improved methodology for N_2O_5 is potentially valuable for numerous applications in synthetic nitro chemistry, including a variety of useful transformations developed recently by researchers in the British Defence Research Agency (RARDE, Fort Halstead) (*1*). These transformations, many leading to energetic materials of defense interest, include C- and N-nitrations or nitrolyses, oxacyclic ring cleavages leading to vicinal dinitrate esters, and selective O-nitrations without affecting sensitive ring structures. Since the latter transformations require nitrations conducted under relatively mild (i.e., non-acidic) conditions, improvements in methodology for acid-free N_2O_5 should be particularly beneficial.

[1]Current address: Research and Technology Division, Weapons Division, Naval Air Warfare Center, Code 474220D, China Lake, CA 93555–6001

There are two general, industrially feasible preparative routes to N_2O_5 (1): (1) ozonolysis of N_2O_4 (2–5); and (2) electrolytic oxidation of N_2O_4 solutions in nitric acid (6–11). Although the latter electrochemical reaction was reported as early as 1948 (12), its potential for industrial-scale production of energetic materials was not recognized and exploited until Lawrence Livermore National Laboratory researchers developed an "industrial" process based on it in the early 1980s (6–8). The two half-reactions involved in this electrolytic process are oxidation of N_2O_4 to N_2O_5 (anodic reaction) and reduction of HNO_3 to N_2O_4 (cathodic reaction):

$$N_2O_4 + 2NO_3^- \leftrightarrow 2N_2O_5 + 2e^-$$
$$2e^- + 4HNO_3 \leftrightarrow N_2O_4 + 2NO_3^- + 2H_2O$$
$$\overline{4HNO_3 \leftrightarrow 2N_2O_5 + 2H_2O}$$

The electrolytic process has since been developed extensively by British researchers (9, 11, 13), and its application to the improved production of energetic materials has been reviewed (1). Modifications of the electrolytic oxidation of N_2O_4 have also been reported recently: Bloom, Fleischmann, and Mellor (University of Southampton) oxidized a 0.25 M solution of N_2O_4 in dichloromethane to N_2O_5 (0.15 M, 80% current efficiency at 75% conversion) using tetrabutylammonium nitrate and tetrabutylammonium tetrafluoroborate as electrolytes (14). The product solution was successfully used for organic nitrations.

A relevant example of extraction technology applied to the N_2O_5–HNO_3 system was reported about 50 years ago by Caesar and co-workers at Stein, Hall & Co. (Long Island City, NY) for the purpose of preparing nitrate esters of carbohydrates (15–17). The objective of that group's development of this process modification is summarized by the opening paragraph of their patent (16): "This invention relates to acid absorption, and more particularly to a process for the absorption or elimination of acids from a reaction which liberates acids. The invention includes especially the application of the process for the selective removal of nitric acid from a non-aqueous solution thereof with its anhydride nitrogen pentoxide."

The basis of this system is simply the affinity of fluoride ions for protons, forming bifluoride ion. In the specific systems utilizing sodium fluoride in the presence of nitric acid, the process is simply the metathetical reaction of the acid with sodium fluoride to make sodium nitrate and hydrogen fluoride; the HF subsequently combines with sodium fluoride to make the bifluoride, which is insoluble in typical chlorinated organic solvents:

$$HNO_3 + 2NaF \xrightarrow[0\,°C]{CHCl_3} NaHF_2 \downarrow + NaNO_3 \downarrow$$

As an example of the absorptive capacity of NaF for HNO_3, a solution of chloroform containing 51.5 g/L HNO_3 was agitated with sodium fluoride corresponding to 20 g

NaF per liter of solution. After removal of acidified NaF (NaHF$_2$), residual acid corresponded to 17 g/L (*16*), a reduction equivalent to 1.72× the weight of NaF used. Absorptive capacity is therefore greater than suggested by the stoichiometry of the reaction above. As an example of the extent of nitric acid absorption possible with this system, a chloroform solution containing 50.6 g/L HNO$_3$ (0.80 *M*) was agitated at 10–15 °C with sodium fluoride corresponding to 100 g NaF per liter of solution. After removal of acidified NaF (NaHF$_2$), residual acid corresponded to only 0.9 g/L (0.014 *M*), a removal of >98% of the acid present (*15*). The efficiency of removal achievable with this system generally correlates with the quantity of sodium fluoride used. The degradation of N$_2$O$_5$ during such typical treatments was found not to be significant, and several examples of nitrations of acid-sensitive substrates such as starches and dextrins were demonstrated using the N$_2$O$_5$ solutions.

Technical Approach

Samples of production-grade N$_2$O$_5$/HNO$_3$ solution, prepared by electrolysis of N$_2$O$_4$/HNO$_3$, were procured from Thiokol, Inc. (Longhorn Division, Marshall, TX). The Thiokol samples were nominally 25% N$_2$O$_5$ in nitric acid. For certain experiments, N$_2$O$_5$/HNO$_3$ samples were prepared on site at TPL, Inc. by ozonolysis of N$_2$O$_4$/HNO$_3$ solutions. Initially, ^{14}N NMR spectroscopy was considered as a possibly convenient means of characterizing N$_2$O$_5$/HNO$_3$ solutions with respect to N$_2$O$_5$ content: ^{14}N NMR's relatively large dispersion produces signals at δ 0 for NO$_3^-$, δ –126 for NO$_2^+$, $\delta \sim$ –60 for N$_2$O$_5$ in inert solvents (CCl$_4$, CH$_3$NO$_2$), and δ –43 for 100% HNO$_3$. Solutions of N$_2$O$_5$ in HNO$_3$ in theory should exhibit one signal between δ –43 and δ –60, depending on N$_2$O$_5$ content. Although a solution of N$_2$O$_5$/HNO$_3$ was reported (as an unpublished result) to exhibit a shift of δ –47.5 (*18*), when these authors later published collective values of ^{14}N shifts for anhydrous nitrates (including N$_2$O$_5$ in other solvents), this value was missing (*19*). Consistent with the discrepancies noted here, Ogg and Ray reported an ^{14}N NMR spectrum for an N$_2$O$_5$/HNO$_3$ solution that is virtually indistinguishable from that of 100% HNO$_3$ (*20*).

Therefore, we adopted *for solutions containing significant nitric acid* a reported ^{1}H NMR technique in which chemical shift differences between N$_2$O$_5$/HNO$_3$ solutions and 100% HNO$_3$ are measured, which have been tabulated by Harrar and Pearson (*6*). This technique is based on the precedent observation by Happe and Whittaker of a concentration-dependent chemical shift of the proton in nitric acid (*21*). Harrar and Pearson report ^{1}H chemical shift *differences* between quantified N$_2$O$_5$ solutions and "anhydrous" nitric acid, but they do not report absolute chemical shifts of either sample. An absolute chemical shift for 100% HNO$_3$ was reported by Happe and Whittaker as δ 11.853. They also report that a transformation of the nitric acid concentration via $P(X) = 3X/(2 - X)$, where X is the mole fraction of nitric acid, results in a function, P, that is approximately linear with respect to the ^{1}H chemical shift, δ_H. This same transformation applied to the N$_2$O$_5$–HNO$_3$ system (but defining X as the mole fraction of N_2O_5 and with Harrar and Pearson's chemical shift differences corrected by δ 11.853 as the reference value for anhydrous nitric acid) also produces a linear function (equation 1) with a squared linear correlation coefficient $r^2 = 0.9996$.

$$P(N_2O_5) = \frac{3X_{N_2O_5}}{2 - X_{N_2O_5}} = 0.13108\delta_H - 1.5563 \tag{1}$$

This calibration equation allows an estimation of N_2O_5 content based on the absolute chemical shift measured for the protons in these solutions. (Initially, we attempted this estimation by comparison to a commercial sample of "100%" nitric acid, similarly to Harrar and Pearson's procedure. However, due to the difficulty of maintaining anhydrous nitric acid at this purity, chemical shifts were observed for this reference material that were more consistent with material of ~98% purity, according to the chemical shift calibration reported by Happe and Whittaker. Thus, one measured shift of δ 11.668 corresponds to a nitric acid concentration of only 98.04 mol% HNO_3. We therefore adopted the approach of referencing to the reported absolute shift for 100% HNO_3.)

Furthermore, because of the known effects of paramagnetic solutes on shifting NMR absorption peaks, the N_2O_5/HNO_3 samples from Thiokol had to be redistilled. These solutions were otherwise dark-colored, bordering on black, from dissolved or particulate metal from the industrial vessels used for the electrolysis. (There was also concern about possible detrimental effects of the metal and metal salt impurities upon isolation of pure N_2O_5 via other techniques investigated in the course of this study.) Distillations were conducted at low temperature under static vacuum by bulb-to-bulb transfer from a ~0 °C pot to a liquid-nitrogen-cooled receiving flask. The distilled solutions were light yellow, as expected for N_2O_5/HNO_3. After storage and redistillation, the N_2O_5/HNO_3 sample used in the acid extraction experiments analyzed as 17.63% N_2O_5 by the ^1H NMR technique described above.

Our initial experiments toward an acid-extraction process—utilizing oven-dried reagent-grade sodium fluoride as the acid scavenger in less-polar solvents—followed the general procedure described by Caesar (*15–17*), in which chloroform is used as the predominant solvent for N_2O_5 from which acid is to be removed. These systems varied several conditions: N_2O_5/HNO_3 samples (two homemade batches plus a Thiokol sample); N_2O_5/HNO_3 concentrations in chloroform (5–25 w/v%); ratios of sodium fluoride to acid (ranging from 1~13 weight-equivalents); temperatures (ranging from –20 to 0 °C, though most were run at 0 °C); and contact times (with or without manual agitation) ranging from ~15 min to >2 h. Transfer of liquids was accomplished under a flow of dry nitrogen or argon or in a glove bag; and manual agitation by swirling the reaction flask was usually performed throughout the reaction (i.e., the contact time). It may be expected that surface area of the acid absorbent (i.e., sodium fluoride) and agitation of the heterogeneous reaction would influence its efficiency.

Results and Discussion

A typical result from the chloroform experiments was that significant residual acid remained after short contact times (~15 min), especially when low sodium fluoride/acid ratios were employed. When longer contact times were allowed (>2 h) at high sodium fluoride/acid ratios, absorption of essentially all of the nitric acid seemed to be achieved (according to initial analyses by ^{1}H NMR)—but at the expense of apparently sorbing (absorbing or adsorbing) essentially all of the N_2O_5 as well.

A serendipitous variation in conditions for this acid extraction approach involved a change in solvent to acetonitrile, which was also expected to be convenient for practical nitrations by N_2O_5. The first experiment in this solvent gave the first indication that nitric acid and nitrogen pentoxide were removed *at different rates* in systems employing this acid absorption principle. In particular, ~85% of the original nitric acid content was removed in this experiment (according to ^{1}H NMR) while the majority (≥98%) of the expected N_2O_5 content was still present according to quantitative nitration of benzene following the method described by Bloom et al. (*14*). This result dictated a more careful analysis of the course of the reaction as acid is absorbed. Thus, a repetition of the conditions of that extraction experiment next entailed careful monitoring of the progress of the sorptions via ^{1}H and ^{14}N NMR spectrometry. In both NMR analyses, the acetonitrile solvent signals were used as internal standards for quantification of the relative amount of nitric acid (by ^{1}H NMR) or total nitro content (by ^{14}N NMR).

The results were initially analyzed, in the case of the ^{14}N data, by plotting the simple ratio of integrals of the nitro peak (δ –47 to –64) and the acetonitrile peak (δ –139)—corresponding to $(2[N_2O_5]+[HNO_3])/[CH_3CN]$—versus time. The best-characterized run was continued up to a reaction time (i.e., contact time between N_2O_5/HNO_3 and sodium fluoride) of 269 min at time intervals of ~30 min. The experiment utilized 9.79 g of N_2O_5/HNO_3 solution (measured as 17.63% N_2O_5) plus 41.40 g of sodium fluoride (NaF/HNO_3 mole ratio = 7.7) in 40 mL acetonitrile, maintained at 0 °C during monitoring of the reaction. After this duration of 4.5 h, real-time inspection of the NMR data indicated that residual HNO_3 and N_2O_5 contents were both quite small. Relevant NMR data from this experiment are given in Table I.

The nitro content kinetics from ^{14}N NMR data are shown in Figure 1. The trend exhibited by the data is also interesting: after an initial exponential decay attributable to absorption of nitric acid by sodium fluoride, the onset of a faster removal of nitro species (presumably N_2O_5) occurs at about 150–180 min. The trend of the later removal also looks like a decaying exponential, so the observed trend may be a convolution of two phenomena behaving in this manner. A further consideration of natural phenomena may suggest, however, that events such as the sorption of chemical species tend not to start instantaneously after absolute inactivity for long periods (in the absence of obvious extrinsic factors, such as the addition of reactants). On this

Table I. NMR Data From Sodium Fluoride Acid Absorption (Acetonitrile)

Time from NaF addn (min)	δ_{1H} (HNO$_3$) (^{1}H)	HNO$_3$ integration[a] (^{1}H)	CH$_3$CN integration[a] (^{1}H)	δ_{14N} (N$_2$O$_5$)	HNO$_3$ + N$_2$O$_5$ integration[a] (^{14}N)	CH$_3$CN integration[a] (^{14}N)
0	13.81	0.666	13.898	−47.16	13.01	65.00
29	11.40	0.185	7.089	−50.23	5.86	66.11
59	10.88	0.121	7.427	−53.90	4.20	72.99
89	10.26	0.0720	7.494	−57.62	1.47	36.29
117	9.73	0.0252	7.651	−60.51	1.13	36.85
151	9.43	0.0164	7.680	−62.79	1.04	37.75
179	9.11	1.00	616.55	−63.39	13.86	528.50
210	8.49	0.473	313.096	−63.57	0.647	37.775
240	8.05	0.175	304.017	−64.02	0.339	37.209
269	7.86	1.027	3570.985	−64.23	0.924	572.621

[a] Arbitrary units.

basis, it should be presumed that the later phenomenon is not another simple exponential decay with an absolute onset at a time several hours into the system reaction. Rather, another physically realistic mathematical function that a chemical phenomenon such as N$_2$O$_5$ sorption may be expected to follow is the *logistic sigmoid*, which is followed by autocatalytic second-order reactions, for example (22). This function has the general form:

$$C = C_{min} + \frac{C_{max} - C_{min}}{1 + e^{-k(t-t_{50})}} \quad (2)$$

in which (for example) a concentration, C, ranging from a minimum value, C_{min}, to a maximum value, C_{max}, is expressed as a function of time; t_{50} is the time at which C reaches the midpoint between C_{min} and C_{max}. This function initially has low values but increases at a rate that approximates an exponential function until reaching the midpoint; the maximum value is then approached at a rate that approximates an inverted exponential decay.

A physical phenomenon that may exhibit such behavior is the gradual onset of an event (e.g., a new chemical reaction) caused by the intermediate formation of a reactive species. The following speculation may be offered about the chemical nature of this phenomenon. The ^{1}H NMR method for determination of N$_2$O$_5$ in nitric acid is based on the observed changes in chemical shifts believed to be caused by hydrogen bonding between covalent nitric acid and nitrate ion (due to NO$_2^+$NO$_3^-$) to produce a solvated nitrate ion (*6, 21*). This hydrogen bonding may be sufficiently strong to allow sorption (hypothetically, chemical adsorption) of N_2O_5 onto the acidified solid acid-absorbent which had started as sodium fluoride. Until the concentration of sodium bifluoride ($NaHF_2$) had appreciably built up via absorption of HNO_3, there was no significant affinity of sodium *fluoride* for N_2O_5, consistent with the results of Caesar (*16*). When the solid acid-absorbent eventually contained a high surface concentration of acidic sites, these may then assume an affinity for sorption of N_2O_5 via hydrogen bonding similar to that causing NMR chemical shift changes.

The data points of Figure 1 (derivable from Table I data) were therefore fit to a convolution of a simple exponential decay with finite offset followed by another removal of nitro species following a logistic sigmoid function. The convolution assumes that the apparent offset (*C*) of the simple exponential decay constitutes the experimental maximum value of the logistic decay. The overall behavior thus has the form:

$$\frac{2[N_2O_5] + [HNO_3]}{[CH_3CN]} = \left(ae^{-k_A t} + C\right) - \left(\frac{C}{1 + e^{-k_B(t-t_{50})}}\right) \quad (3)$$

In this convolution, k_A and k_B are characteristic kinetic rate constants for disappearance of ^{14}N due to chemical absorption of HNO_3 by NaF (k_A) and autocatalytic sorption of N_2O_5 by $NaHF_2$ (k_B, analogous to k in equation 2). The fit of the complete data set to this function is quite good, as shown by the regression curve in Figure 1. The regression equation (with *t* in minutes hereafter) is:

$$\frac{2[N_2O_5] + [HNO_3]}{[CH_3CN]} = \left[0.16766 \cdot e^{-0.035315 t} + 0.031898\right] - \left[\frac{0.031898}{1 + e^{-0.036777(t-212.40)}}\right] \quad (4)$$

Another transformation of the dependent variable may make the meaning of the kinetic data more straightforward in appearance: $(2[N_2O_5]+[HNO_3])/[CH_3CN]$ was normalized to its initial value at time zero, $(a + C)$, predicted by the decay function of equation 3. The *relative* removal of nitro species content is then easily envisioned in Figure 2, which plots the normalized function

$$\frac{\left(\frac{2[N_2O_5] + [HNO_3]}{[CH_3CN]}\right)}{\left(\frac{2[N_2O_5] + [HNO_3]}{[CH_3CN]}\right)_0} = \left[(1 - 0.15984) \cdot e^{-0.035315 t} + 0.15984\right] - \left[\frac{0.15984}{1 + e^{-0.036777(t-212.40)}}\right] \quad (5)$$

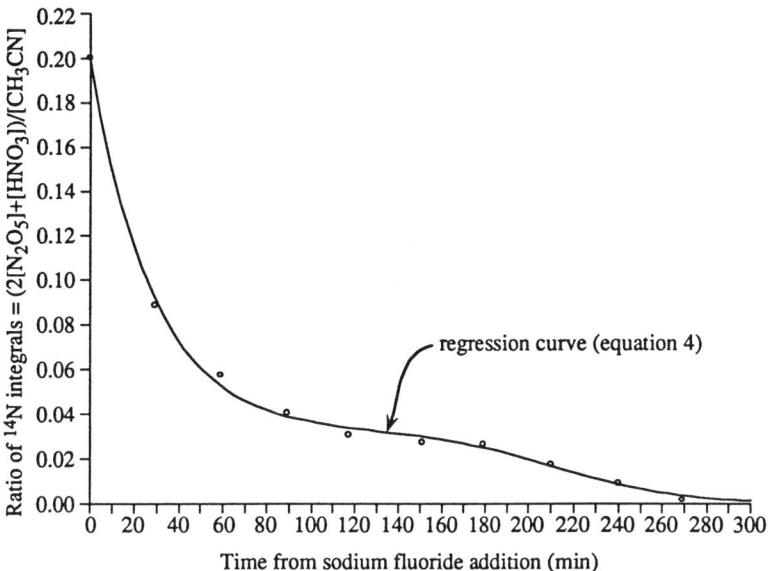

Figure 1. Concentration data analysis, by ^{14}N NMR, of nitric acid absorption (N_2O_5/HNO_3 in acetonitrile, 0 °C)

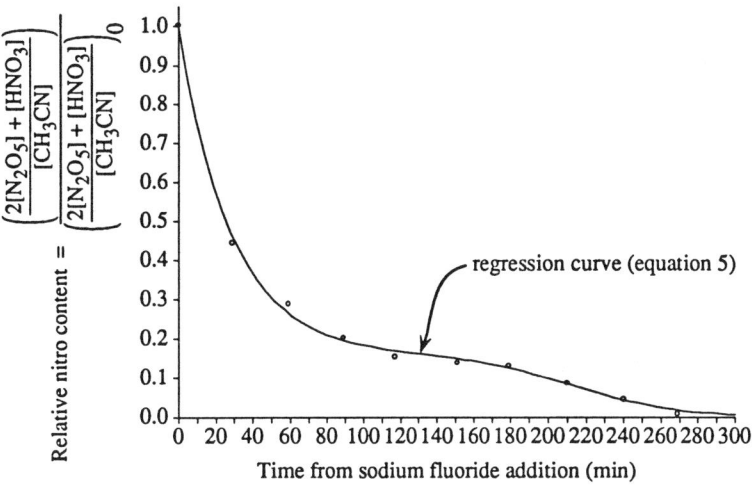

Figure 2. Concentration data analysis by ^{14}N NMR (normalized) (N_2O_5/HNO_3 in acetonitrile, 0 °C)

In contrast, a plot of the ^{1}H NMR data from this experiment (Figure 3) shows no unexpected delayed behavior in the absorption of nitric acid. Fitting the data initially to a decaying exponential with a finite offset yielded a calculated offset of 0.000078 ± 0.000558, so the model was reverted to a simple exponential decay to zero. In this case, the regression curve was calculated as:

$$[HNO_3]/3[CH_3CN] = 0.047666 \cdot \exp(-0.019365t) \qquad (6)$$

As with the ^{14}N data, this dependent variable can be normalized to the fitted intercept value at time zero, yielding equation 7, plotted in Figure 4:

$$\frac{([HNO_3]/3[CH_3CN])}{([HNO_3]/3[CH_3CN])_0} = \exp(-0.019365t) \qquad (7)$$

A useful comparison—for the assessment of the utility of this technique as a practical method for preparation of "acid-free" N_2O_5 solutions—is that of the relative extents of removal of nitric acid and of N_2O_5 throughout this process. The time dependence of total nitro species content was given by equation 5 (Figure 2), and that of nitric acid by equation 7 (Figure 4). It would be instructive, however, to plot the relative residual N_2O_5 content as a function of relative residual HNO_3 content. In order to estimate relative residual N_2O_5 content, the contribution of N_2O_5 to the initial ratio of integrals was calculated from the measured N_2O_5 concentration in nitric acid for this quantity of this N_2O_5/HNO_3 sample: 9.79 g of 17.63 wt% N_2O_5. In terms of mole quantities:

$$\left(\frac{2m_{N_2O_5}}{m_{HNO_3} + 2m_{N_2O_5}}\right)_{theor} = \frac{2(0.015980)}{0.12797 + 2(0.015980)} = 0.19983 \qquad (8)$$

To assess the validity of the ^{14}N NMR integrations in measuring N_2O_5 content, the *theoretical initial* value of the integral ratio (nitro vs. CH_3CN) can be calculated similarly:

$$\left(\frac{m_{HNO_3} + 2m_{N_2O_5}}{m_{CH_3CN}}\right)_0 = \frac{0.12797 + 2(0.015980)}{0.76097} = 0.21030 \qquad (9)$$

The measured value of the initial ^{14}N integration ratio is 0.200, a deviation of only –4.77%, typical of the accuracy of NMR integrations. Substituting the concentration-dependent expression for t given by equation 7 into equation 5, defining the expressions for relative N_2O_5 content and relative nitric acid content as N_{rel} and H, respectively, and normalizing to the theoretical ratio 0.19983 yields the desired dependence:

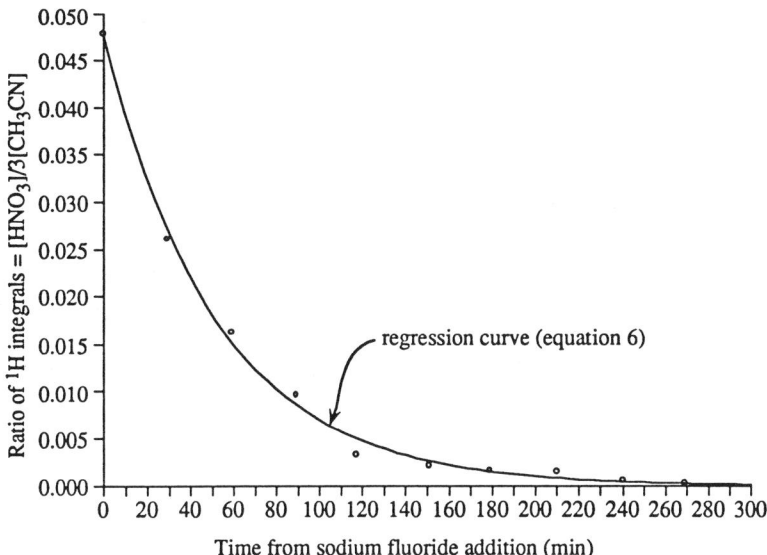

Figure 3. Concentration data analysis, by ^{1}H NMR, of nitric acid absorption (N_2O_5/HNO_3 in acetonitrile, 0 °C)

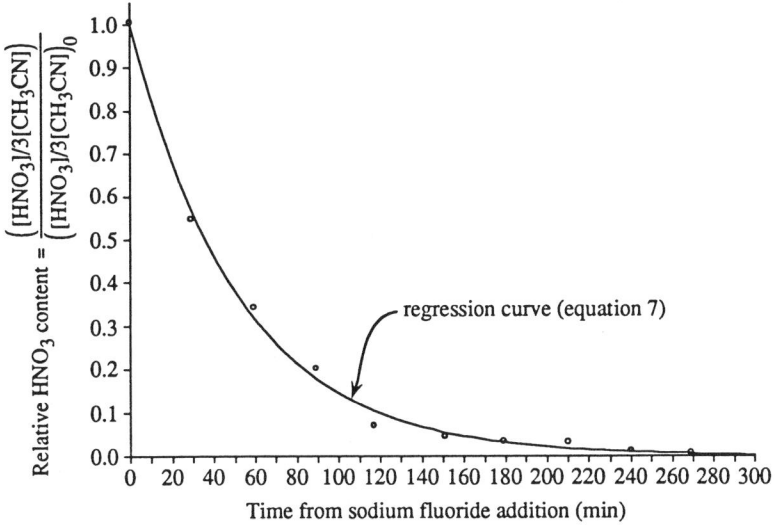

Figure 4. Concentration data analysis by ^{1}H NMR (normalized) (N_2O_5/HNO_3 in acetonitrile, 0 °C)

$$N_{\text{rel}} = \frac{\left[(1 - 0.15984) \cdot e^{1.8237 \cdot \ln(H)} + 0.15984\right] - \left[\dfrac{0.15984}{1 + e^{1.8991 \cdot \ln(H) + 7.8115}}\right]}{0.19983} \quad (10)$$

The plot of this function (Figure 5) shows the N_2O_5 content (relative to theory) as a function of relative residual nitric acid content. It shows a gradual but distinct fall-off in N_2O_5 content as nitric acid content is reduced completely to zero. The rate of N_2O_5 removal increases as relative nitric acid content falls to very low values (<2 rel%). This clearly indicates a significant tradeoff in the yield of usable N_2O_5 as a function of nitric acid removal. Nevertheless, the results represented by Figure 5 demonstrate an advancement in technology for conveniently removing a great majority of excess nitric acid from N_2O_5/HNO_3 solutions in order to produce substantially acid-free N_2O_5 in organic solvents suitable for conducting nitrations sensitive to excess acid.

As a result of the discovery of the competing sorption of N_2O_5 upon prolonged contact with sodium bifluoride, a simple variation of this process was envisioned that would allow removal of the majority of nitric acid prior to the onset of sorption of the contained N_2O_5. Then, transfer of the solution from contact with the excess of sodium fluoride solid to a new container would allow continuation of the acid absorption process with a smaller quantity of sodium fluoride (more closely simulating the conditions employed by Caesar et al.) and obviating the sorption of N_2O_5 by the excess solids (sodium fluoride, bifluoride, and nitrate).

In a modification of the previous experiment in acetonitrile, approximately the same initial conditions were employed: 10.16 g of N_2O_5/HNO_3 solution (measured as 17.87% N_2O_5) plus 42.12 g sodium fluoride (NaF/HNO_3 mole ratio = 7.6) in 40 mL acetonitrile, maintained in the temperature range of –8 to 0 °C. After a reaction (contact) time of 150 min under these conditions, the solution was filtered via a Schlenk-type filter (Kontes #21550) into a different flask. It is estimated from the quantitative trends of the previously characterized reaction that the system at this time contained essentially all of the contained N_2O_5 (~1.82 g) and ~10% (~0.97 g) of the original HNO_3. The solution was filtered into another flask containing 1.58 g of sodium fluoride (NaF/HNO_3 mole ratio = 2.4), which was theoretically sufficient to chemically absorb the remaining nitric acid.

Analysis of this system by ^{14}N NMR, as in the previous experiment, yields a plot (Figure 6) of nitro content that is qualitatively similar to the previous experiment. In this variation, however, the trend even up to 4.5 h during regular monitoring of the reaction indicated a distinctly positive and appreciable value for ultimate nitro content. After a prolonged reaction time of 46 h (by leaving the reaction solution in a freezer), the residual nitro content ($\mathcal{R}$) was confirmed as >10% of the initial value: $\mathcal{R}_\infty$ = 0.021807 (by regression) vs. $\mathcal{R}_0 \equiv (a + C) = 0.21253$. (The measured value of $\mathcal{R}$ at 46 h was 0.0220.) The identity of residual nitro species was confirmed as N_2O_5 by quantitative nitration of benzene, according to the method of Bloom et al. (*14*), and

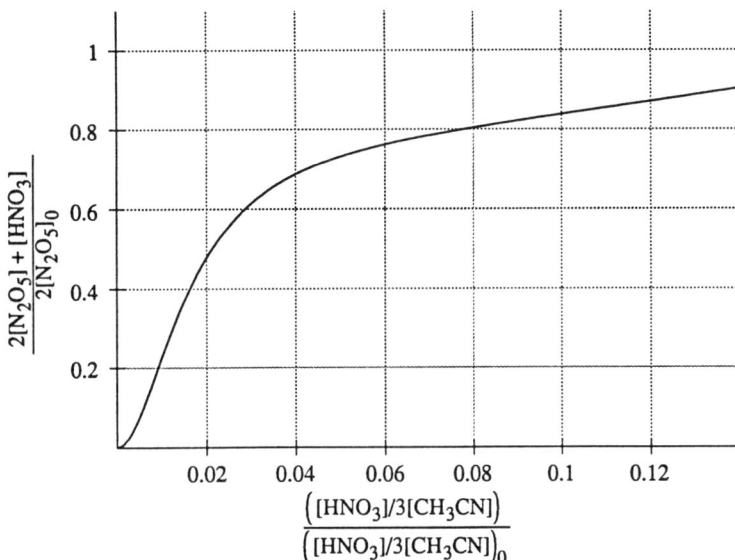

Figure 5. Relative N_2O_5 content dependence on relative nitric acid content

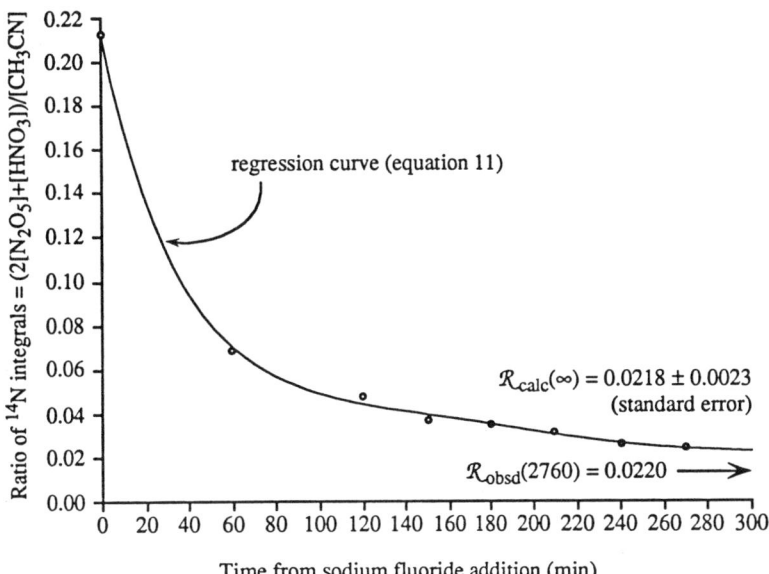

Figure 6. Concentration data analysis by ^{14}N NMR (modified procedure by solution transfer at 150 min, N_2O_5/HNO_3 in acetonitrile, −8 to 0 °C)

by the ^{1}H NMR spectrum at this time, in which no residual nitric acid was detected. The regression equation expressing total nitro content as a function of reaction time is:

$$\frac{2[N_2O_5] + [HNO_3]}{[CH_3CN]} = \left[0.17257 \cdot e^{-0.029004\,t} + 0.039964\right] - \left[\frac{0.039964 - 0.021807}{1 + e^{-0.032058(t-204.00)}}\right] \quad (11)$$

The regression equation expressing relative nitric acid content as a function of reaction time prior to the solution transfer at 150 min is:

$$[HNO_3]/3[CH_3CN] = 0.057624 \cdot \exp(-0.016651t) \quad (12)$$

The regression equation expressing relative nitric acid content as a function of reaction time after the solution transfer at 150 min is:

$$[HNO_3]/3[CH_3CN] = 0.0063662 \cdot \exp[(-0.0087676(t-150)] \quad (13)$$

As previously seen in equation 7, the latter two simple exponential decays can be normalized to the initial value by neglect of the pre-exponential factor. The acid absorption trend indicated by ^{1}H NMR (Figure 7) is qualitatively similar to that previously observed, except that the kinetics of the system instantaneously change upon transfer of the N_2O_5/HNO_3 solution into a new quantity of sodium fluoride reactant.

Figure 8 again shows the trend in residual N_2O_5 content as a function of the residual nitric acid content, this time in the modified reaction system.

Despite this modification of conditions, only a 49–50% recovered yield of N_2O_5 was realized: the theoretical final value of the ^{14}N NMR integral ratio is $(2m_{N_2O_5}/m_{CH_3CN})_\infty = 2(0.0168)/0.761 = 0.0442$ compared to the observed value of 0.0218. Speculation involving our earlier assumption of hydrogen bonding between N_2O_5 and sodium bifluoride would suggest that the inefficiency in this particular system is still due to a significant quantity of sodium bifluoride relative to N_2O_5; in this case, the molar ratio is still ~0.92. In contrast, the systems described by Caesar et al. (*15–17*) employed ratios of *sodium fluoride* to N_2O_5 in the range of 0.21–1.14. Although quantification of the acid content of their N_2O_5 was not reported, it may be assumed that the quantity of sodium bifluoride formed from residual nitric acid in their generated N_2O_5 is much less than the relative amount (~0.92) in our transferred solution.

The possibility of a metathetical reaction yielding nitryl fluoride in the current system via the reaction of N_2O_5 "product" with excess sodium fluoride was suggested by the observation by Ogg and Ray of such a reaction between the neat reactants at 35 °C (*23*):

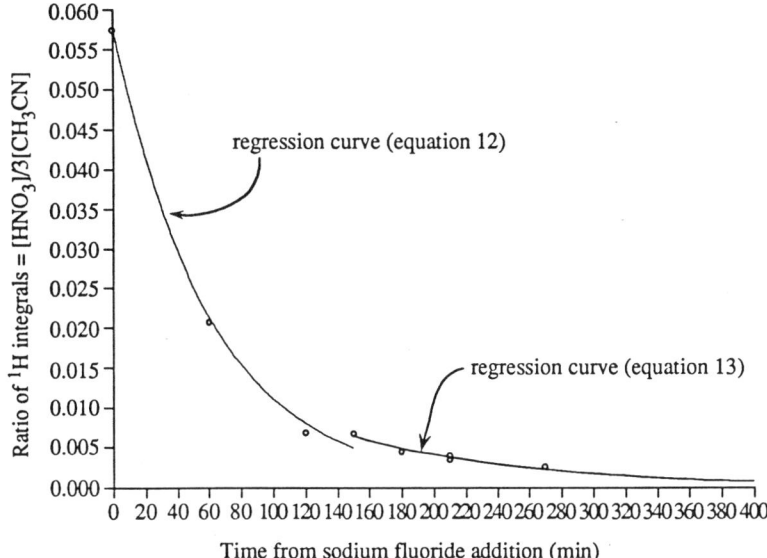

Figure 7. Concentration data analysis by ^{1}H NMR (modified procedure by solution transfer at 150 min, N_2O_5/HNO_3 in acetonitrile, −8 to 0 °C)

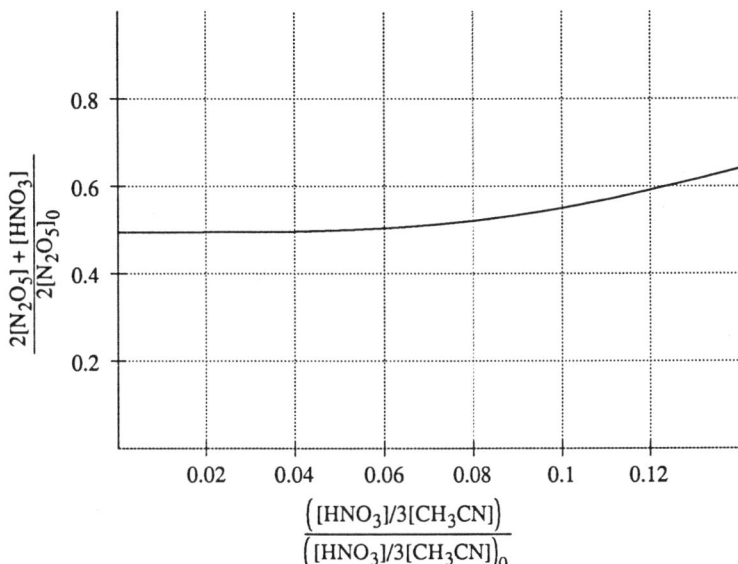

Figure 8. Relative N_2O_5 content dependence on relative nitric acid content (modified procedure by solution transfer at 150 min)

$$NaF + N_2O_5 \rightarrow NO_2F + NaNO_3$$

In the acid-absorption reactions conducted in our study, a search for NO_2F by ^{19}F NMR analysis was conducted on aliquots taken at 3 h and 4 h reaction time (in a system similar to that producing the data of Table I). Although a small trace but distinct ^{19}F absorption found around δ +399 (vs. $CFCl_3$) is at variance with one report (24) of the ^{19}F chemical shift of NO_2F as δ +221 (vs. $CFCl_3$), it is in a range bounded by shifts given in two other reports (25): δ +393 (26) and δ +401 (27). As Wilson and Christe claimed essentially the reverse reaction at 0 °C

$$NO_2F + LiNO_3 \rightarrow N_2O_5 + LiF$$

as a new preparative method for N_2O_5 (28), such metathetical reactions may involve reversible equilibria. Then NO_2F was preparable by Ogg and Ray (23) via its removal as the most volatile component of the system. In our acid-absorption experiments, a small amount of soluble, acidic fluoride was apparent during the course of the reaction in acetonitrile, evidenced by its ^{19}F absorption around δ −182. Ultimately, no acidic fluoride is detectable by ^1H NMR in acetonitrile.

NMR Chemical Shift Analyses. In addition to the conclusions about absorption efficiency derivable from the concentration data presented above, other interesting data are available from the NMR analytical approach used in this experiment. In particular, NMR chemical shifts of the monitored species (collective nitro species by ^{14}N NMR, nitric acid by ^1H NMR) follow a distinct trend as the acid absorption process progresses. The ^{14}N NMR chemical shifts (referenced to external nitromethane-d_3) from the absorption experiment following the original procedure (Table I) are plotted as a function of reaction time in Figure 9. The resulting data set clearly shows another example of logistic sigmoid behavior. In this case, the asymptotes exhibited by the experimental data also correspond very well to reported NMR data for the N_2O_5/HNO_3 system transitioning from high nitric acid concentration to pure N_2O_5 in inert solvents (such as acetonitrile used here). The regression curve through these points is:

$$\delta_{14N}(N_2O_5) = -64.22 + \frac{(-43.59) - (-64.22)}{1 + e^{0.026122(t-58.879)}} \tag{14}$$

Thus, the theoretical maximum asymptote found here (δ −43.59 ± 0.73) corresponds to the reported shift for 100% nitric acid (δ −43 ± 0.5) (18), while the theoretical minimum asymptote (δ −64.22 ± 0.13)—achieved upon complete removal of nitric acid—is reasonably close to that reported for N_2O_5 in inert solvents (δ −62 ± 1 in chloroform) (19). Also, this qualitative trend of logistic sigmoid behavior is similar to one reported by Seel et al. (29), by whom the ^{14}N NMR chemical shift of nitric acid was determined as a function of excess sulfuric acid content, which progressively protonated and dehydrated the acid to generate nitronium ion *in situ*.

Upon determination of a suitable calibration for the N_2O_5–HNO_3 system in a particular solvent and concentration, the ^{14}N chemical shift may after all constitute a convenient diagnostic for the extent of acid removal when the acid content is relatively low.

A similar analysis of the ^1H NMR data revealed only one discrepancy: the initial chemical shift for the N_2O_5/HNO_3 solution prior to sodium fluoride addition was much greater than that predicted by the trend of the data *after* commencement of the acid absorption (Figure 10). This is hypothesized by us to be due to the drastic changes undergone by the HNO_3–NO_3^- interaction (i.e., hydrogen bonding as described above) when acid is effectively irreversibly absorbed by the sodium fluoride. For the sake of this analysis, therefore, the initial chemical shift (δ 13.81) was assumed to be the asymptotic limit, and the fitted logistic sigmoid curve was constrained to this maximum value. The remaining parameters fit to the experimental data then take the form:

$$\delta_{1H}(HNO_3) = 6.62 + \frac{13.81 - 6.62}{1 + e^{0.0092462(t-97.713)}} \qquad (15)$$

Although the trend is not quite as well defined as that of the ^{14}N NMR data, and the chemical shift undergoes change throughout a narrower range, the ^1H chemical shift may also constitute a complementary convenient diagnostic for residual acid content.

Conclusions and Recommendations

We have demonstrated a technical innovation in the form of a convenient process for the separation of nitrogen pentoxide from its solutions in nitric acid, as commercially prepared by electrolysis. This process involves the chemical absorption of the nitric acid by sodium fluoride (producing sodium bifluoride) in inert organic solvents, such as acetonitrile, suitable for use on many nitrolyzable substrates. Although the concept of absorption of HNO_3 from solutions of N_2O_5 in inert solvents has been previously described (*17*), the prior systems in which efficiency of acid removal was sufficient for nitration of acid-sensitive substrates (e.g., starches) used N_2O_5 solutions containing only minor, adventitious concentrations of nitric acid. The innovation of the current process is that nitric acid may comprise a large majority of the N_2O_5/HNO_3 solution and still be effectively removed from it.

Simple variations of the process may allow even more efficient removal of nitric acid without the complication of N_2O_5 sorption. One demonstrated modification, which involved the transfer of the N_2O_5 solution away from the acid absorbent after removal of the great majority of nitric acid but prior to commencement of the N_2O_5

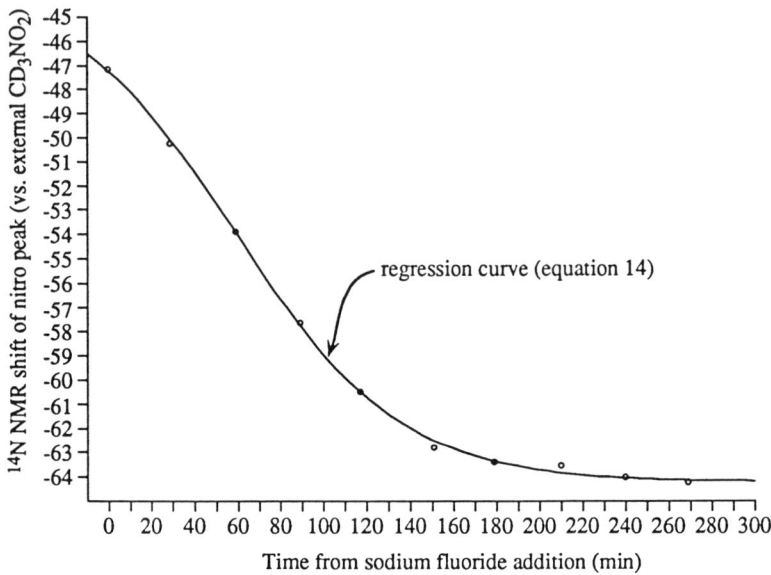

Figure 9. Nitro species ^{14}N NMR chemical shift dependence on nitric acid absorption

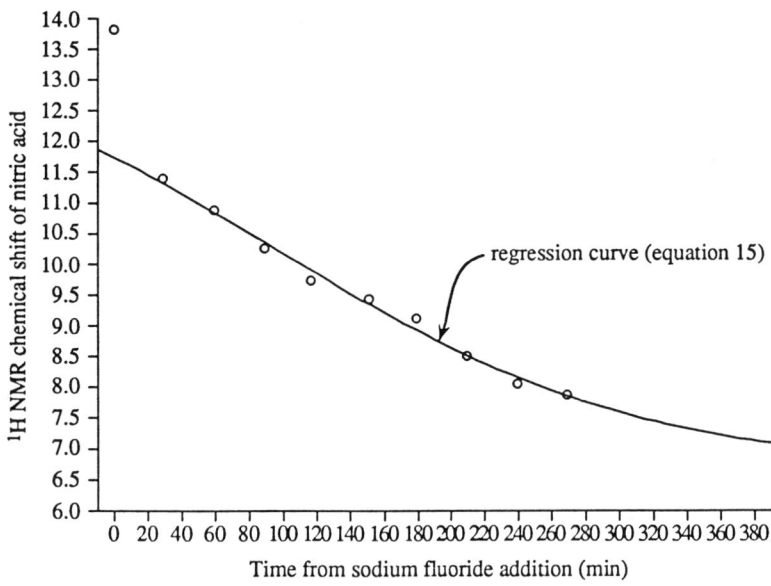

Figure 10. Nitric acid ^{1}H NMR chemical shift dependence on nitric acid absorption

sorption, showed the feasibility of the concept for isolation of practical quantities of acid-free N_2O_5. The success of this modification makes further process improvements apparent: a continuous flow system of N_2O_5/HNO_3 solution through solid sodium fluoride would allow the relatively fast NaF–HNO_3 reaction to occur while continuously removing N_2O_5 from contact with the sodium bifluoride that complicates the absorption process.

In comparing costs of various technologies for N_2O_5 production, the low cost of sodium fluoride (currently on the order of 60¢/lb ≈ $0.0555/mol) may make the described process economically competitive with alternative methods if its efficiency can ultimately be improved to achieve approximately stoichiometric absorption of nitric acid. (Also, technology is apparent for re-converting the product salts to the original reactants, nitric acid and sodium fluoride.) Obvious process parameters that require optimization include: the minimum sodium fluoride/acid ratio necessary for efficient acid removal; contact time between the reactants (especially in a flow system); quantitative concentration of the N_2O_5/HNO_3 solution in suitable organic solvents; the chemical nature of solvents in which the process is most effective; recycling methodology to recover spent acid absorbent.

Acknowledgments

The financial support of this work by the Naval Surface Warfare Center (White Oak), under Small Business Innovation Research contract N60921-93-C-0117, is gratefully acknowledged. We also thank: Dr. William Koppes (Naval Surface Warfare Center, White Oak and Indian Head) for helpful technical discussions; Dr. Richard A. O'Brien (TPL, Inc.) for assistance with the procurement of materials and preliminary analysis of N_2O_5/HNO_3 solutions early in this project; and Mr. Ken Poush (Thiokol, Inc., Longhorn Division, Marshall, TX) for data and techniques for characterization of N_2O_5/HNO_3 samples.

Literature Cited

(1) Millar, R. W.; Colclough, M. E.; Golding, P.; Honey, P. J.; Paul, N. C.; Sanderson, A. J.; Stewart, M. J. *Philos. Trans. R. Soc. London, Ser. A* **1992**, *339*(1654), 305-319.
(2) Guye, P. A. U.S. Patent 1 348 873, 1920.
(3) Wulf, O. R.; Daniels, F.; Karrer, S. *J. Am. Chem. Soc.* **1922**, *44*, 2398-2401.
(4) Gibson, G.; Beintema, C. D.; Katz, J. J. *J. Inorg. Nucl. Chem.* **1960**, *15*, 110-114.
(5) Harris, A. D.; Trebellas, J. C.; Jonassen, H. B. *Inorg. Synth.* **1967**, *9*, 83-88.
(6) Harrar, J. E.; Pearson, R. K. *J. Electrochem. Soc.* **1983**, *130*, 108-112.

(7) McGuire, R. R.; Coon, C. L.; Harrar, J. E.; Pearson, R. K. U.S. Patent 4 432 902, 1984.
(8) McGuire, R. R.; Coon, C. L.; Harrar, J. E.; Pearson, R. K. U.S. Patent 4 525 252, 1985.
(9) Marshall, R. J.; Shiffrin, D. J.; Walsh, F. C.; Bagg, G. E. G. Eur. Pat. Appl. 295 878 A1, 21 Dec 1988.
(10) Foller, P. C. Brit. UK Pat. Appl. 222 3031, 28 Mar 1990.
(11) Bagg, G. E. G. PCT Int. Appl. WO 90/07 020, 28 Jun 1990.
(12) Zawadski, J.; Bańkowski, Z. *Rocz. Chem.* **1948**, *22*, 233-247.
(13) Rodgers, M.; Swinton, P. F. Presented at the 209th National Meeting of the American Chemical Society, Anaheim, CA, April 1995; paper I&EC 35.
(14) Bloom, A. J.; Fleischmann, M.; Mellor, J. M. *Electrochim. Acta* **1987**, *32*, 785-790.
(15) Caesar, G. V.; Goldfrank, M. *J. Am. Chem. Soc.* **1946**, *68*, 372-375.
(16) Caesar, G. V. U.S. Patent 2 432 280, 1947.
(17) Gruenhut, N. S.; Cushing, M. L.; Caesar, G. V. *J. Am. Chem. Soc.* **1948**, *70*, 424-425.
(18) Chew, K. F.; Derbyshire, W.; Logan, N. unpublished results reported by: Logan, N. In *Nitrogen NMR*; Witanowski, M.; Webb, G. A., Eds.; Plenum Press: New York, 1973; p. 331.
(19) Chew, K. F.; Healy, M. A.; Khalil, M. I.; Logan, N.; Derbyshire, W. *J. Chem. Soc. Dalton Trans.* **1975**, 1315-1318.
(20) Ogg, R. A., Jr.; Ray, J. D. *J. Chem. Phys.* **1956**, *25*, 1285-1286.
(21) Happe, J. A.; Whittaker, A. G. *J. Chem. Phys.* **1959**, *30*, 417-421.
(22) Frost, A. A.; Pearson, R. G. *Kinetics and Mechanism*; John Wiley: New York, 1953; pp. 19–20.
(23) Ogg, R. A., Jr.; Ray, J. D. *J. Chem. Phys.* **1956**, *25*, 797-798.
(24) Mason, J. B.; van Bronswijk, W. *Chem. Commun.* **1969**, 357-358.
(25) We thank a reviewer for pointing out Gmelin's tabulation of ^{19}F NMR shifts for NO_2F, which notes confusion among the literature reports: Gmelin-Institut für Anorganische Chemie, *Gmelin Handbook of Inorganic Chemistry*, 8th ed.; Springer-Verlag: Berlin, 1987; Fluorine Supplement Vol. 5, p. 164.
(26) Fox, W. B.; Wamser, C. A.; Eibeck, R.; Huggins, D. K.; MacKenzie, J. S.; Juurik, R. *Inorg. Chem.* **1969**, *8*, 1247-1249.
(27) Solomon, I. J. "Advanced Oxidizer Chemistry" *U.S. Clearinghouse Fed. Sci. Tech. Inform.* **1967**, AD 670531; *Chem. Abs.* **1969**, *70*, 96098g.
(28) Wilson, W. W.; Christe, K. O. *Inorg. Chem.* **1987**, *26*, 1631-1633.
(29) Seel, F.; Hartmann, V.; Gombler, W. *Z. Naturforsch.* **1972**, *27b*, 325-326.

RECEIVED October 24, 1995

Chapter 10

Nitrated Hydroxy-Terminated Polybutadiene: Synthesis and Properties

M. E. Colclough and N. C. Paul

Defence Research Agency, Fort Halstead, Sevenoaks, Kent TN14 7BP, United Kingdom

This paper reports the synthesis and properties of nitrated hydroxy-terminated polybutadiene (NHTPB), an energetic binder which combines the excellent elastomeric properties of hydroxy-terminated polybutadiene (HTPB) with the energy associated with nitrate-ester groups, and is potentially useful in propellant, explosive and pyrotechnic systems.

Attempts to nitrate pre-formed polymers using conventional nitrating agents invariably result in either chain scission of the polymer chains or cross linking reactions leading to intractable materials. Dinitrogen pentoxide (N_2O_5) in an inert solvent, although a powerful nitrating agent, can be used to carry out nitrations in a less destructive environment. Commercial HTPB polymer reacted with N_2O_5 was found to add to the double bonds to form *vicinal* nitro-nitrato groupings along the polymer backbone. An alternative strategy, producing superior materials, was to convert some of the double bonds to oxirane groups and then to *vicinal* dinitrate ester groups by reaction with N_2O_5.

NHTPB is a liquid polymer, the viscosity of which can be varied by changing the percentage conversion of double bonds to dinitrate ester groups. The glass transition temperature is low (down to -58°C) and is again dependent on the amount of substitution. The product is miscible with energetic plasticisers (unlike HTPB) and undergoes an isocyanate cure to give an energetic rubber.

Energetic compounds are the basis of all explosive and propellant compositions and the majority of such compounds contain nitro groups. These may be C-Nitro (eg trinitrotoluene), O-Nitro (Nitrate Esters; eg nitroglycerine) or N-Nitro (Nitramine; eg RDX). Nitration is therefore an important part of the synthesis of energetic compounds. Energetic compositions contain an oxidiser and a fuel element which, in the case of high explosives, are contained within each molecule, whereas, in propellants, these elements are generally provided by a mixture of compounds. Binders are added to compositions to improve mechanical properties and provide safer compositions which are less vulnerable to accidental stimuli.

Initially, natural products such as waxes were used as binders but most binders are now polymeric materials where the energetic material particles are embedded in a rubbery matrix which can thus absorb mechanical shock. Currently, 'inert' binders, such as the hydroxy terminated polybutadienes, whilst conferring excellent properties, 'dilute' the available energy of the energetic component. The potential rewards in both performance and vulnerability of compositions that can be achieved by using an energetic binder have been reported elsewhere (*1*).

There are two possible approaches to the synthesis of energetic polymers and these are; the polymerisation of an energetic monomer and the introduction of energetic groups to a pre-formed inert polymer. The former approach has been applied to the synthesis of energetic polyoxetanes and polyoxiranes (*1*) whilst the latter has been applied to the synthesis of NHTPB, the subject of this paper.

Previous workers (2) have prepared a nitrated HTPB via a nitromercuration demercuration route; however the desirednitration was accompanied by a number of side reactions which caused both degradation of the polybutadiene backbone and crosslinking of the polymer with the formation of insoluble materials .

Generally, attempts to introduce energetic groups into polymers by conventional nitration procedures, such as nitration with nitric acid or mixed acids, results in either degradation of the polymer backbone or crosslinking of the polymer with the formation of insoluble materials.

A preliminary nitration study using dinitrogen pentoxide (N_2O_5) showed that the HTPB could be nitrated directly through addition of the N_2O_5 molecule across the C=C double bonds in HTPB but leaving the polymer backbone intact. This reaction is shown below in Equation 1.

$$\text{-}(CH_2\text{-}CH=CH\text{-}CH_2)\text{-} \xrightarrow{N_2O_5/CH_2Cl_2} \text{-}(CH_2\text{-}\underset{ONO_2}{\underset{|}{CH}}\text{-}\underset{}{\overset{NO_2}{\overset{|}{CH}}}\text{-}CH_2)\text{-} \quad [1]$$

The amount of nitration could be controlled by varying the molar ratios of N_2O_5/polymer such that up to about 90% of the double bonds could be reacted. However, with high degrees of addition (above about 50%) the material produced was found to be intractable, precipitating from the reaction as a gelled solid. Removal of the swelling solvent under vacuum resulted in a yellow powdery material and the rubbery properties were lost.

More significantly, with even small nitration values, of the order of 10 to 20%, there were serious concerns over the inherent thermal instability of the products obtained.

From our concurrent work on the reactions of N_2O_5 with various substrates(*3*), it was known that N_2O_5 reacted cleanly with oxiranes to produce the*vicinal* dinitrate ester grouping under mild conditions (Equation 2). Compounds with *vicinal* dinitrate ester groupings have been shown to have acceptable stability.

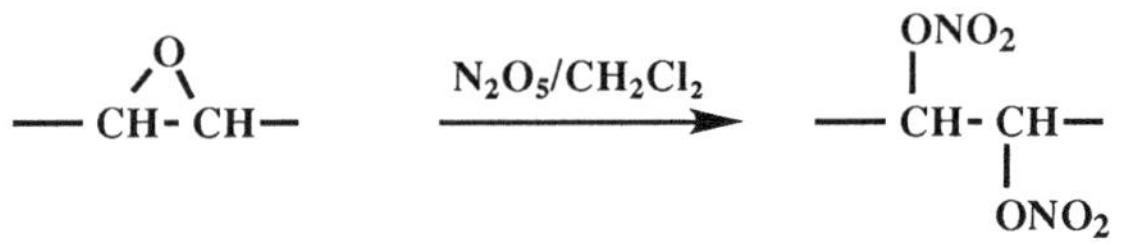

$$\text{—}\overset{O}{\overset{/\backslash}{CH\text{-}CH}}\text{—} \xrightarrow{N_2O_5/CH_2Cl_2} \text{—}\underset{ONO_2}{\underset{|}{CH}}\text{-}\underset{}{\overset{ONO_2}{\overset{|}{CH}}}\text{—} \quad [2]$$

Therefore, in order to introduce nitrate ester groups into a polybutadiene using this N_2O_5/epoxide reaction, it was first necessary to convert a proportion of the C=C double bonds in the backbone into epoxide groups.

This conversion is readily achieved through by reaction with peracetic acid in the presence of an ion exchange resin as catalyst, to give a polymers with a percentage of the double bonds converted to epoxide groups.

The epoxide groups so formed can then be reacted with N_2O_5 to introduce *vicinal* dinitrate ester groups on the polymer backbone. The sequence of reactions is shown in (Scheme 1) below.

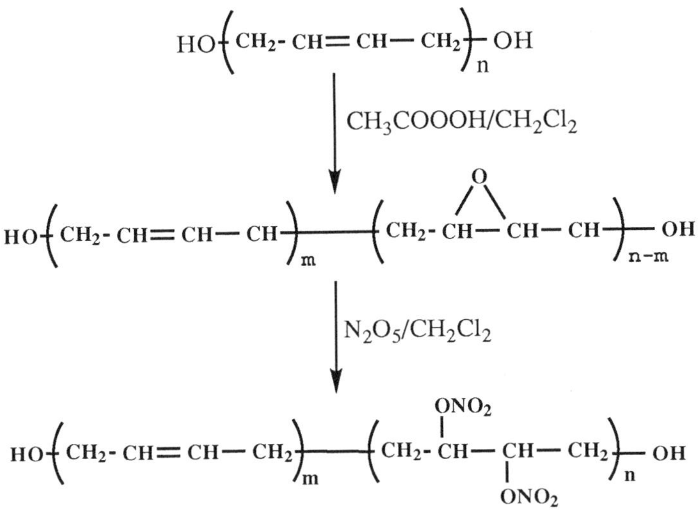

$$HO\text{-}(CH_2\text{-}CH=CH\text{-}CH_2)_n\text{-}OH$$

$$\downarrow CH_3COOOH/CH_2Cl_2$$

$$HO\text{-}(CH_2\text{-}CH=CH\text{-}CH_2)_m\text{-}(CH_2\text{-}\overset{O}{\overset{\triangle}{CH\text{-}CH}}\text{-}CH)_{n-m}\text{-}OH$$

$$\downarrow N_2O_5/CH_2Cl_2$$

$$HO\text{-}(CH_2\text{-}CH=CH\text{-}CH_2)_m\text{-}(CH_2\text{-}\underset{ONO_2}{\overset{ONO_2}{CH\text{-}CH}}\text{-}CH_2)_n\text{-}OH$$

Scheme 1.

Experimental Details

The procedures described herein are general procedures for the epoxidation of 13% of the double bonds and their subsequent nitration (m=0.87n). To obtain different levels of epoxidation, and hence nitration, a variation in stoichiometry of reagents is required, as described in the discussion.

Materials. All materials were used as received from the commercial suppliers unless otherwise stated. The polybutadiene used was poly-bd ®R20LM resin supplied by Atochem.(US), a low molecular weight hydroxy-terminated polybutadiene. Amberlite IR-120(Na form) ion exchange resin was supplied by BDH Chemicals and was washed with hydrochloric acid (5M) before use. Acetic anhydride and hydrogen peroxide solution were used as supplied by BDH and dichloromethane was dried with calcium hydride and distilled before use. N_2O_5 was made by gas phase reaction of dinitrogen tetroxide (N_2O_4) with ozone (O_3) ,trapping at -78°C, (4, 5) and subsequent storage in a stoppered flask at -60°C . Magnesium sulphate, sodium bicarbonate and ferrous sulphate were used as anhydrous reagents supplied by BDH.

Epoxidation. Acetic anhydride (75ml) was dissolved in dichloromethane (75ml) and then stirred with hydrogen peroxide solution (60% w/v;61ml) for 15 minutes. Amberlite resin (32.4g) was then added and the mixture stirred for 30 minutes at 30-35°C. The mixture was cooled in an ice bath to 5°C and a solution of R20LM (100g) in dichloromethane (300ml) added at such a rate that the temperature did not exceed 15°C. (It is important to keep the temperature below this figure since side reactions, causing the polymer to have carbonyl contamination, occur above 15°C). When the addition was complete the reaction was stirred for a further 5 minutes and then filtered to remove the ion exchange resin.

The clear dichloromethane solution was separated from the yellow acid layer and washed continuously with water until the water washings were free from acid. Stirring with wet ferrous sulphate then destroyed any residual peroxide by reducing it to acid, which was removed by stirring with excess sodium hydrogen carbonate. The solution was then dried with magnesium sulphate, filtered and the solvent evaporated to leave the pure epoxidised polymer. (NB It is important that the product solution is not concentrated until all the peroxide has been removed. This can be checked by the absence of peaks in the infrared spectrum for the carbonyl groups in diacetyl peroxide at 1780 and 1820 cm-1.

The purity of the product was checked by ^{1}H nmr, FTIR and sec, and the epoxide content determined by titration based on a literature method (6).

Nitration. A dichloromethane solution of the epoxidised polymer was made to a concentration of ca 80g of polymer in 200ml of dichloromethane and stirred under nitrogen for 16 hours with calcium hydride. The solution was filtered into a dry, nitrogen flushed round bottomed flask and cooled to -30°C in an acetone/dry ice bath. N_2O_5 (20.3g, for 80g of 13% epoxidised polymer) in dichloromethane (100ml) was added and the cold bath removed after the initial exotherm had finished. When the temperature reached 5°C the reaction mixture was poured into sodium bicarbonate (excess) and water (5ml), and stirred until no acid remained. The solution was then filtered, dried with magnesium sulphate and the solvent evaporated to leave a brown liquid polymer in ca. 95% yield.

Results

NMR. ^{1}H nmr spectra were recorded on a Varian Associates EM 360A nmr spectrometer at 60 MHz. Chemical shifts are reported in ppm downfield from the tetramethylsilane (TMS) reference.

^{13}C nmr spectra were run on a Jeol FX-90Q pulse fourier transform spectrometer operating at 22.5 MHz. Chemical shifts are reported in ppm from the signal due to TMS as internal reference.

The ^{1}H nmr spectrum did not provide much useful information since the signals due to the proton α- to the secondary nitrate ester group should appear around δ5.0-5.5, which is the area occupied by signals of the olefinic protons of the polymer, so the signals are not directly discernable.

The ^{13}C nmr spectrum showed signals in the 70-80ppm range consistent with carbon atoms bearing nitrate ester groups, but the spectrum was too complex to be able to accurately determine the nitrate ester content.

Epoxide Analysis. The residual epoxide content of NHTPB, determined as for the epoxidised polymer (6), was about 3%, implying incomplete reaction with N_2O_5. This was probably due to some reaction of the N_2O_5 with the double bonds directly, to produce vicinal nitro-nitrate groups (Equation.1). This was confirmed by FTIR.

It is possible to remove residual epoxide during work-up of the product by reaction of the epoxide groups with nitric acid to give hydroxy-nitrate. However this causes an increase in the hydroxyl content and the viscosity of the product which can have a deleterious effect on the physical properties.

FTIR. Infra-red spectral measurements were carried out using either a Nicolet 5SX Fourier transform I.R.(FTIR) spectrometer operating in transmittance mode equipped with DTGS detector, 1280 data processor and Zeta 8 plotter, or a Perkin-Elmer 157G IR spectrometer with a resolution of 4 cm^{-1}. Polymer samples were recorded on liquid films between two KBr disks.

The FTIR spectrum shows the presence of nitrate ester (C-ONO$_2$) groups by the peaks at 1277 and 1633cm^{-1}, nitro groups (C-NO$_2$) by the peak at 1551cm^{-1} and the presence of hydroxyl groups by the broad peak at 3450cm^{-1}. The (C-NO$_2$) peak was not very pronounced and arises from the small degree of the slower competing reaction of the C=C double bonds in the nitration.

SEC. Size exclusion chromatography measurements were carried out on a Waters 840 series chromatograph fitted with refractive index and ultra-violet (254 nm) detectors in series. Four PLgel columns (Polymer Laboratories) of porosities 10^5Å, 10^4Å, 10^3Å, and 10^2Å were used throughout. All molecular weights were determined from a universal polystyrene calibration and are quoted in polystyrene equivalents.

SEC shows a broad molecular weight distribution and the approximate molecular weights (taken from an average of a number of analyses) are Mw=8000, Mn=3000(polystyrene equivalents), polydispersity=2.7. The starting material also showed a broad molecular weight distribution with Mw=7000 and Mn=2500, poly dispersity 2.6 (from an average number of batches). The increase in Mw and Mn are consistent with the additional groupings introduced.

Cure Chemistry. The hydroxyl content of the polymer was determined by reaction with acetic anhydride, in the presence of pyridine as catalyst, followed hydrolysis of excess acetic anhydride and titration of the total acetic acid (7·). Over the number of batches produced this was generally found to be ca 1.5 mequiv/g (cf. 1.7 for the starting material)

NHTPB can be cured with a variety of isocyanates to give rubbers, the hardness of which depends on the isocyanate used and the extent to which the double bonds have been nitrated. Using Suprasec DNR (functionality 2.7), a cure ratio (OH/NCO) of 1.2 and temperature of 60°C, gelation occurred after only 4 hours, and after full cure (4 days) a good soft rubber was produced. Using 20% nitrated HTPB (Scheme 1, m=0.80n) the rubber produced under identical conditions was very hard.

When MDI is used as crosslinking agent for NHTPB the time to gelation is 4 hours and a soft rubber is produced after a cure time of 24 hours. Using IPDI as the isocyanate the time to gelation was 24 hours and after a cure time of 3 days a soft rubber is again produced.

Viscometry. Viscometry was carried out using a Brookfield rotational cylinder viscometer with a variable temperature accessory

The viscosity of the HTPB starting material was 14 poise at 25°C On epoxidation followed by nitration there was a significant increase in viscosity to 128 poise at 25°C and 14 poise at 60°C for a 10% nitrated material. The viscosity of a sample of a 20% nitrated HTPB was also measured to be 2000 poise (upper limit of machine parameters) at 25°C.

DSC. Differential scanning calorimetry (DSC) was carried out on a Stanton-Redcroft DSC 700 over the range -100 to100°C at a scan rate of 10°C/minute for the lower temperature measurements (glass transition) and over the range 100-250°C at a scan rate of 20°C/Minute for the higher temperature measurements (decomposition).

The glass transition temperature (Tg) of NHTPB was measured to be -58°C. At high temperature the onset of exotherm was 156°C and the exotherm maximum was 209°C. The Tg of 20% nitrated HTPB was found to be -22°C

Vacuum Stability Test. NHTPB was subjected to a vacuum stability test on a 5 gram sample at 100°C for 40 hours. At the end of this time 2.5cm^3 of permanent gases had been produced.

Preliminary Hazard Assessment. NHTPB was found to have a temperature of ignition (production of smoke) of 173°C. Using a liquid impact test (Rotter) the Figure of Insensitiveness (F of I) was out of range, and there was no response in a mallet friction test to a steel mallet on a steel anvil. There was no propagation of detonation in a BAM 50/60 steel tube test and a Koenen steel tube test gave an event at a critical diameter of 3.5mm.

Miscibility with Energetic Plasticiser. Energetic plasticisers are added to polymer bound compositions to improve low temperature physical properties and to add additional energy to the sytem. HTPB is, however, immiscible with most of the energetic plasticisers in current use whereas NHTPB has been shown to be miscible with energetic plasticisers in current use.

Samples of NHTPB were mixed with a number of plasticisers at plasticiser - polymer ratios from 0.1:1 up to 1:1. Plasticisers tested were; nitroglycerine (NG), diethylene glycol dinitrate (DEGDN), triethylene glycol dinitrate (TEGDN), metriol trinitrate (MTN), K10 and bis-dinitropropyl acetal/formal (BDNPA/F) and in all cases homogenous mixes were obtained with no evidence of phase separation.

Discussion

The desirable features of using liquid curable elastomers in propellants, pyrotechnics and other energetic materials have been documented previously (8). In particular the polymeric binder wets the solid oxidiser to provide a void free matrix, which causes a decrease in burning rate, enhanced mechanical and safety properties, and the facility to be cast into large or irregular cases. HTPB is a widely used binder and this suggests that NHTPB may find application in the equivalent fields.

NHTPB has some properties which are not dissimilar to HTPB. The low viscosity at ambient temperature enables ease of handling in a laboratory or processing environment. It has favourable cure properties in that it has cured with commercially available aliphatic and aromatic isocyanates with the former being the better due to the reasonable rate of cure. The glass transition temperature is slightly higher than that of HTPB, as one might expect, due to the introduction of bulky groups onto the polymer backbone. However it is still very good from the point of view of retaining elastomeric properties at the temperatures required.

As far as energy is concerned polybutadiene binders are known to have comparatively favourable heats of formation (+100 to -100 cal/g) because of their unsaturation. High energy compositions are available due to the capability of the polybutadienes to achieve high solids loadings. This therefore implies that NHTPB could be especially useful since although the presence of nitrate ester groups may lead to a slight decrease in the solids loading capability, their presence enables the energy of the system to be increased by use of an energetic plasticiser, with which, as described earlier, NHTPB is completely miscible (unlike HTPB).

The high energy release associated with the nitrate ester groups combined with a high solids loading and an energetic binder would therefore lead to a very high energy composition. Alternatively, the use of NHTPB, with or without an energetic plasticiser could lead to better mechanical and safety properties in that equivalent energy could be achieved using a lower solids loading. A visualisation of the energy associated with NHTPB is shown by the vigorous way in which a cured sample of NHTPB burns whereas a cured HTPB sample will not support a flame.

The percentage of double bonds in HTPB converted to dinitrate ester groups is important in determining the physical properties of the product. As described earlier, the viscosity and the glass transition temperature are both increased with increasing level of nitration. The product on which most attention has been concentrated is 10% nitrated, and this appears to be a reasonable compromise between retention of low viscosity and glass transition temperature, energy and the presence of sufficient nitrate ester groups to ensure miscibility with energetic plasticisers. However the potential to change the properties is an important point since different users may have different requirements eg. more or less energy, higher or lower viscosity, higher or lower Tg, all of which are controllable by variation in the extent of nitration.

Conclusions

N_2O_5 in an inert organic solvent can be used to directly nitrate polybutadiene in a clean manner; without degradation of the polymer chain. The material produced by this direct nitration, although energetic, was unsuitable as a binder in that the *vicinal* C-nitro-nitrate ester grouping was inherently unstable. NHTPB, containing *vicinal* dinitrate ester groupings is an energetic prepolymer made by, first epoxidising, then nitrating 10% or more of the double bonds in commercially produced HTPB.

The physical properties and stability of NHTPB were acceptable for the use envisaged. The nitration step highlights the selectivity of N_2O_5 in organic solvents giving preferential reaction with the oxirane rings in the presence of the less reactive double bonds. This nitrating medium is also milder as evidenced by the absence of chain scission of the polymer backbone.

A very important aspect of this material is that, unlike HTPB, it is miscible with the majority of energetic plasticisers which enhances the low temperature properties at the same time as increasing the energy of a formulation.

The cure chemistry, thermal stability and hazard properties of NHTPB have been examined in preliminary studies and all tests have produced encouraging results. There is scope for synthesis of a range of products with varying nitrate-ester content, viscosity, and Tg to suit the requirements of the user.

Literature Cited

1. Colclough, M.E *et al.*; *Polymers for Advanced Technologies* **1994**, *5* , 554-560.
2. Chien, J.W.C., *et al* ;*J Polym Sci* , **1980**, 7051.
3. Golding, P , Millar, R.W., Paul, N.C., Richards, D.H., *Tetrahedron*, **1993**, 7051.
4. Golding, P., Millar, R.W., Paul, N.C., Richards, D.H., *Tetrahed. Letts.*, **1988**,*29*, 2731.
5. Harris, A.D., Trebellas, J.C., Jonassen, H.B., *Inorg. Synth.,* 9, McGraw-Hill, NY, 1967; Vol. 9, 83-88.
6. Dijkstra, R. and Dahmen, E.A.M.F., *Analyt. Chim. Acta*, **1964**, *31* , 38.
7. Mann, F. G., Saunders, B. C., *Practical Organic Chemistry* ;4th Edition, Longman: London, 1974; 450.
8. Shaw, G.C., Reed, R., Munson, W.D., Roberts, J.A., *Proceedings of the Second International Pyrotechnics Seminar,* **1970**, *55* .

© Crown Copyright DRA 1995

RECEIVED January 9, 1996

Chapter 11

Novel Syntheses of Energetic Materials Using Dinitrogen Pentoxide

R. W. Millar, M. E. Colclough, H. Desai, P. Golding[1], P. J. Honey, N. C. Paul, A. J. Sanderson[2], and M. J. Stewart[3]

Defence Research Agency, Fort Halstead, Sevenoaks, Kent TN14 7BP, United Kingdom

An overview of the utility of dinitrogen pentoxide (N_2O_5) in the synthesis of energetic materials is presented. New nitration methodologies based on dinitrogen pentoxide have been developed which overcome many of the drawbacks of conventional mixed acid (HNO_3-H_2SO_4) media, especially when dealing with sensitive substrates. Two principal nitration systems have been developed:- i) N_2O_5 in pure nitric acid, which possesses strength similar to mixed acid systems, and ii) N_2O_5 in organic solvents, mainly chlorinated hydrocarbons, which can accomplish nitration of acid-sensitive substrates and selective nitrations. Advantages of these novel systems over conventional media are highlighted both in the preparation of N_2O_5 and its handling and use in aromatic and heterocyclic nitrations, particularly the novel ring-opening nitration of strained-ring heterocycles, as well as in the synthesis of energetic polymers and their precursors.

C-Nitro (**1**), nitramine (**2**) and nitrate ester (**3**) functionalities are the building blocks for energetic materials (*1*) whether established or novel. They are introduced into precursor or substrate molecules by a process called nitration (*2,3*) which classically has employed either pure nitric acid or nitric-sulphuric acid mixtures. In the present work we aim to show how a novel nitrating agent, dinitrogen pentoxide or N_2O_5, can complement or even replace some of these classical procedures.

Nitrations can be broadly divided into three classes for the purposes of this review:- i) nitration on carbon; ii) nitration on heteroatoms (principally nitrogen and

$-\overset{|}{\underset{|}{C}}-NO_2$ $\overset{\diagdown}{\underset{\diagup}{N}}-NO_2$ $-O-NO_2$

    1   2   3

[1]Current address: AWE plc, Aldermaston, Berkshire RG7 4PR, United Kingdom
[2]Current address: NATO Insensitive Munitions Information Center, NATO Headquarters, B–1110 Brussels, Belgium
[3]Current address: Defence Research Agency, Farnborough, Hantshire GU14 6TD, United Kingdom

oxygen), and iii) selective nitrations, where other acid-sensitive functionalities are present in the substrate molecule. Some of these reactions are exemplified in Table I, with particular reference to the use of N_2O_5. As discussed below, certain other reagents are capable of effecting these transformations, but we shall see that few offer the scope and versatility that N_2O_5 exhibits. The utility of the various classes of product, C-nitro, nitramine or nitrate ester, will be illustrated throughout this paper and summarised at the end.

Table I. Summary of N_2O_5 Nitrations

Reaction Type	Conditions*	Examples	Product Name
aromatic nitration	A / O	benzene → nitrobenzene (NO_2)	(C-) nitro compound
nitrolysis	A / O	$R_2N-COCH_3 \longrightarrow R_2N-NO_2$	nitramine
ring cleavage	O	H_2C-CH_2 (epoxide) $\longrightarrow O_2NO(CH_2)_2ONO_2$	nitrate ester or nitramine-nitrate
selective nitration	O	H_3C, CH_2OH strained ring $\longrightarrow$ H_3C, CH_2ONO_2 strained ring	nitrate ester derivative of strained-ring compound

* A = N_2O_5 in anhyd. HNO_3; O = N_2O_5 in halogenated solvent (e.g. CH_2Cl_2).

Nitration Potential of N_2O_5 and its Applications to Energetic Materials Synthesis

N_2O_5 has the capability of introducing nitro groups into substrate molecules, in other words it is a nitrating agent (4,5). Although it had been prepared as early as 1849 (6), it was largely neglected as a nitrating agent until the 1920s when the first systematic study of this type was undertaken (7). Presumably this neglect was a result of the difficulty in obtaining the reagent in a pure form as well as problems in storage resulting from its poor thermal stability (4), aspects which will be covered in more detail presently. The current resurgence of interest within this research group and elsewhere has demonstrated the versatility of N_2O_5, which can be used to generate all three classes of energetic grouping from suitable precursors in clean, specific and frequently high-yielding reactions.

Nitration reactions, of any of the types mentioned here, are believed to proceed via the nitronium ion, NO_2^+ (8), and the activity of a nitrating agent is believed to depend more or less directly on the concentration of nitronium ion. To illustrate this point, the nitrating potential of a selection of nitration systems commonly used both in the laboratory and in production is summarised in Table II. Thus, for a nitronium compound NO_2Y, the nitrating ability generally follows a well-established trend which correlates with the base strength or nucleofugacity of Y (8): hence NO_2BF_4 ~ HNO_3-H_2SO_4 > N_2O_5-HNO_3 > N_2O_5-halogenated solvent ~ HNO_3-Ac_2O > pure HNO_3 > $C(NO_2)_4$.

Table II. Selected Nitration Media and their Properties

System	Application				By-products	Features
	Arom[a]	N'am[b]	N'est[c]	Sel.[d]		
Pure HNO$_3$	+	++	+	-	Dilute HNO$_3$, acyl nitrates	Low nitrating strength, limited substrate choice, disposal/recycling problems
HNO$_3$-H$_2$SO$_4$	++	++	++	-	Dilute mixed acid, acyl nitrates	High nitrating strength, wide substrate choice, severe disposal problems
NO$_2^+$BF$_4^-$	+	++	(+)	-	H$^+$BF$_4^-$, acyl tetrafluoroborate	Very high nitrating strength, limited substrate choice, high expense & very corrosive by-products.
HNO$_3$-Ac$_2$O	(+)	++	++	+	Acetyl nitrate (xs), AcOH	Moderate nitrating strength, limited substrate choice, <u>high hazard</u> - detonable
C(NO$_2$)$_4$	+	?	?	+	Nitroform + lower nitromethanes	Low nitrating strength, selectivity possible but extremely hazardous
N$_2$O$_5$/HNO$_3$	++	++	++	-	Strong HNO$_3$, acyl nitrates	High nitrating strength, wide substrate choice, easy recyclability
N$_2$O$_5$/halogen-ated solvent	+	++	++	++	<u>Potentially none</u> (with ring-opening nitrations)	Moderate nitrating strength, wide substrate choice, by-product disposal problems <u>virtually eliminated</u>

[a]Arom. = aromatic nitrations; [b]N'am = nitramine syntheses; [c]N'est = nitrate ester syntheses; [d]Sel. = selective nitrations. ++ = highly suitable; + = less suitable but possible; - = not possible

It is apparent that N_2O_5 exhibits a dichotomy of behaviour according to the medium selected, on the one hand 100% nitric acid, giving a medium with high nitronium ion concentration on account of the high degree of dissociation of N_2O_5 arising from the high polarity of the solvent (9,10), and on the other the organic solvents, typically chlorinated hydrocarbons, where a low nitronium ion concentration is found owing to the essentially undissociated nature of N_2O_5 in this environment (11). Thus, the first medium provides a potent, unselective nitration system akin to mixed acid but with some advantages over the classical system, while the second enables gentler conditions to be achieved, which are essentially non-acidic and hence useful for nitrating acid-sensitive substrates or for performing selective nitrations. Comparison with other systems shows that none is capable of achieving such a range of effects, particularly with sensitive substrates, where systems such as acetyl nitrate or tetranitromethane are precluded from scale-up work on account of their excessive hazard.

As was shown in Table I, the harsher nitric acid system is more suitable for aromatic nitrations or nitrolyses to yield nitramines from their N-acyl precursors (although, under suitable conditions, these transformations can sometimes also be effected in appropriate organic solvent media), while on the other hand, the organic media, particularly chlorinated hydrocarbons or freons, are essential to enable the ring cleavage or selective nitration reactions to be carried out. With some substrates possessing both strained rings and labile groups, the reaction can be "fine-tuned" to enable selective nitration of these compounds to be attained - this important aspect of N_2O_5 chemistry will be covered in more detail later. It should be emphasised that, whilst reaction in the first two categories can sometimes be effected under organic solvent conditions as well, the latter two categories require exclusively organic solvent conditions, also preferably free from adventitious acid in certain cases.

Preparation of N_2O_5

Two main routes are currently in use for the preparation of N_2O_5, both in our laboratories and industrially. The first, by electrolysis of nitric acid in the presence of N_2O_4 (equation 1), is based on laboratory studies carried out at the Lawrence

$$2 HNO_3 \xrightarrow[-2e^-]{N_2O_4} N_2O_5 + H_2O \quad \text{cell membrane} \quad (1)$$

Livermore National Laboratory (12) and further developed by the DRA in the U.K. to a fully commercial process (13). This generates a solution of N_2O_5 of some 15-35% (wt./wt.) strength in anhydrous nitric acid according to the requirement. Rates of production upwards of several kg per day are currently available and scale-up to higher production levels poses no serious technical problems (14); the electrolytic cell employed is shown in Figure 1. N_2O_5 generated in this way finds application mainly in the first two types of nitration reaction - aromatic nitration and especially nitrolysis to generate HMX (see below).

The second route, ozonation of N_2O_4 (equation 2), although known for some decades (15), was not, to our knowledge, developed into a viable <u>large-scale</u> synthetic process until the inception of work in our laboratories. The gas-phase

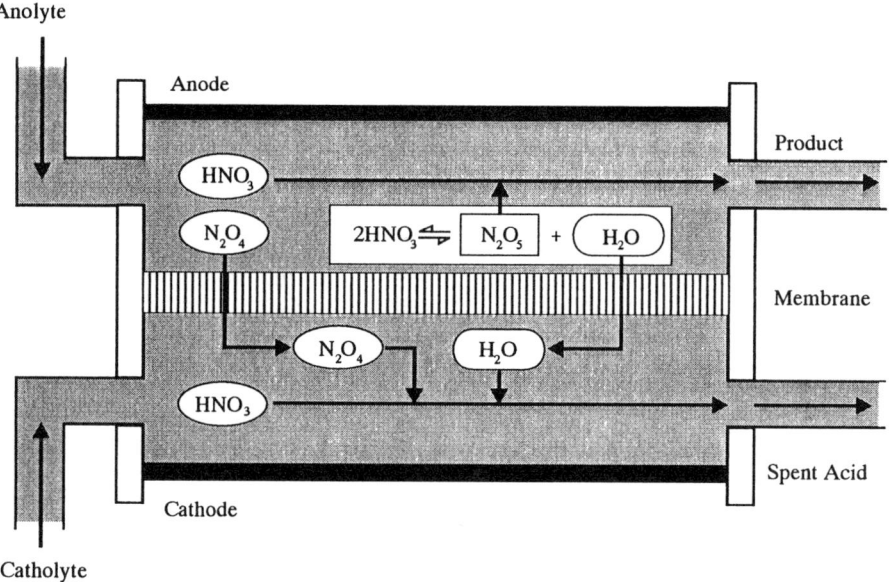

Figure 1. Schematic diagram of electrolytic cell for generating N_2O_5 in nitric acid.

$$N_2O_4 \xrightarrow{O_3} N_2O_5 \qquad (2)$$

reaction of ozone (as a 5-12% mixture with oxygen) with N_2O_4 generates essentially acid-free N_2O_5 which is trapped as a solid and can be stored at low temperature for lengthy periods until required. On the larger scale, reaction can occur in the liquid phase yielding up to 10% (wt./wt.) N_2O_5/dichloromethane solution directly, and with high capacity ozonisers production rates of up to 20 tonnes per annum should be feasible.

N_2O_5- Nitric Acid Nitrations

Safety Note: In these nitrations, and others described later, energetic materials are produced which may be hazardous, and appropriate precautions should be taken in their preparation and handling. Also, many of the starting materials, reagents and products are toxic, corrosive or present other hazards and only suitably trained personnel should attempt the chemical transformations described throughout this article.

The utility of N_2O_5 nitrations in energetic materials chemistry is now illustrated, taking firstly reactions in N_2O_5-nitric acid medium. With aromatic nitrations: compounds **5, 7a & 7b** are formed cleanly and in essentially quantitative yield and these products are directly applicable in explosive/plasticiser technology (equations 3 & 4) (*16-19*). Furthermore, polynitrofluorenes (e.g. **11**), which are novel thermally-stable explosives, have been made for the first time (*16*) (equations 5 & 6). Interestingly, in this reaction N_2O_5 effects the introduction of a gem-dinitro moiety.

A remarkable phenomenon has been discovered in N_2O_5-nitric acid nitrations of aromatic substrates, namely an unexpected rate enhancement (Figure 2) of up to 30 times at high N_2O_5 concentration over that which would be expected from nitric acid - sulphuric acid systems of similar concentration (*17-19*). Such a discontinuity in the rate profile would appear to indicate a change in the active nitrating species; one such

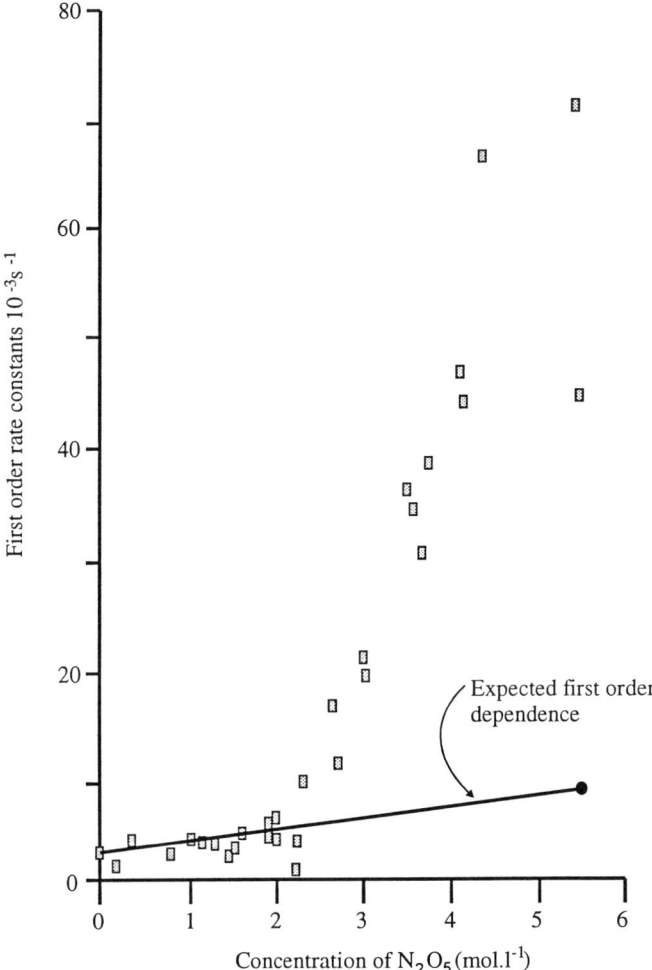

Figure 2. Rate profile for the nitration of phenyltrimethylammonium perchlorate by N_2O_5/HNO_3.

candidate, as yet unverified, might be the NO_3· radical - indeed recent Japanese work has suggested the intermediacy of this species in certain N_2O_5 nitrations (*20*). Such features give N_2O_5-nitric acid solutions unique nitration potential which will be exploited in future studies.

$$8 \xrightarrow{\substack{N_2O_5\text{-enriched }HNO_3 \\ 46 \text{ mins, } 50°C}} 9 \quad 88\% \tag{5}$$

$$10 \xrightarrow{\substack{N_2O_5 \text{ in } CH_2Cl_2 \\ 13 \text{ mins.} \\ \text{room temp.}}} 11 \quad 98\% \tag{6}$$

Before turning to nitrolyses, a very recent development in aromatic nitration using N_2O_5 should be mentioned which, although not carried out in nitric acid medium, nevertheless uses a strongly polar medium to enhance the reactivity of the N_2O_5. The nitration of pyridine (**12**) and its homologues is described (*21*) where high (up to 60%) yields of mononitropyridines (e.g. 3-nitropyridine (**13**, equation 7) are obtained, a massive improvement upon yields hitherto reported for such direct nitrations (*ca* 4.5% for conversion of **12** to **13** (*22*)). The problem however lies in the aggressive medium required - the reaction necessitates the use of sulphur dioxide as solvent, thus reducing the potential for scale-up on account of the hazard.

$$12 \xrightarrow{\substack{N_2O_5/SO_2 \\ -10°C}} 13 \tag{7}$$

Nitrolysis reactions to form nitramines again offer improvements over existing methodology, largely due to the instability of these products in media containing sulphuric acid (*23*). The route of greatest practical utility so far is, of course, the production of HMX (**16**) via DADN (**15**) (*24*), which was under consideration for large scale production in the USA (equation 8).

Further uses of N_2O_5 chemistry in the synthesis of i) nitramines, by a novel synthetic method (nitrodesilylation), and ii) gem-dinitro compounds (Ponzio reaction), will be detailed in subsequent papers.

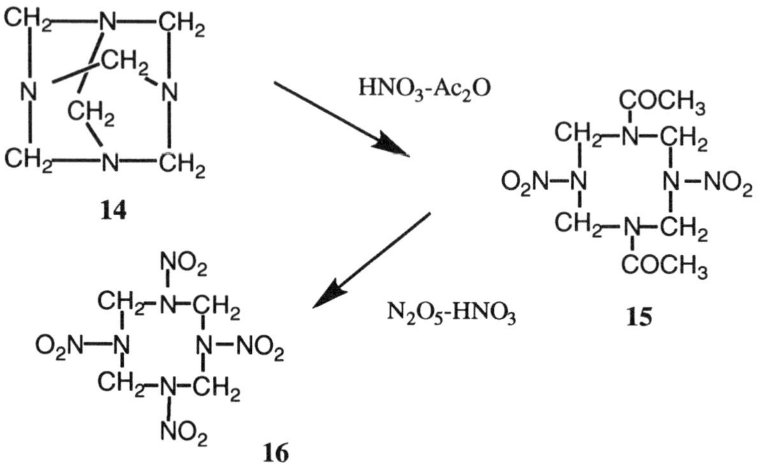

[Structural diagram: Compound **14** (hexamethylenetetramine-like cage) reacts with HNO₃-Ac₂O to give **15** (diacetyl dinitro compound), and with N₂O₅-HNO₃ to give **16** (HMX-like tetranitro compound).] (8)

N$_2$O$_5$ - Organic Solvent Nitrations

Safety Note: See notes at previous section heading (N$_2$O$_5$ - Nitric Acid Nitrations).

These types of reaction embrace two distinct but related areas of chemical synthesis, namely ring-cleavage reactions and selective nitrations, and each will be dealt with in turn.

The ring cleavage reactions (equation 9) constitute an entire family of novel reactions discovered by the Synthetic Chemistry Section at DRA (formerly MOD(PE)) in the early 1980s (25-33), and enable the preparation of multifarious nitrated compounds of high energy content, viz. **18a, 18b, 19a, 19b**, which are applicable in all fields of energetic materials technology except, perhaps, those areas requiring ultimate thermal stability.

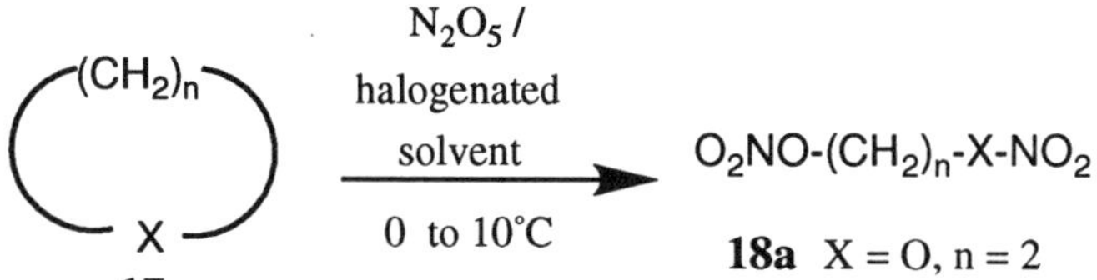

$$\underset{\mathbf{17}}{\overset{(CH_2)_n}{\underset{X}{\bigcirc}}} \xrightarrow[\text{0 to 10°C}]{\text{N}_2\text{O}_5 \text{ / halogenated solvent}} O_2NO\text{-}(CH_2)_n\text{-}X\text{-}NO_2 \qquad (9)$$

n = 2 or 3
X = O or NR (R = alkyl, etc)

18a X = O, n = 2
18b "   , n = 3
19a X = NR, n = 2
19b "   , n = 3

The substrates (**17**) are strained-ring heterocycles with three- or four-membered rings containing either oxygen or nitrogen heteroatoms, making a total of four possible sets of reactions. It should be noted that these reactions possess a common feature, namely, the simultaneous introduction of two energetic groups, either similar - nitrate in the case of the oxygen heterocycles, or dissimilar - nitramine-nitrates in the case of the nitrogen heterocycles. Thus the reaction is a new type of nitration, namely an addition reaction, which features complete utilisation of the nitrating agent, unlike the more usual substitutive nitration where, with N$_2$O$_5$, half of the nitrating agent would be wasted. These reactions are generally high yielding and encompass a wide range of compounds, the main limitation being that R is not equal to H or certain other groups for the nitrogen heterocycles (see below).

As a new nitration method, these reaction sequences possess in common several advantages over conventional nitration routes using mixed acids:-
1) absence of waste acids for disposal
2) simple product separation - evaporation of solvent often suffices
3) ease of temperature control - reactions are essentially non-exothermic
4) high selectivity in position of attack with polyfunctional substrates.

Yields are frequently high (80-98%) and product contamination from by-products is correspondingly low. In addition, the first two features suggest important environmental advantages, since waste liquors from conventional nitrators require tedious post-reaction treatment in order to prevent pollution and indeed to avoid hazard, containing labile chemical constituents such as, for instance, acetyl nitrate. The N_2O_5-strained-ring reactions, on the other hand, avoid these problems and exhibit the further advantages mentioned above. The last feature (selectivity in position of attack) is of relevance primarily to energetic binder synthesis, but before turning to this topic the ring-cleavage nitration reactions will be surveyed.

Oxygen Heterocycles. Strained-ring oxygen heterocycles (**17**, X = O) with 3-membered (epoxide, n = 2) or 4-membered (oxetane, n = 3) rings are cleaved cleanly and in high yield to give the corresponding dinitrate esters **18a** & **18b**, 1,2- or 1,3-dinitrates according to the substrate (equations 10-13) (*25,30,31*). The reaction is quite general and fails to give yields exceeding *ca* 70% in only a few cases, notably heavily-substituted epoxides or oxetanes substituted at the 2-position; in the latter case oxidative cleavage reactions are believed to supervene. Reaction rates may be accelerated in certain cases by catalysis, e.g. addition of Lewis acids such as $AlCl_3$. Applications to the synthesis of energetic plasticiser compounds are shown (equations 10-13), giving rise to both known (nitroglycerine **21** or trimethylolethane trinitrate

$$CH_2\text{-}CH\text{-}CH_2OH \quad \xrightarrow[\text{0 to 10°C}]{\substack{N_2O_5 > 2\text{ mol} \\ AlCl_3 \\ 4h}} \quad O_2NO\text{-}CH_2\text{-}CH\text{-}CH_2\text{-}ONO_2 \quad \underset{ONO_2}{|} \quad (10)$$

20   **21**   73%

$$\underset{\textbf{22}}{H_3C\diagup\diagdown CH_2OH} \quad \xrightarrow[\text{0 to 10°C}]{\substack{N_2O_5 > 2\text{ mol} \\ 28h}} \quad H_3C\text{-}C\diagup\diagdown\underset{CH_2ONO_2}{\overset{CH_2ONO_2}{CH_2ONO_2}} \quad (11)$$

22   **23**   88%

23) and novel compounds (octanetetrol tetranitrate **25** or erythritol tetranitrate **27**). The reactions appear to be unaffected by even moderate amounts of nitric acid, formed either from atmospheric moisture or by reaction with other functionalities in the substrate molecule (e.g. in the cases of glycerol **20** or 3-hydroxymethyl-3-methyloxetane **22**); in this respect these nitrations contrast sharply with the nitrogen heterocycle reactions, particularly those of the aziridines (see below).

The possession of ring strain in these oxygen heterocycles has been shown to be a prerequisite for these reactions to proceed in high yield - substrates with 5-, 6- and 7-membered rings, i.e. **28a, 29a** & **30a** (Scheme 1) react sluggishly yielding generally less than 5% of the corresponding dinitrates, unless other functionalities (particularly

$$\underset{24}{CH_2\text{-}CH\text{-}(CH_2)_4\text{-}CH\text{-}CH_2} \atop \underset{O}{\diagdown}\underset{}{\diagup}\underset{O}{\diagdown}\underset{}{\diagup}} \xrightarrow[\substack{0 \text{ to } 10°C \\ 6h}]{N_2O_5 > 2 \text{ mol}}$$

$$O_2NOCH_2\text{-}\underset{\underset{ONO_2}{|}}{CH}\text{-}(CH_2)_4\text{-}\underset{\underset{}{|}}{\overset{\overset{ONO_2}{|}}{CH}}\text{-}CH_2ONO_2 \quad 91\%$$
$$ 25 (12)$$

$$\underset{26}{CH_2\text{-}CH\text{-}CH\text{-}CH_2} \atop \underset{O}{\diagdown}\underset{}{\diagup}\underset{O}{\diagdown}\underset{}{\diagup}} \xrightarrow[\substack{0 \text{ to } 10°C \\ 4h}]{\substack{N_2O_5 > 2 \text{ mol} \\ AlCl_3}}$$

$$O_2NOCH_2\text{-}\underset{\underset{ONO_2}{|}}{CH}\text{-}CH\text{-}CH_2\text{-}ONO_2 \quad 55\%$$
$$ 27 (13)$$

a second oxygen atom in the 3-position) are present. However, in such cases (compounds **28b, 29b & 30b**), products formed by competing reaction pathways are also found, e.g. formate-nitrates **33**, as well as the expected hemiformal nitrates **32**; the former are believed to originate from hydride abstraction, proceeding via stable dioxolenium cation intermediates (27).

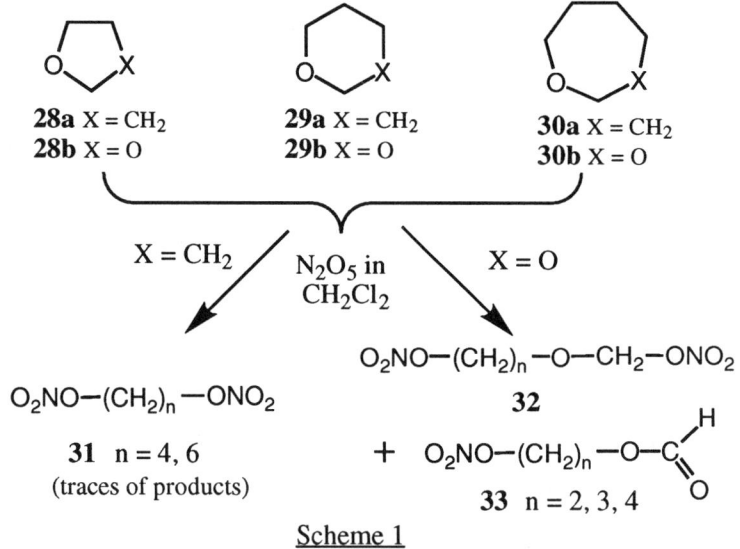

Scheme 1

Finally, some studies (30) of product stereo- and regiochemistry in the reaction of N_2O_5 with two epoxides, cyclopentene oxide (**34**) and cyclo-octene oxide (**35**), have indicated that the reactions appear to proceed in a stepwise rather than in a concerted fashion, suggesting the intermediacy of ionic species: thus with **34** the *trans*-dinitrate (**36**) is isolated, rather than the *cis* product, and with **35** the 1,4-

11. MILLAR ET AL. *Novel Syntheses of Energetic Materials*

$$\text{34} \xrightarrow{N_2O_5/CH_2Cl_2} \text{36} \quad (14)$$

$$\text{35} \xrightarrow{N_2O_5/CH_2Cl_2} \text{37} \quad (15)$$

dinitrate (**37**), formed by a Cope-type rearrangement (*34*) is the sole isolated product (equations 14 & 15). The role of ionic intermediates in the reactions of epoxides with N_2O_5 have been confirmed in other studies (*35*).

Nitrogen Heterocycles. With the nitrogen heterocycles (**17**, X = NR), the 3-membered (aziridine, n = 2) and 4-membered (azetidine, n = 3) compounds give rise to the corresponding nitramine-nitrate products **19a** & **19b** in many cases in good to excellent yields (equation 9) (*26,28,32,33*). As mentioned earlier, the aziridine reactions are particularly susceptible to acid resulting from either adventitious moisture or from hydroxylic species within the substrate molecule, and in such cases only low yields of the desired nitramine-nitrates are obtained (*32*), with homopolymerisation of the aziridine supervening (*36*).

The products find application as melt-castable explosives or plasticisers; thus the explosive materials pentryl **39** (equation 16) and Tris-X **41** (equation 17) feature a high oxygen balance and the latter is an entirely novel compound (*29*), inaccessible by other routes, illustrating the utility of N_2O_5-ring cleavage reactions. Both of these products are highly energetic and show other favourable physical properties. These strained-ring/ N_2O_5 reactions can also yield energetic plasticisers, starting from nitrogen heterocycles (equations 18 & 19): Bu-NENA (**44**), for instance, is a component of some LOVA propellant compositions, whilst nitrocarbamates such as **46** make excellent energetic plasticisers (*37*).

Finally, the cleavage reactions of nitrogen heterocycles have been found to be not quite as general as those of the oxygen series, with the course of the reaction being affected by the nature of the exocyclic nitrogen substituent (R) as well as the ring size (Scheme 2 & equation 20). Thus when R is hydrogen no nitramine-nitrates

$$\text{38} \xrightarrow{N_2O_5/\ CHCl_3} \text{39} \quad 82\% \quad (16)$$

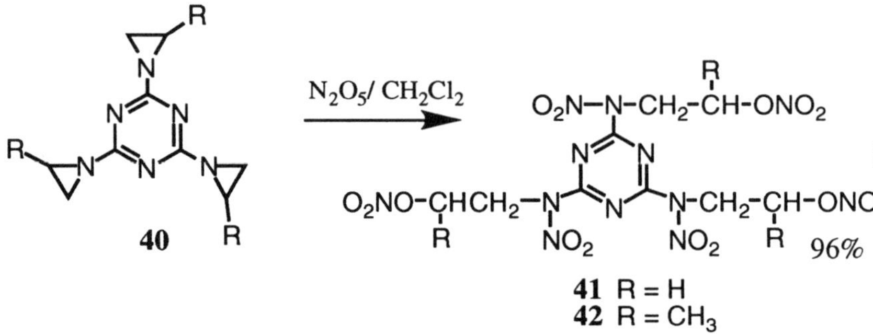

are obtained: aziridines such as **49** yield 1,2-dinitrates **51** by a rearrangement reaction with expulsion of N_2O, whilst azetidine **54** gives the cyclic nitramine, N-nitroazetidine **55**, by a substitutive nitration (28).

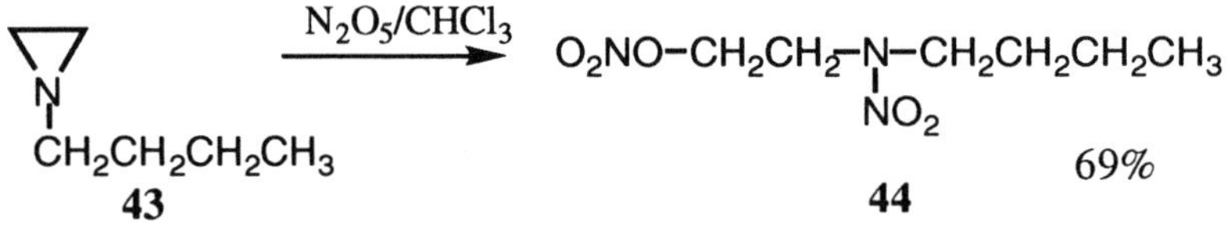

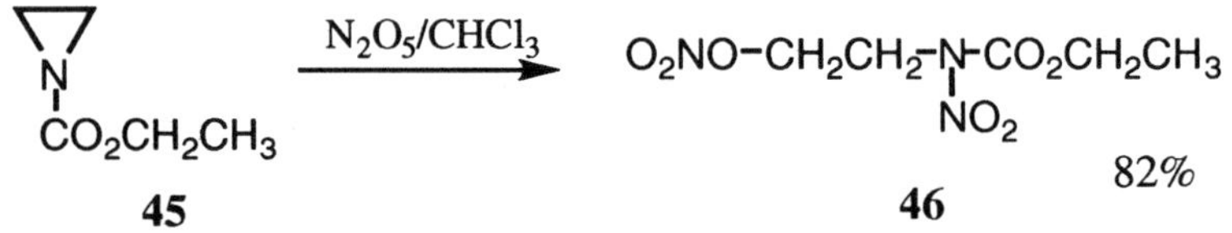

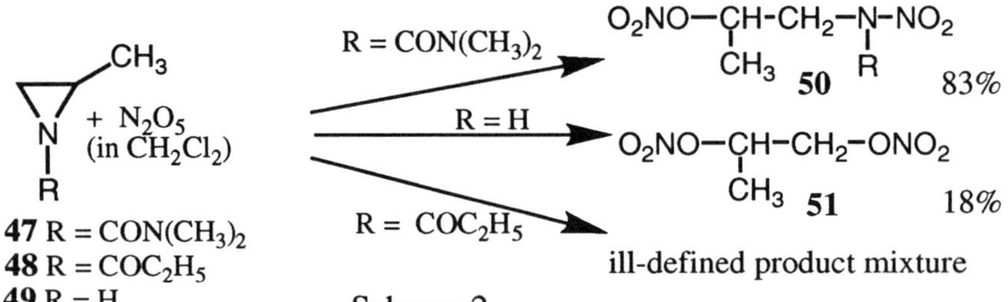

Scheme 2

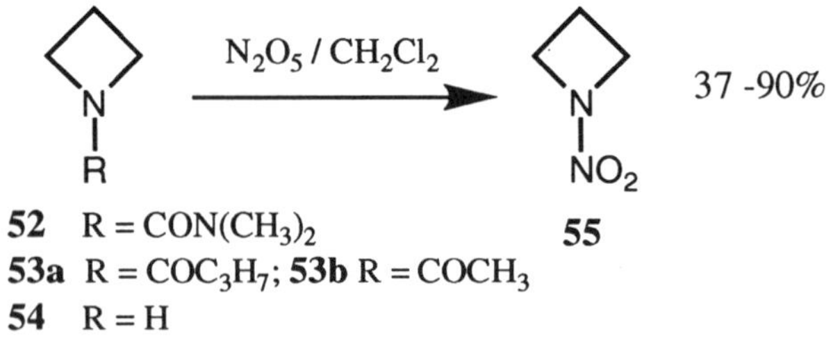

47 R = $CON(CH_3)_2$
48 R = COC_2H_5
49 R = H

52 R = $CON(CH_3)_2$
53a R = COC_3H_7; **53b** R = $COCH_3$
54 R = H

The behaviour becomes more complex with other N-acyl substituents, some aziridines (e.g. the carbamyl compound **47**) showing predominantly ring-opening behaviour giving nitramine-nitrate products (e.g. **50**) whilst others (e.g. the propionyl compound **48**) yield mixtures, indicating competing reaction pathways (*32*). The azetidines, on the other hand, tend toward deacylative nitration (nitrolysis), yielding N-nitroazetidine **55** with groups such as carbamyl (e.g. **52**), butyryl (**53a**) or acetyl (**53b**) (*28,33*). Such deacylations are, of course, well known in nitration chemistry (*38,39*) and reflect the lower ring strain in azetidines which permits alternative reactions to occur.

Selective Nitrations and Energetic Binders

Selective nitrations, as suggested earlier, are feasible in organic solvent medium and, owing to the mild nitrating power of N_2O_5 under these conditions, it is possible to allow a nitration to proceed only partially to completion where two or more functional groups of widely differing lability are present in the same molecule. Such an approach, of course, would be quite impossible with conventional strongly acid media, e.g. nitric-sulphuric acid mixtures, whilst other known mild reagents, e.g. acetyl nitrate or tetranitromethane, are impractical on large scale on account of excessive hazard and expense.

Two approaches to energetic binder macromolecules by selective nitration chemistry using N_2O_5 are illustrated by the following reactions (*40*):- firstly, the epoxy groups in an epoxidised hydroxy-terminated polybutadiene **57** can be cleaved by N_2O_5 to yield an energetic polybutadiene containing vicinal dinitrate groups (equation 21). This material, known as NHTPB (nitrated hydroxy-terminated polybutadiene, **58**) is currently undergoing scale-up and evaluation in collaboration with industry, since it is readily preparable on large scale from cheap precursors and shows desirable properties for binder applications (*41*). Incidentally, the reactivity of the chain double bonds appears sufficiently low to be negligible in this application; were this not so, formation of 1,2-nitro-nitrates (**61**) would occur (*42*) (equation 22) which would have a deleterious effect on the properties of the NHTPB, since this is known to be an unstable grouping. Further work is underway to assess the scope of the N_2O_5-alkene reaction (*43*).

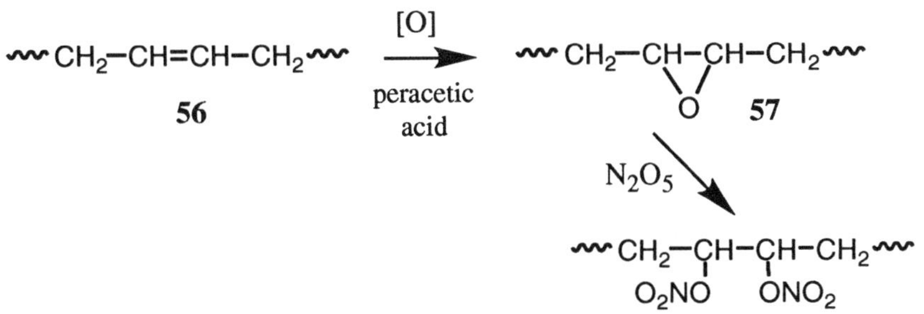

$$\sim CH_2-CH=CH-CH_2\sim \xrightarrow[\text{peracetic acid}]{[O]} \sim CH_2-\underset{\underset{O}{\diagdown\diagup}}{CH-CH}-CH_2\sim$$
$$\mathbf{56} \qquad\qquad\qquad\qquad \mathbf{57}$$

$$\xrightarrow{N_2O_5} \sim CH_2-\underset{O_2NO}{CH}-\underset{ONO_2}{CH}-CH_2\sim$$
$$\mathbf{58} \qquad (21)$$

$$\left(CH_2-\underset{O_2NO}{CH}-\underset{ONO_2}{CH}-CH_2\right)_n \left(CH_2-CH=CH-CH_2\right)_{1-n}$$
$$\mathbf{59}$$

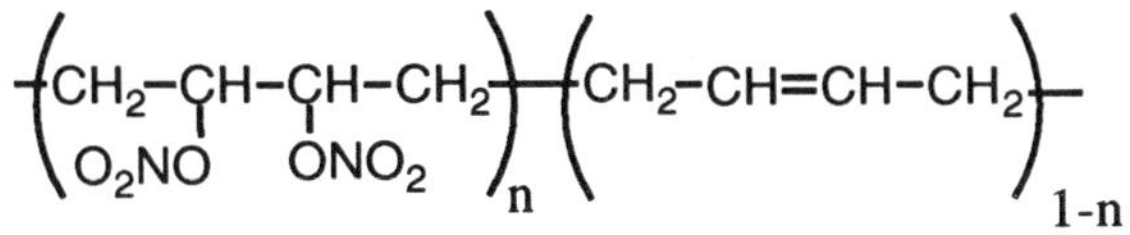

The second approach involves the generation of energetic monomers by selective nitration: candidate molecules such as 3-hydroxymethyl-3-methyloxetane (**22**) and

$$\underset{60}{\diagup\!\!\!=\!\!\!\diagdown} \xrightarrow{N_2O_5} \underset{61}{O_2N-\overset{|}{\underset{|}{C}}-\overset{|}{\underset{|}{C}}-ONO_2} \quad (22)$$

glycidol (**20**) have been found to be susceptible to nitration on the hydroxyl function whilst leaving the strained ring unchanged (equations 23 & 24), and essentially quantitative yields in the nitration step have been achieved under suitable conditions (*14,41,44,45*).

$$\underset{\mathbf{22}}{\text{H}_3\text{C}\diagdown\!\!\diagup\text{CH}_2\text{OH oxetane}} \xrightarrow[-5°C\ 45\ min.]{N_2O_5,\ 1\ mol} \underset{\mathbf{62}}{\text{H}_3\text{C}\diagdown\!\!\diagup\text{CH}_2\text{ONO}_2\ oxetane} \quad 95\% \quad (23)$$

$$\underset{\mathbf{20}}{\text{CH}_2\text{—CH—CH}_2\text{OH epoxide}} \xrightarrow[-25°C\ 10\ min.]{N_2O_5,\ 1\ mol} \underset{\mathbf{63}}{\text{CH}_2\text{—CH—CH}_2\text{ONO}_2\ epoxide} \quad 85\% \quad (24)$$

The nitrated strained-ring monomers, 3-hydroxymethyl-3-methyl-oxetane nitrate (**62**) and glycidyl nitrate (**63**), are then polymerised cationically to their respective polymers resulting in materials with the desired molecular weight range and suitable hydroxyl functionality, enabling subsequent cross-linking to a polyurethane rubber to be carried out (equation 25). The resulting energetic rubbers constitute a significant new class of energetic materials with wide application in propellant and explosive technology, notably nabling viable high performance low vulnerability propellant formulations to be manufactured for the first time (*46*), and such materials are undergoing scale-up to tonnage levels in collaboration with industry.

$$\underset{\substack{\mathbf{62}\ R=CH_3,\\ R'=CH_2ONO_2}}{\text{oxetane with R, R'}} \xrightarrow{\text{Cationic polymerisation}} \underset{\mathbf{64}}{\left(O-CH_2-\underset{R'}{\overset{R}{C}}-CH_2\right)_n} \quad (25)$$

Conclusions

N_2O_5 nitrations are versatile and have much to offer the energetic materials chemist/technologist. The areas in which this chemistry has made an impact are summarised in Table III; it can be seen that further applications await, particularly on the right hand column, where further types of energetic binder precursors possessing nitramine or C-nitro functionality are possible candidates.

Table III. Applications of N_2O_5 Nitrations (in Nitric Acid or Organic Solvent)

	Product Type	Plasticisers[a]	Crystalline H.E.[b]	Thermally-stable H.E.[c]	Polymer precursors[d]
Aromatic nitrations	$C-NO_2$	+		+	
Nitrolysis	$N-NO_2$		+	+	(+)
Ring cleavage	$N-NO_2$ $O-NO_2$	+	+	+	
Selective nitrations	$O-NO_2$ ($N-NO_2$)				+

[a]Nitrate esters/nitroaromatics; [b]Nitramines/nitramine-nitrates; [c]Nitramines/nitroaromatics; [d]Nitrate esters (nitramines);

"+" indicates application implemented, in parentheses yet to be implemented.

To conclude, it has been established that two essentially different but complementary systems are available:-
1) N_2O_5 in nitric acid, and
2) N_2O_5 in organic solvents.

The former system, now available on pilot plant scale, can effect nitrations of the aromatic and nitrolysis type, useful for generating C-nitro and nitramine products respectively. It is a potent nitration system which has already shown advantage over mixed acid systems, particularly in the preparation of nitramines such as HMX.

The N_2O_5 organic solvent system, on the other hand, although a milder nitration medium can nevertheless effect nitration of a wide variety of substrates, in particular the selective nitration of polyfunctional substrates and also nitration of polymers such as epoxidised HTPB. These products are of great importance industrially and have demonstrated that the future of N_2O_5 in the field of energetic materials chemistry is assured.

References

1. Urbanski, T. *Chemistry & Technology of Explosives*, Pergamon Press: Oxford, UK, 1967; Vols. 1-3, & 1984; Vol. 4.
2. Schofield, K. *Aromatic Nitration*, Cambridge University Press: Cambridge, UK, 1980.
3. Olah, G. A.; Malhotra, R.; Narang, S. C. In *Nitration: Methods & Mechanisms*; Feuer, H.; Ed.; Organic Nitro Chemistry Series; VCH: New York, 1989.
4. Addison, C. C.; Logan, N. In *The Chemistry of Dinitrogen Pentoxide*; Colburn, C. B., Ed.; Developments in Inorganic Nitrogen Chemistry; Elsevier: Amsterdam, 1973, Ch. 2.
5. Fisher, J. W. In *Nitro Compounds: Recent Advances in Synthesis & Chemistry*; Feuer, H., Nielsen, A. T., Eds.; Organic Nitro Chemistry Series; VCH: New York, 1990, Ch. 3.
6. Deville, M. H. *Compt. Rend. Acad. Sci. Paris* **1849**, *28*, 257-260.
7. Haines, L. B.; Adkins H. *J. Am. Chem. Soc.* **1925**, *47*, 1419-1426.
8. Ingold, C. K. *Structure and Mechanism in Organic Chemistry*, 2nd Edition; Bell & Sons: London, 1969; Ch. 6; see also ref. 3, Ch. 2.

9. Ingold, C. K.; Millen, D. J. *J. Chem. Soc.* **1950**, 2612-2619.
10. Odokienko, S. S.; Latypov, N. V.; Shokhor, I. N.; Fedorov Y. A.; Vishnevskii, E. N. *J. Appl. Chem. USSR* **1978**, *51* Part 2, 666-668; *Chem. Abstr. 88* 159376r.
11. Chedin, J. *Compt. Rend. Acad. Sci. Paris* **1935**, *201*, 552-554.
12. Harrar, J. E.; Pearson, R. K. *J. Electrochem. Soc.* **1983**, *130*, 108-112.
13. Bagg, G., U.S. Patent 5181996 (publ. 28 June 1990); U.K. Patents 2229449 & 2245003 (publ. 20 Feb. 1991 & 9 Sept. 1992 resp.).
14. Bagg, G., Stewart, M. J., Leeming, W. B. H., Swinton, P. F., "Manufacture of Energetic Binders using N_2O_5" *Proc. Joint International Symp. on Compatibility of Plastics & Other Materials with Explosives, Propellants, Pyrotechnics & Processing of Explosives Propellants & Ingredients, San Diego, CA. 22-24 April 1991*, p. 108, American Defense Preparedness Assoc., Arlington, VA.
15. Harris, A. D.; Trebellas, J. C.; Jonassen, H. B. In *Inorg. Synth.*, McGraw-Hill: New York, 1967; Vol. 9, pp 83-88.
16. Honey, P. J., M.Phil. Thesis, Hatfield Polytechnic, Hatfield, UK (1991).
17. Moodie, R. B.; Stephens, R. J., *J. Chem. Soc. Perkin Trans. 2* **1987**, 1059-1064.
18. Moodie, R. B.; Sanderson A. J.; Willmer, R. *J. Chem. Soc. Perkin Trans. 2* **1990**, 833-836.
19. Moodie, R. B.; Sanderson A. J.; Willmer, R. *J. Chem. Soc. Perkin Trans. 2* **1991**, 645-650.
20. Suzuki, H.; Mori, T. *J. Chem. Soc. Perkin Trans. 2* **1995**, 41-44 and references contained therein.
21. Bakke, J. M.; Hegbom, I. *Acta Chem. Scand.* **1994**, *48*, 181-182.
22. Acheson, R. M. *An Introduction to the Chemistry of Heterocyclic Compounds, 3rd Edition*; J. Wiley & Sons: New York, 1976; p. 237.
23. Wright, G. F. In *Methods of Formation of the Nitramine Group, its Properties & Reactions*; Feuer, H., Ed.; The Chemistry of the Nitro and Nitroso Groups, Part 1; Interscience: New York, 1969; Ch. 9.
24. Siele, V. I.; Warman, M.; Leccacorvi, J.; Hutchinson R. W.; Motto, R.; Gilbert, E. E.; Benzinger, T. M.; Coburn, M. D.; Rohwer, R. K.; Davey, R. K. *Propellants & Explosives* **1981**, *6*, 67-73.
25. Golding, P.; Millar, R. W.; Paul, N. C.; Richards, D. H. *Tetrahed. Letts.* **1988**, *29(22)* 2731-2734.
26. Golding, P.; Millar, R. W.; Paul, N. C.; Richards, D. H. *Tetrahed. Letts.* **1988**, *29(22)* 2735-2736.
27. Golding, P.; Millar, R. W.; Paul, N. C.; Richards, D. H. *Tetrahed. Letts.* **1989**, *30(46)* 6431-6434.
28. Golding, P.; Millar, R. W.; Paul, N. C.; Richards, D. H., *Tetrahed. Letts.* **1991**, *32(37)* 4985-4988.
29. Millar, R. W.; Paul, N. C.; Richards, D. H.; Bunyan, P.; Golding, P.; Rowley, J. A., *Propellants, Explosives & Pyrotechnics*, **1993**, *18* 55-61.
30. Golding, P.; Millar, R. W.; Paul, N. C.; Richards, D. H. *Tetrahedron* **1993**, *49(32)* 7037-7050.
31. Golding, P.; Millar, R. W.; Paul, N. C.; Richards, D. H. *Tetrahedron* **1993**, *49(32)* 7051-7062.
32. Golding, P.; Millar, R. W.; Paul, N. C.; Richards, D. H. *Tetrahedron* **1993**, *49(32)* 7063-7076.
33. Golding, P.; Millar, R. W.; Paul, N. C.; Richards, D. H. *Tetrahedron* **1995**, *51(17)* 5073-5082.
34. Rosowsky, A. In *Ethylene Oxides*; Weissberger, A., Ed.; The Chemistry of Heterocyclic Compounds; Wiley-Interscience: New York, 1964; Vol. 19, Part 1, pp. 284, 378.
35. Dormer, J., Moodie, R. B., *J Chem. Soc. Perkin Trans. 2* **1994**, 1195-1200.

36. Dermer, O. C., Ham, G. E. *Ethyleneimine & Other Aziridines*; Academic Press: New York & London, 1969.
37. Olsen, R. E.; Fisch, D. W.; Hamel, E. E. In *Nitrations by Nitronium Tetrafluoroborate;* Gould, R. F., Ed.; Advanced Propellant Chemistry (Adv. Chem. Ser. No. 54); Washington DC: American Chemical Society (1966); Ch. 6.
38. Smith, P. A. S. *Open Chain Nitrogen Compounds*; Benjamin: New York, 1966; Vol. 2, Ch. 15.
39. Gilbert, E. E., Leccacorvi J. R., Warman, M. In *The Preparation of RDX from 1,3,5-Triacylhexahydro-s-triazines*; Albright, L. F.; Hanson, C., Eds. Industrial & Laboratory Nitrations (A.C.S. Symp. Ser. No. 22); American Chemical Society: Washington DC , 1975; Ch. 23.
40. Colclough, M. E., Desai, H., Millar, R. W., Paul, N. C., Stewart, M. J., Golding, P., *Polymers for Advanced Technologies* **1994**, *5*, 554-560.
41. Stewart, M., Arber, A., Bagg, G., Colclough, E., Desai, H., Millar, R., Paul, N., Salter, D., "Novel Energetic Monomers & Polymers prepared using Dinitrogen Pentoxide Chemistry" *Proc. 21st Annual Conference of I.C.T. on Technology of Polymer Compounds & Energetic Materials, 3-6 July 1990.* Fraunhofer Institut für Chemische Technologie, Karlsruhe, FRG.
42. Stevens, T. E., Emmons, W. D., *J. Am. Chem. Soc.* **1957**, *79*, 6008.
43. Lewis, R. J., Moodie, R. B., "Mechanistic and Kinetic Studies of the Nitration of Alkenes", poster presented at Zeneca FCMO Organic Reactivity Meeting, Huddersfield, UK, July 1994.
44. Golding, P., Millar, R. W., Paul, N. C., U.S. Patent 5145974 (publ. 8 Feb. 1990); U.K. Patent 2240799 (publ. 8 Jan. 1992).
45. Leeming, W., Paterson, D. H., Paul, N., Desai, H., "Scale-up of Polyglycidyl Nitrate Manufacture; Process Development and Assembly", *Proc. Joint International Symp. on Energetic Materials Technology, New Orleans, LA 4-7 Oct. 1992:* American Defence Preparedness Assoc., Arlington VA.
46. Debenham, D., Leeming, W. B. H., Marshall E. J., "New Energetic Formulations containing Nitropolyethers", as ref. 14, p. 119.

© British Crown Copyright 1995/ DRA Farnborough Hants. U.K.
Published with the permission of the Controller of Her Britannic Majesty's Stationery Office.

RECEIVED January 16, 1996

Chapter 12

A New Route to Nitramines in Nonacidic Media

R. W. Millar

Defence Research Agency, Fort Halstead, Sevenoaks, Kent TN14 7BP, United Kingdom

The novel synthesis of nitramines and nitramides by nitrolysis of the corresponding N-trialkylsilyl compounds using dinitrogen pentoxide (N_2O_5) is described. In seventeen examples the yields are generally in the range 70 to over 90%, falling below the lower figure only if alkylsilyl groups with chain lengths greater than two are employed. The reactions are characterised by their cleanliness, and the co-products, trialkylsilyl nitrates, are relatively stable and volatile, facilitating isolation of the nitrated products. Furthermore, these trialkylsilyl nitrates, unlike the acyl nitrates produced in conventional nitrolyses, are isolable and can be used to nitrate further substrates, thus eliminating problems of disposal of spent liquors from conventional reactions. The process is both mild and versatile, enabling nitramine functions to be introduced into a variety of molecular environments, and two notable cases are exemplified, namely N-nitroaziridines and N,N'-dinitroaminals.

Nitramines are a class of compound finding widespread application in propellant and explosive technology, and their chemistry has been reviewed (*1-5*). They are commonly prepared by the reaction of secondary amides with nitric acid in dehydrating media such as acetic anhydride (*3*), although other routes are possible, for instance by the addition of nitrate salts of secondary amines to acetic anhydride in the presence of a catalyst, e.g. chloride ion (*6*), by direct interaction of an amine with dinitrogen pentoxide, N_2O_5 (*5,7*), or by nitrolysis of gem-diamines with nitric acid-acetic anhydride (*2,3,8*). More recently developed methods include the reaction of N,N-dialkylamides with nitronium tetrafluoroborate (*9*), the reaction of *tert.*-butylamines with nitric acid or N_2O_5(*10*), and the action of nitric acid-acetic anhydride on *tert.* amines with *in situ* oxidation of the resulting nitrosamines with peracetic acid (*11*). (Routes involving oxidation of <u>isolated</u> nitrosamines have been disregarded owing to the high toxicity of these compounds.)

Many of these routes have disadvantages such as contamination of the product by nitrosamines which are awkward to remove (*12*), the use of reagents which are not available cheaply on an industrial scale (e.g. NO_2BF_4), or the production of co-products which are difficult to dispose of, notably acyl nitrates. Further problems may arise from inaccessibility of substrates, for instance in the direct nitration of amines (7), certain categories of amine either do not form the nitramine (particularly highly basic amines), or may not be preparable in their unsubstituted form (e.g. hexahydropyrimidines - see later). Such shortcomings limit the scope and utility of existing routes for the synthesis of nitramines.

The problems outlined above are exemplified in one of the most commonly used routes for nitramine synthesis, namely the reaction of secondary amides with nitric acid under dehydrating conditions (equation 1), where the cleavage of the N-acyl

$$\begin{array}{c} R \\ \diagdown \\ R' \end{array} N-H \xrightarrow{a} \begin{array}{c} R \\ \diagdown \\ R' \end{array} N-\underset{\underset{O}{\|}}{C}-R'' \xrightarrow{b} \begin{array}{c} R \\ \diagdown \\ R' \end{array} N-NO_2 \quad \mathbf{I} \quad (1)$$

amine

amide (R" = alkyl)
carbamate (R" = alkoxy)

nitramine

+

O_2NOCOR'' **II**

acyl nitrate
(R" = alkyl)

[Reagents:- a) acylating agent (eg R"COCl)
b) nitrating agent, esp. pure HNO_3, $NO_2^+BF_4^-$ or N_2O_5]

bond results in formation of the desired nitramine (**I**), but an acyl nitrate co-product (**II**) is also formed during the reaction (termed a nitrolysis (3)). The disposal of these acyl nitrates is awkward and also poses safety problems in certain circumstances, for instance in the synthesis of HMX from DADN (13). A further drawback of the nitrolysis of acylamines is that cleavage of N-C bonds *other* than the acyl linkage may occur, resulting in competing reaction pathways and hence lower yields and product contamination, and in extreme cases little of the desired product may be formed (e.g. N,N-dimethylurethane (**III**) yields ethyl N-methyl-N-nitrocarbamate (**IV**) instead of N,N-dimethylnitramine (4)). Finally, some acyl derivatives of polycyclic polyamines (e.g. the precursor of bicyclo-HMX, **V**) are completely inert to nitrolysis (14).

III   **IV**   **V**

In an attempt to overcome these twin problems of controlling the direction of nitrolysis reactions and forming more easily handlable co-products, the replacement of acyl functions by other readily nitrolysable groups was considered. It was felt that these problems stemmed largely from the inertness of the nitrogen atom towards electrophilic attack as a result of the electron-withdrawing acyl function, and therefore employment of substituents with the opposite inductive effect, i.e. electron-*donating* substituents, would be beneficial. With this rationale in mind, obvious candidate elements for consideration would be the group IV metalloids, and it was already known that stannylamines could be nitrolysed to yield nitramines (Nielsen, A. T., NWC China Lake, Calif., personal communication, 1988). Furthermore, publications in the mid-1980s had indicated that C-silyl compounds could be cleaved by reagents such as nitronium tetrafluoroborate to yield C-nitro compounds (*15-17*). However, as no reports were known of the nitrolysis of the corresponding N-silyl compounds, silylamines (**VI**), this therefore seemed an obvious class of compound to examine.

In the subsequent discussion, the N-silyl substrates are divided into two categories - i) dialkyl and cycloalkyl silylamines, the largest category with thirteen examples, and ii) silylamides (including ureas and carbamates) with four examples. Because of their different chemistries, both in the preparation and handling of the substrates as well as their nitration chemistry, this subdivision will be maintained throughout the paper.

Discussion

Silylamines. The silylamines (**VI**) were derived from the corresponding secondary amines, formed *in situ* where necessary (e.g. **VIm**, see below). Reaction with dinitrogen pentoxide (N_2O_5) in halogenated solvents such as dichloromethane generated the nitramines (**I**) cleanly and in good to excellent yield (equation 2 and Table I). The reaction was found to be general for a range of alkyl substituents both

$$\underset{\text{amine}}{\overset{R}{\underset{R'}{\diagdown}}N-H} \quad \xrightarrow{a} \quad \underset{\text{silylamine}}{\overset{R}{\underset{R'}{\diagdown}}N-\overset{R''}{\underset{R''}{\overset{|}{Si}}}-R''} \quad \xrightarrow{b} \quad \underset{\text{nitramine}}{\overset{R}{\underset{R'}{\diagdown}}N-NO_2} \quad (2)$$

$$\text{VI} \qquad\qquad\qquad \text{I}$$

$$+$$

$$\underset{\text{silyl nitrate}}{O_2NOSiR''_3} \quad \text{VII}$$

[Reagents:- a) silylating agent (R''_3SiX where X is a leaving group such as halogen, dialkylamino etc)

b) nitrating agent, esp. N_2O_5, also NO_2BF_4]

on nitrogen (R & R') and silicon (R"), with the highest yields being obtained with trimethylsilyl derivatives (R = CH_3, see Table I). The reaction is applicable to cases which have proved troublesome in the past, for instance sterically hindered nitramines such as **If**, and yields were in many cases improved, sometimes markedly, upon those hitherto obtained.

Cyclic dinitramines (**Il** & **Im**) were likewise preparable without difficulty from the corresponding disilyl precursors. It is notable that the precursor to **Im**, 1,3-bis(trimethylsilyl)-hexahydropyrimidine (**VIm**, Table I) is derived from an unstable diamine (hexahydropyrimidine, **VIII**) and highlights an intrinsic advantage of the novel nitration over other methods which require the use of the free amine, which may be unavailable. Furthermore, the dinitramine product (**Im**), which contains a

Table I: Nitrodesilylation Reactions using N_2O_5

1. Monosilylamines $R^1R^2N-Si(R^3)_2R^4$

VI	R^1	R^2	R^3	R^4	Rn. Time (hr)	Rn. Temp. (°C)	Yield of Nitramine
a	$-(CH_2)_2O(CH_2)_2-$		CH_3	CH_3	2	0±2	80%
b	$-(CH_2)_5-$		CH_3	CH_3	0.75	-5±2	81%
c	$-(CH_2)_4-$		CH_3	CH_3	0.5	-7 to -1	76%
d	CH_3	CH_3	CH_3	CH_3	0.75	-5±2	78%
e	C_2H_5	C_2H_5	CH_3	CH_3	0.75	-5±2	84%
f	$i-C_4H_9$ [a]	$i-C_4H_9$	CH_3	CH_3	0.75	-5 to 0	87%
g	$-(CH_2)_2O(CH_2)_2-$		CH_3	$t-C_4H_9$	2.25	0 to +5	37%
g					6	+5 to +10	40%[b]
h	$-(CH_2)_2O(CH_2)_2-$		$n-C_4H_9$	$n-C_4H_9$	1.5	-5 to +5	39%
i	$-(CH_2)_2O(CH_2)_2-$		C_2H_5	C_2H_5	0.75	0 to +5	61%[b]
j	$i-C_4H_9$ [a]	$i-C_4H_9$	C_2H_5	C_2H_5	1	0 to +5	70%
k	$-CH_2-CH(CH_3)-$		CH_3	CH_3	10 min.	0±5	-[c]

2. Disilylamines

		Rn. Time (hr)	Rn. Temp. (°C)	Yield of Nitramine
l	$(CH_3)_3Si-N\underset{}{\overbrace{}}N-Si(CH_3)_3$ (piperazine)	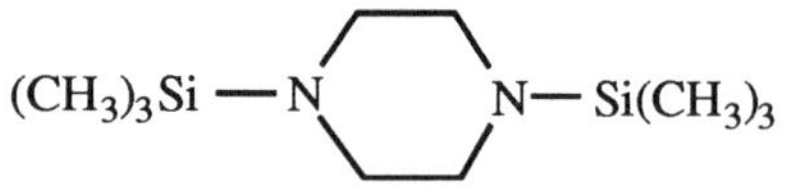 1	-8 to 0	91%
m	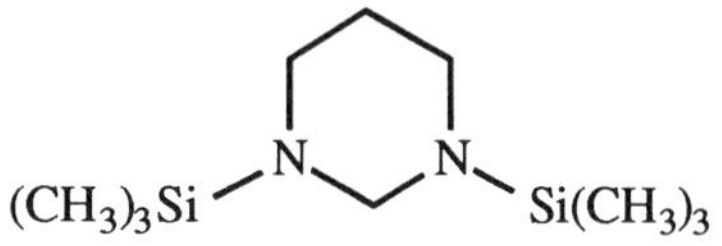 $(CH_3)_3Si-N\underset{}{\overbrace{}}N-Si(CH_3)_3$ (hexahydropyrimidine)	1	-5 to +5	69%[d]

Continued on next page

Table I: Nitrodesilylation Reactions using N_2O_5 (Contd.)

3. Silylamides

VI	Starting Material	Product	Rn. Time (hr)	Rn. Temp. (°C)	Yield of Nitramine
n	Me–C(=O)–N(Me)(SiMe₃)	Me–C(=O)–N(Me)(NO₂)	0.75	-5	79%
o	2-oxazolidinone-N-SiMe₃	2-oxazolidinone-N-NO₂	0.7	0	80%
p	Me–N(SiMe₃)–C(=O)–N(SiMe₃)–Pr^n	Me–N(NO₂)–C(=O)–N(NO₂)–Pr^n	0.5	-10 to -5	75%
q	Me₃Si–N,N'–SiMe₃ cyclic urea	O_2N–N,N'–NO_2 cyclic urea	0.75	-5	82%

[a] i-C_4H_9 = $(CH_3)_2CHCH_2$-
[b] Larger excesses of N_2O_5 (50 and 100% resp.) used
[c] N-Nitroaziridine not isolated - reacted further *in situ* (see text)
[d] Mode of addition reversed (N_2O_5 added to silylamine)

geminal dinitramine moiety which is a substructural fragment found in the RDX and HMX molecules, is preparable in a yield (69%) twice that reported in the hitherto best method (by nitro-denitrosation of the 1,3-dinitroso compound **IX** (*18*)). This hints at the potential of this reaction, and its viability is subject only to the availability of

VIII: HN–NH cyclic (1,3-diazinane)

IX: ON–N,N'–NO (1,3-dinitroso-1,3-diazinane)

suitable silylated precursors; in this respect, few limitations have been encountered, one of the few groups which is incompatible with silylating agents being the nitrile group.

The behaviour of one substrate, the N-trimethylsilylaziridine **VIk** was notable. Upon reaction with 1 mol N_2O_5 the N-nitroaziridine **X** was formed *in situ* in *ca* 80% yield, and further reaction with excess of the reagent resulted in the formation of the N,N-dinitramine-nitrate **XI** (equation 3), which was characterised spectroscopically,

$$\underset{\textbf{VIk}}{\triangle\!\!\!\!\!N\text{-SiR''}_3} \xrightarrow[\text{1 mol}]{N_2O_5} \underset{\textbf{X}}{\triangle\!\!\!\!\!N\text{-NO}_2} \xrightarrow[\text{excess}]{N_2O_5} \underset{\textbf{XI}}{CH_3\text{-CH(ONO}_2)\text{-CH}_2\text{-N(NO}_2)_2} \quad (3)$$

the nitramine asymmetric stretching band in the i.r. being observed at 1607 cm^{-1}, in line with previous findings (*19*). The formation of the N-nitroaziridine **X** constitutes the second reported synthesis of this class of compound (*20,21*) and the first by direct electrophilic substitution, although a N-nitroaziridine intermediate was postulated in earlier work on the nitration of propyleneimine by N_2O_5 (*22*). Also, the N,N-dinitramine-nitrate **XI** was contaminated with some propane-1,2-diol dinitrate (**XII**); such compounds are known to be decomposition products of N,N-dinitramines (Coon, C. L., LLNL Livermore, Calif., personal communication, 1992). Therefore the behaviour of this silylamine opens the door to some novel chemistry by affording classes of compound which are only otherwise obtainable with extreme difficulty; furthermore by the ring-opening nitration yields high-energy compounds such as **XI** ($C_3H_6N_4O_7$) which possesses a similar oxygen balance to nitroglycerine.

Silylamides. As mentioned above, three representative classes of acyl substrates were investigated: one amide (**VIn**), one carbamate (**VIo**) and twoureas - **VIp** (acyclic) and **VIq** (cyclic); see Table I and equation 4. All gave the corresponding nitramine derivatives in good yields (75-82%), which in two cases (**VIn** & **VIo**; equations 4a & 4b) were significant improvements upon those obtained previously (27% and 53% respectively (*23*)). The acyclic dinitrourea (**Ip**) was a new compound, prepared from the known silyl precursor **VIp** (*24*), while preparation of the cyclic dinitrourea (**Iq**), although feasible by direct nitration of trimethyleneurea (**XIII**) in 87% yield (*25*) was nevertheless facilitated by the nitrodesilylation route by the aforementioned advantages of simplified workup and absence of contaminating by-products (equation 4c).

Conclusions

Nitrodesilylations of silylamines and silylamides by N_2O_5 afford the corresponding N-nitro compounds, nitramines and nitramides respectively, in good to excellent yields. The reaction is of wide applicability, and several products bearing 1,3-bis-(N-nitro) functions have been prepared, notably compounds **Im, Ip** & **Iq**. The success of the method with these materials is significant and augurs well for the extension to polynitramines; for instance, RDX should be available from the corresponding tris-(N-trimethylsilyl) precursor (**XIV**), and indeed the starting materials for this synthesis have already been described in the literature (*26,27*). The conditions necessary to

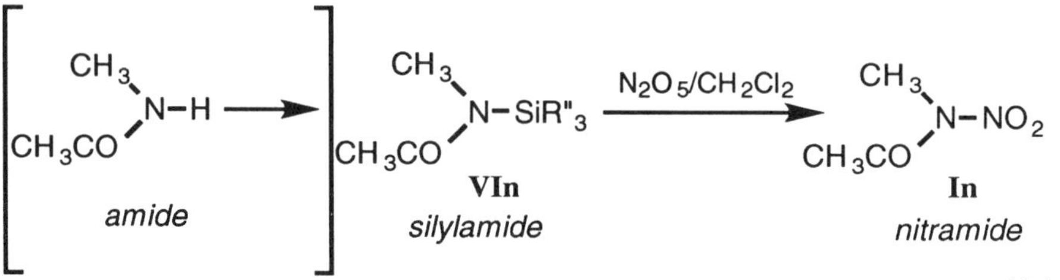

$$\left[\begin{array}{c} CH_3 \\ CH_3CO \end{array} N-H \right] \longrightarrow \left[\begin{array}{c} CH_3 \\ CH_3CO \end{array} N-SiR''_3 \right] \xrightarrow{N_2O_5/CH_2Cl_2} \begin{array}{c} CH_3 \\ CH_3CO \end{array} N-NO_2$$

amide — silylamide **VIn** — nitramide **In**

(4a)

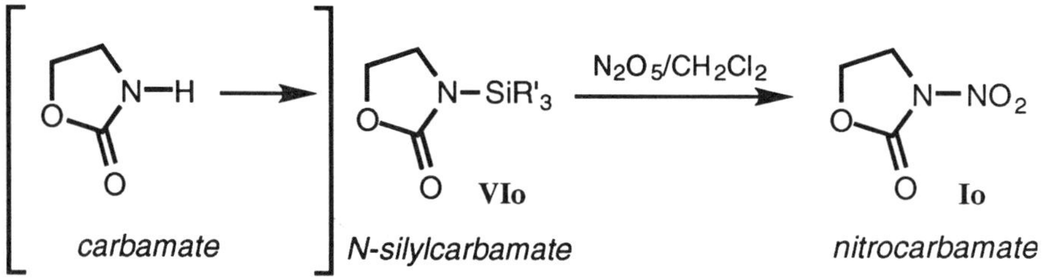

carbamate — N-silylcarbamate **VIo** — nitrocarbamate **Io**

(4b)

urea — N,N'-disilylurea — N,N'-dinitrourea

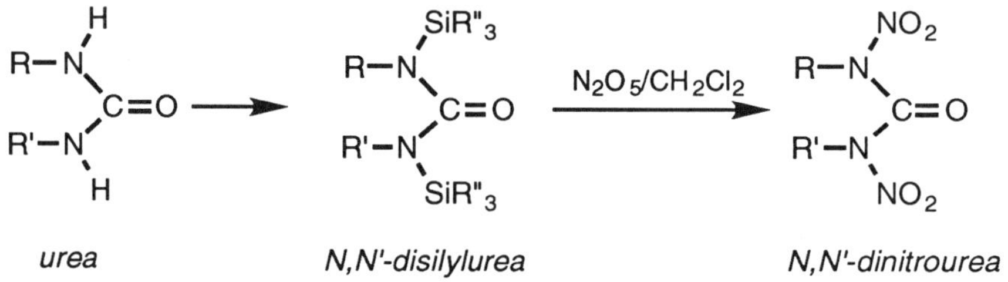

XIII: R = R' = -(CH$_2$)$_3$- **VIp:** R = CH$_3$, R' = n-C$_3$H$_7$ **Ip:** R = CH$_3$, R' = n-C$_3$H$_7$
VIq: R,R' = -(CH$_2$)$_3$- **Iq:** R,R' = -(CH$_2$)$_3$-

(4c)

effect the cleavage of the N-Si bond are much milder than, for instance, those required to cleave N-acyl substrates (i.e. amides) and suggest applications in those areas where cleavage of acetamides have failed to yield nitramine products, e.g. polycyclic nitramines such as bicyclo-HMX (14).

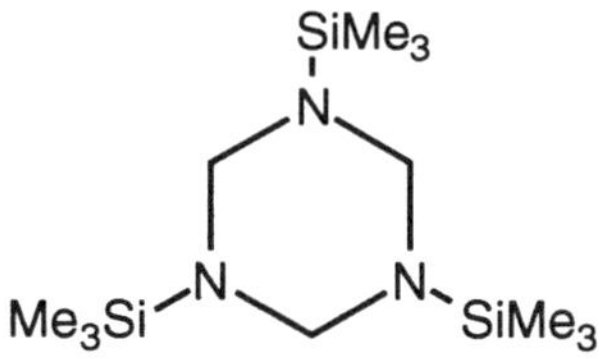

XIV

The by-product from the nitration reaction, the volatile silyl nitrate $(CH_3)_3SiONO_2$ (**VII**, R" = CH_3, equation 2), is a nitrating agent in its own right (*17*) and could easily be collected by distillation and used profitably to carry out other nitrations, such as toluene to dinitrotoluene. The formation of silyl nitrates such as **VII** is also preferable to the acidic by-products (viz. BF_3 or HBF_4) which would arise from the corresponding reactions with NO_2BF_4. Another advantage is the non-acidic nature of the reaction medium which suggests applications involving acid-sensitive substrates hitherto precluded from study in conventional nitration media.

Finally, concerning the novelty of the nitrodesilylations, although some mention has been made of the formation of N-nitroheteroaromatics by the reaction of the corresponding N-silyl compounds with NO_2BF_4 (*17,28*), the products are principally of use as transfer nitrating reagents. Furthermore, silylamines of primary amines have been shown to behave differently upon attempted nitration with oxides of nitrogen such as N_2O_4, undergoing deamination (*29*). Likewise, previous investigations of the reactions of some simple silylamines with N_2O_5 did not result in a general nitramine synthesis (*30*), although the recently reported (*31,32*) action of nitronium tetrafluoroborate on N-trimethylsilylamines has come close to achieving this goal. Hence the work reported here constitutes the first detailed description of nitrodesilylation by N_2O_5 of a comprehensive range of substrates, comprising secondary amines, amides, carbamates and ureas.

Experimental

Safety Note: In the reactions described herein energetic materials are produced which may be hazardous, and appropriate precautions should be taken in their preparation and handling. Also, many of the starting materials, reagents and products are toxic, corrosive or present other hazards and only suitably trained personnel should attempt the chemical transformations described here.

Materials and Apparatus. The following silylamines were purchased from Aldrich Chemical Co.: 4-(trimethylsilyl)morpholine (**VIa**), 1-(trimethylsilyl)pyrrolidine (**VIc**), N-trimethylsilyldimethylamine (**VId**) and N-trimethylsilyldiethylamine (**VIe**); two silylamides, N-methyl-N-(trimethylsilyl)acetamide (**VIn**) and N-(trimethylsilyl)-oxazolidin-2-one (**VIo**) were purchased from Fluka Chemicals. The remaining silylamines were prepared as follows:- n-alkylsilylamines derived from strong bases by direct reaction with chlorotrimethylsilane in the presence of a proton acceptor (triethylamine), either as described in the literature: 1-(trimethylsilyl)piperidine (**VIb**) (*33*) and 1,4-bis(trimethylsilyl)piperazine (**VII**) (*34*), or by modified literature methods: 4-(tri-(n-butyl)silyl)morpholine (**VIh**) and 4-(triethylsilyl)morpholine (**VIi**). Trimethylsilylamines derived from weaker bases, or silylamines bearing branched alkyl substituents on Si were prepared by lithiating the amine using n-butyllithium prior to reaction with the chlorosilane (*35,36*): N-(trimethylsilyl)di-isobutylamine (**VIf**), 4-(*tert.*-butyldimethylsilyl)morpholine (**VIg**) and N-(triethylsilyl)di-isobutylamine (**VIj**). 1-Trimethylsilyl-2-methylaziridine (**VIk**) was prepared by the reaction of propyleneimine with chlorotrimethylsilane in n-pentane (*37*).

Physical data of these materials are as follows: **VIf**: b.pt. 72°C/10 mm; ^1H nmr, δ: -0.05 (s,9); 0.75 (d,12); 1.70 (sp,2); 2.45 (d,4) ppm; i.r. ν_{max} (liquid film): 1468 (m), 1386 (m), 1367 (m), 1318 (w), 1260 (m), 1248 (s) cm^{-1}; **VIg**: b.pt. 120-130°C/15 mm; ^1H nmr, δ: 0.05 (s,6); 0.90 (s,9), 2.91 (t,4), 3.55 (t,4); i.r. ν_{max} (liquid film): 1471 (w), 1386 (m), 1372 (m), 1258 (s), 1159 (m), 1258 (s) cm^{-1}; **VIh**: b.pt. 160-163°C/9 mm; ^1H nmr, δ: 1.1 (m,29); 2.83 (t,4); 3.55 (t,4) ppm; i.r. ν_{max} (liquid film): 1457 (w), 1374 (m), 1255 (m), 1160 (w), 1116 (s), 1080 (m) cm^{-1}; **VIi**: b.pt.

155-160°C/ 47 mm; ^{1}H nmr, δ: 0.6 (m,15); 2.80 (t,4); 3.50 (t,4); i.r. ν_{max} (liquid film): 1456 (w), 1371 (w), 1256 (m), 1161 (m), 1115 (s) cm^{-1}; **VIj**: b.pt. 125-130°C/ 1.5 mm; ^{1}H nmr, δ: 0.8 (m,27); 1.72 (sp,2); 2.45 (d,4); i.r. ν_{max} (liquid film): 1466 (m), 1385 (m), 1366 (w), 1163 (m), 1029 (s) cm^{-1}.

The remaining silylamine, 1,3-bis(trimethylsilyl)hexahydropyrimidine (**VIm**) was prepared by the following novel procedure: hexahydropyrimidine (3.0 g) in anhydrous ether solution (150 ml), prepared as described in the literature (*38*), to which triethylamine (7.05 g) had been added, was treated with chlorotrimethylsilane (7.6 g) at 15 to 20°C - further anhyd. ether was added to facilitate stirring. After 20 min. at this temperature, the mixture was filtered and concentrated (Rotavapor) to give a cloudy oil which was distilled (bulb-to-bulb) to give a mobile oil (1.9 g), b.pt. 90-5°C/ 3 mm). A second batch distilled on the Spaltrohr had b.pt. 80-2°C/ 6.5 mm; ^{1}H nmr, δ: 0.05 (s,18); 1.31 (m,2); 2.97 (t,4); 3.92 (s,2) ppm; i.r. ν_{max} (liquid film): 1452 (m), 1385 (m), 1365 (m), 1250 (s), 1200 (m) cm^{-1}.

N,N'-Bis(trimethylsilyl)-N-methyl-N'-(n-propyl)urea (**VIp**) was prepared by a modified literature procedure from n-propyl isocyanate and heptamethyldisilazane (*24*); 1,3-bis(trimethylsilyl)-3,4,5,6-tetrahydro-2(*1H*)-pyrimidinone (**VIq**) was prepared by silylation of the trimethyleneurea using N-trimethylsilyltrifluoroacetamide as follows:- N,N'-Trimethyleneurea (3,4,5,6-tetrahydro-2(*1H*)-pyrimidinone) (5.80 g, 50 mmol) and N-methyl-N-(trimethylsilyl)trifluoroacetamide (24.9 g, 125 mmol) in acetonitrile (30 ml, dried over 4A molecular sieve) were heated under reflux for 18 hr under an atmosphere of dry nitrogen. The mixture was cooled and, after removal of the solvent under reduced pressure, bulb-to-bulb distillation of the residue gave a fore-run of N-methyltrifluoroacetamide, b.pt. <100°C/ 1.0 mm, followed by the main fraction, b.pt. 110-120°C/ 0.8 mm, 3.0 g (22%), identified as 1,3-bis(trimethylsilyl)-3,4,5,6-tetrahydro-2(*1H*)-pyrimidinone (**VIq**) from its ^{1}H nmr spectrum and m.pt. of 78-79°C (lit. (*39*) m.pt. 78-80°C); ^{1}H nmr, δ: 0.20 (s,18); 1.70 (m,2); 3.15 (t,4) ppm; i.r. ν_{max} (liq. film): 1692, 1666, 1438, 1306, 1249, 845 cm^{-1}.

N_2O_5 was prepared by ozonation of N_2O_4 as described previously (*40*) and stored at -70°C before use. Dichloromethane was dried by passage through a column of chromatographic silica gel (BDH) before use.

^{1}H nmr spectra were recorded on a Varian Associates EM 360A nmr spectrometer equipped with an EM3630 homonuclear lock-decoupler operating at 60 MHz. Chemical shifts are reported in ppm downfield from TMS used as an internal standard. Infra-red spectral measurements were carried out using a Nicolet 5SX FTIR spectrometer equipped with a DTGS detector. Melting points were determined in open capillaries on a Büchi 510 apparatus and are uncorrected.

Nitrations

General method: Silylamines VIa-j and VII &m. The silylamine (20 mmol) was dissolved in dichloromethane (10-15 ml) and added dropwise with stirring and cooling (at -5±2°C) to a solution of N_2O_5 (22 mmol) in the same solvent (20-40 ml) [disilylamines **VII** and **VIm** were reacted with 44 mmol N_2O_5]. After addition was complete the mixture was stirred for the period shown at the temperature shown in Table I. The reaction mixture was then drowned in saturated NaHCO$_3$ solution (30-40 ml) and the organic layer separated. The aqueous layer was extracted with dichloromethane and the combined extracts were washed further with saturated sodium bicarbonate solution, dried over anhydrous MgSO$_4$ and evaporated under water-pump vacuum below 30°C. The nitramine products, **Ia-lm** (see Table I), were collected and characterised by m.pt., i.r. and nmr spectra; the majority were isolated as colourless solids (after trituration (ethanol) if necessary), except for **Ib** and **Ie**

which were oils. In the case of **Ia**, the m.pt. of the product (N-nitromorpholine) was not depressed on admixture with an authentic sample, prepared from morpholine and N_2O_5 by the method of Emmons *et al* (*7*). In certain cases (**VIg, VIi** and **VIm**), larger excesses of N_2O_5 (100, 50 and 100% resp.) were used, and owing to the instability of nitramine **Im** to acid, the mode of addition of the reagents was reversed [otherwise a highly impure product was obtained].

4-Nitromorpholine (Ia), m.pt. 50-51°C (lit. (*7*) 50-52°C); ^{1}H nmr, δ: 3.80 (s) ppm; i.r. v_{max} (mull): 1524, 1512 (-NO_2 asymm.), 1315, 1254 (-NO_2 symm.) cm^{-1}.

1-Nitropiperidine (Ib), oil, ^{1}H nmr, δ: 1.65 (m,6), 3.85 (m,4) ppm; i.r. v_{max} (liq. film): 1514 (-NO_2 asymm.), 1329, 1279, 1242 (-NO_2 symm.) cm^{-1}.

1-nitropyrrolidine (Ic), m.pt. 55.5-56.5°C (lit. (*44,7*) 56°C, 58-59°C resp.); ^{1}H nmr, δ: 2.00 (m,4), 3.70 (m,4) ppm; i.r. v_{max} (mull): 1500 (-NO_2 asymm.), 1307, 1280 (-NO_2 symm.) cm^{-1}.

Dimethylnitramine (Id), m.pt. 54.5-55°C (lit (*41*) 58°C); ^{1}H nmr, δ: 3.43 (m) ppm; i.r. v_{max} (mull): 1500, 1450 (-NO_2 asymm.), 1333, 1290, 1260 (-NO_2 symm.) cm^{-1}.

Diethylnitramine (Ie), oil, ^{1}H nmr, δ: 1.25 (t,6), 3.80 (qr,4) ppm; i.r. v_{max} (liq. film): 1509, 1470 (-NO_2 asymm.), 1377, 1282 (-NO_2 symm.) cm^{-1}.

Di-isobutylnitramine (If), m.pt. 79-80°C (lit. (*42,43*) 76-77°C, 81.5-82.5°C resp.); ^{1}H nmr, δ: 0.85 (d,6), 2.20 (sp,1), 3.53 (s,2) ppm; i.r. v_{max} (mull): 1526, 1492, 1462 (-NO_2 asymm.), 1327, 1270 (-NO_2 symm.) cm^{-1}.

1,4-Dinitropiperazine (Il), m.pt. 202°C(dec.) (lit. (*43*) 215°C); ^{1}H nmr, δ(D_6-DMSO): 4.03 (s) ppm; i.r. v_{max} (mull): 1545 (-NO_2 asymm.), 1335, 1285, 1241 (-NO_2 symm.) cm^{-1}.

1,3-Dinitrohexahydropyrimidine (Im), m.pt. 82-83°C (lit. (*18*) 84°C); ^{1}H nmr, δ: 1.92 (m,2), 3.95 (t,4), 5.75 (s,2) ppm; i.r. v_{max} (mull): 1556, 1538 (-NO_2 asymm.),1275 (-NO_2 symm.) cm^{-1}.

Preparation of 1-(N,N-dinitramino)propan-2-ol nitrate (XI). N-(Trimethyl-silyl)propyleneimine (**VIk**) (2.32 g, 18.0 mmol) in dichloromethane (6 ml) was added over 12 min. at -15°C to a solution of N_2O_5 (2.0 g, 18.5 mmol) in the same solvent (*ca* 10 ml). The resulting mixture was stirred for 10 min. whilst allowing to warm to 0°C (yellow colour develops), then it was pipetted rapidly into a solution containing excess N_2O_5 (5.0 g, 46.3 mmol) in dichloromethane (*ca* 15 ml) and the mixture was monitored by HPLC (RP18 column, acetonitrile-water 60:40 vol./vol. eluant, monitor at 210 nm) for disappearance of the N-nitroaziridine. After *ca* 90 min. at 0±5°C the N-nitroaziridine peak (shortest retention time) had reached a minimum and the peak assigned as the N,N-dinitramine (longest retention time, λ_{max} *ca* 240 nm) had reached a maximum relative to the propane-1,2-diol dinitrate which was also present in the mixture (intermediate retention time).

The mixture was worked up in the usual manner (wash with saturated sodium bicarbonate solution, dry over anhydrous $MgSO_4$ and evaporate) to give an oil (2.25 g, 62%) which comprised 1-(N,N-dinitramino)propan-2-ol nitrate (**XI**), v_{max} (liquid film) 1647 (-ONO_2 asymm.), 1607 (-N(NO_2)$_2$ asymm.), 1274 (-ONO_2 symm.), 1244 (-N(NO_2)$_2$ symm.), 850 (-ONO_2), 815 (-N(NO_2)$_2$) cm^{-1} in admixture with propane-1,2-diol dinitrate (**XII**).

Reactions of silylamides VIn-q. The silylamide (20 mmol) was dissolved in dichloromethane (10-15 ml) and added dropwise with stirring and cooling (at -5±2°C) to a solution of N_2O_5 (22 mmol) in the same solvent (20-40 ml) [disilylamides **VIp** and **VIq** were reacted with 44 mmol N_2O_5]. After addition was complete the mixture was stirred for the period shown at the temperature shown in Table I. The reaction mixture was then worked up as described above.

N-Nitro-N-methylacetamide (In), oil, ^{1}H nmr, δ: 2.68 (s,3), 3.60 (s,3) ppm, δ(CCl_4): 2.61 (s,3), 3.55 (s,3) ppm (lit. (23) δ(CCl_4): 2.43, 3.31 ppm); i.r. v_{max} (liq. film): 1721 (CO), 1569 (-NO_2 asymm.), 1239 (-NO_2 symm.) cm^{-1} (lit. (23) CCl_4 soln.: 1720, 1575 cm^{-1}).

3-Nitro-1,3-oxazolidin-2-one (Io), m.pt. 106.5-107°C (lit. (23) 108-109.5°C); ^{1}H nmr, δ: 4.42 (s) ppm; i.r. v_{max} (mull): 1786 (CO), 1554 (-NO_2 asymm.), 1282 (-NO_2 symm.) cm^{-1}.

N,N'-Dinitro-N-methyl-N'-(n-propyl)urea (Ip), oil, ^{1}H nmr, δ: 0.98 (t,3), 1.67 (m,2), 3.69 (s,3), 4.08 (t,2) ppm; i.r. v_{max} (liq. film): 1723 (CO), 1590 (-NO_2 asymm.), 1287 (-NO_2 symm.) cm^{-1}.

1,3-dinitro-3,4,5,6-tetrahydro-2(*1H*)-pyrimidinone (N,N'-Dinitro-N,N'-(trimethylene)urea, **Iq**), m.pt. 118°C(dec.) (lit. (25) 121-122°C); ^{1}H nmr, δ: 2.35 (m,2), 4.18 (t,4) ppm; i.r. v_{max} (mull): 1753 (CO), 1571 (-NO_2 asymm.), 1266 (-NO_2 symm.) cm^{-1}.

Literature Cited

1. Wright, G. F. In *Methods of Formation of the Nitramino Group, its Properties and Reactions*; Feuer, H., Ed.; The Chemistry of the Nitro and Nitroso Groups, Part 1 (The Chemistry of Functional Groups Series, Patai, S., Series Ed.); Interscience: New York, 1969, Ch. 9.
2. Smith, P. A. S. *Open Chain Nitrogen Compounds*; Benjamin: New York: 1966, Vol. 2, Ch. 15.
3. Urbanski, T. *Chemistry and Technology of Explosives*; Pergamon Press: Oxford, 1967, Vol. 3.
4. Coombes, R. G. In *Nitration*; Sutherland, I. O., Ed.; Comprehensive Organic Chemistry (Barton, D.; Ollis, W. D., Series Eds.); Pergamon: Oxford, 1979, Vol. 2.
5. Fischer, J. W. In *Nitro Compounds: Recent Advances in Synthesis & Chemistry*; Feuer, H.; Nielsen, A. T., Eds.; Organic Nitro Chemistry Series; VCH: New York, 1990, Ch. 3, pp 338-346.
6. Chute, W. J.; Herring, K. G.; Tombs, L. E.; Wright, G. F. *Canad. J Res.* **1948**, *26B*, 89-103.
7. Emmons, W. D.; Pagano, A. S.; Stevens, T. E. *J Org. Chem.* **1958**, *23*, 311-3.
8. Chapman, F. *J Chem. Soc.* **1949**, 1631.
9. Andreev, S. A.; Novik, L. A.; Lebedeev, B. A.; Tselinskii, I. V.; Gidaspov, B. V. *J. Org. Chem. USSR* **1978**, *14*(2), 221-224.
10. Cichra, D. A.; Adolph, H. G. *J Org. Chem.* **1982**, *47*(12), 2474-2476.
11. Boyer, J. H.; Pillai, T. P.; Ramakrishnan, V. T. *Synthesis,* **1985**, 677-679.
12. Bottaro, J. C.; Schmitt, R. J.; Bedford, C. D. *J. Org. Chem.* **1987**, *52*(11), 2292-2294.
13. Siele, V. I.; Warman, M.; Leccacorvi, J.; Hutchinson, R. W.; Motto, R.; Gilbert, E. E.; Benzinger, T. M.; Coburn, M. D.; Rohwer, R. K.; Davey, R. K. *Propellants & Explosives* **1981**, *6*, 67-73.

14. Koppes, W. M.; Chaykovsky, M.; Adolph, H. G. *J Org. Chem.* **1987**, *52*(6), 1113-1119.
15. Schmitt, R. J.; Bottaro, J. C.; Malhotra, R.; Bedford, C. D. *J Org. Chem.* **1987**, *52*(11), 2294-2297.
16. Olah, G. A.; Rochin, C. *J. Org. Chem.* **1987**, *52*, 701-702.
17. Olah, G. A.; Malhotra, R.; Narang, S. C. In *Nitration: Methods & Mechanisms*; Feuer, H., Ed.; Organic Nitro Chemistry Series; VCH: New York, 1989, Ch. 4.
18. Willer, R. L.; Atkins, R. L. *J Org. Chem.* **1984**, *9*, 5147-5150.
19. Aerojet-General Corp., Brit. Pat. 1126591 (6 Sept. 1963).
20. Haire, M. J.; Boswell, G. A. *J Org. Chem.* **1977**, *42*, 4251-4256.
21. Haire, M. J.; Harlow, R. L. *J Org. Chem.* **1980**, *45*, 2264-2265.
22. Golding, P.; Millar, R. W.; Paul, N. C.; Richards, D. H., *Tetrahed. Letts.* **1991**, *32*(37) 4985-4988.
23. White, E. H.; Chen, M. C.; Dolak, L. A. *J Org. Chem.* **1966**, *31*, 3038-3046.
24. Roesky, H. W.; Lucas, J. *Inorg. Synth.* **1986**, *24*, 120-121.
25. McKay, A. F.; Wright, G. F. *J Am. Chem. Soc.* **1948**, *70*, 3990-3994.
26. Bestmann, H. J.; Wölfel, G. *Angew. Chem. Internat. Ed.* **1984**, *23*, 53.
27. Morimoto, T.; Takahashi, T.; Sekiya, M. *Chem. Comm.* **1984**(12), 794-795.
28. Glass, R. S.; Blount, J. F.; Butler, D.; Perrotta, A.; Oliveto, E. P. *Canad. J. Chem.* **1972**, *50*, 3472-3477.
29. Wudl, F.; Lee, T. B. K. *Chem. Comm.* **1970**, 490-491.
30. Schultheiss, H.; Fluck, E. *Z. Anorg. Allg. Chem.* **1978**, *445*, 20-26.
31. Olah, G. A. In *Methods for Preparing Energetic Nitro-Compounds: Nitration with Superacid Systems, Nitronium Salts and Related Complexes*; Chemistry of Energetic Materials; Olah, G. A.; Squire, D. R., Eds.; Academic Press Inc.: San Diego, 1991, Ch. 7, p.197.
32. Dave, P. R.; Forohar, F.; Axenrod, T.; Bedford, C. D.; Chaykovsky, M.; Rho, M-K.; Gilardi, R.; George, C. *Phosphorus, Sulfur, Silicon Relat. Elem.* **1994**, *90*(1-4), 175-184.
33. Breed, L. W.; Haggerty W. J.; Harvey, J. *J Org. Chem.* **1960**, *25*, 1804-1806.
34. Kakimoto, M. A.; Oishi, Y.; Imai, Y. *Makromol. Chem. Rapid Commun.* **1985**, *6*, 557-562.
35. Rauchschwalbe, G.; Ahlbrechte, H. *Synthesis* **1974**, *9*, 663-665.
36. Sundberg, R. J.; Russell, H. F. *J Org. Chem.* **1973**, *38*, 3324-3330.
37. Scherer, O. J.; Schmidt, M. *Chem. Ber.* **1965**, *98*, 2243.
38. Titherley A. W.; Branch, G. E. K. *J Chem. Soc.* **1913**, 330.
39. Birkhofer, L.; Kühlthau, H. P.; Ritter, A. *Chem. Ber.* **1960**, *93*, 2810-2813.
40. Harris, A. D.; Trebellas, J. C.; Jonassen, H. B. *Inorg. Synth.* **1967**, *9*, 83-88.
41. Lamberton, A. H. *Q Rev.* **1951**, *5*, 75-98.
42. Luk'yanov, O. A.; Seregina, N. M.; Tartakovskii, V. A. *Bull. Acad. Sci. USSR Chem. Ser.* **1976**(1) 220-221; *Chem. Abs.* **1976**, *84*, 135574.
43. Robson, J. H.; Reinhart, J. *J Am. Chem. Soc.* **1955**, *77*, 2453-2457.
44. Suri, S. C.; Chapman, R. D. *Synthesis* **1988**, 743-745.

© British Crown Copyright 1995/ DRA Farnborough Hants. U.K.
Published with the permission of the Controller of Her Britannic Majesty's Stationery Office.

RECEIVED January 16, 1996

Chapter 13

Reinvestigation of the Ponzio Reaction for the Preparation of *gem*-Dinitro Compounds

P. J. Honey[1], R. W. Millar[1], and Robert G. Coombes[2]

[1]Defence Research Agency, Fort Halstead, Sevenoaks, Kent TN14 7BP, United Kingdom
[2]Department of Chemistry, Brunel University, Uxbridge, Middlesex UB8 3PH, United Kingdom

This paper presents some recent work to re-investigate the Ponzio reaction for the conversion of oximes to the gem-dinitro group. The Ponzio reaction is one of a number of less aggressive non-acidic methods used to nitrate aliphatic compounds, which are usually destroyed in the mixtures of nitric and sulphuric acids commonly used to nitrate aromatic compounds. The effect of varying the Ponzio reaction solvent, and of changing the nitrating species from N_2O_4 to N_2O_5 will also be presented. The reaction mechanism will be discussed with reference to ^{15}N CIDNP nmr studies. Where relevant the results will be compared with other aliphatic nitration methods.

A review (1) in 1964 stated that the ONR initiated a programme of work in 1947 on aliphatic polynitro- compounds. The US Government is today still continuing to sponsor research, through the auspices of ONR (2) and ARDEC (3), to find new polynitroaliphatic compounds which would have potential uses as explosive and/or propellant ingredients. The last review of any significance was by Larson (4) in 1969, more recent reviews have restricted themselves to a specific range of products The most comprehensive review on aliphatic nitro compounds is the 460 page review by Padeken et al (5), although this volume is not readily available to most practising chemists. It is therefore worth highlighting some of the various reactions available for producing aliphatic nitro compounds before presenting the investigations undertaken by the authors. These are split into reactions not involving oximes (Table I) and reactions involving oximes (Table II).

This is obviously not a comprehensive list but it helps to illustrate that nitration of aliphatic compounds requires a wide variety of reaction conditions and the selection of the nitration route is often dependent on the structure of the substrate. This can result in low yields which reduces the economic usefulness of the desired product. It was the apparent ease with which the Ponzio reaction products were obtained by Frojmovic and Just (37) which promoted the interest in the reactions synthetic potential for aliphatic polynitro- compounds.

Table I. Aliphatic Nitration Reactions not Involving Oximes

Reaction	Reaction centre	Reagent	Product	Comments	Refs
1a. Victor Meyer	R-CH$_2$-Br	AgNO$_2$	R-CH$_2$-NO$_2$	Best for primary nitroalkanes, polynitro if separated by at least three carbons, poor yields for secondary, other products for tertiary, uses costly silver salts.	6, 7 & 4
1b. Kornblum et al modifications	R-CH$_2$-Br	NaNO$_2$ in DMF or DMSO	R-CH$_2$-NO$_2$	Higher yields for secondary, but still only mono-substitution.	8
2. ter Meer	R-CH(NO$_2$)Cl	NaNO$_2$	RCH(NO$_2$)$_2$	Preferably for short chain primary gem-halonitro, secondary gives other products, some intermediates detonable.	9 & 10
3a. Kaplan Shechter	R$_2$CHNO$_2$	AgNO$_2$	R$_2$C(NO$_2$)$_2$	Best for secondary gem-dinitro's, limited to compounds not containing an electron-withdrawing group, uses costly silver salts.	11, 4 & 12
3b. Matacz et al modifications	R$_2$CHNO$_2$	K$_3$Fe(CN)$_6$	R$_2$C(NO$_2$)$_2$	Successful for secondary, modifications by Kornblum et al gave higher yields/purity for both primary and secondary, some intermediate salts unstable.	13, 14, 2, 3, 15, 16 & 17
4a. Double bonds	R$_2$C=CR$_2$	N$_2$O$_4$	O$_2$N NO$_2$ | | R$_2$C—CR$_2$	Produces vicinal dinitro, best when substituted by other than hydrogen (eg Cl or F), forms stereo isomers, also vicinal dinitro not always sole product.	4 & 18
4b. Active methylene	R$_2$CH	HNO$_3$ or N$_2$O$_4$	R$_2$CHNO$_2$	Mono-substitution, dinitration achieved only when oxygen excluded.	19 & 20

Table II. Aliphatic Nitration Reactions Involving Oximes

Reaction	Reaction centre	Reagent	Product	Comments	Refs
5. Halogen gas	$R_2C=NOH$	Cl_2 or Br_2	$R_2C(NO_2)Cl$	Gem-halonitro produced, would need ter Meer reaction for gem-dinitration.	21, 22, 23 & 24
6. Peroxytrifluoroacetic acid	$R_2C=NOH$	CF_3CO_3H	R_2CHNO_2	Effective for primary and secondary, fails for sterically hindered oximes, mono-substitution.	25, 22, 26 & 2
7a. Halogenation/ Oxidation/ Dehalogenation	$R_2C=NOH$	i KOH/Br_2 ii HNO_3/H_2O_2 iii $KOH/EtOH$	i R_2CBrNO ii R_2CBrNO_2 iii R_2CHNO_2	Best for alicyclic ketoximes, worst for aldoximes and aromatic oximes, unpredictable results for complex alicyclic ketoximes.	27 & 28
7b. Modified to two stages	$R_2C=NOH$	i $NBS/KHCO_3$ dioxane/H_2O ii $NaBH_4$	i R_2CBrNO_2 ii R_2CHNO_2	(i) modified to N-bromosuccinimide (NBS) (eliminates HNO_3 oxidation stage), (iii) modified to sodium borohydride.	29, 21 & 3
8. Pseudonitrole (Scholl)	$R_2C=NOH$	$HNO_3/CH_2Cl_2/$ NH_4NO_3 H_2O_2	$R_2C(NO_2)_2$	Isolation of pseudonitrole possible, few limitations but yields of aliphatic oximes low, other oxygenation methods available.	30, 31, 32, 33, 34, 21 & 3
9. Ponzio Reaction	$R_2C=NOH$	N_2O_4	$R_2C(NO_2)_2$	Applicable for all oximes, best for aromatic ketoximes, yields often poor for aldoximes, N_2O_5 potential reagent.	35, 36 & 37

Synthetic Potential of the Ponzio Reaction.

The main appeal of the Ponzio reaction, for the DRA, is the synthesis of novel polynitro compounds, which are usually highly energetic and often have energy exceeding that of TNT (2,4,6-trinitrotoluene). A study of the published data on the Ponzio reaction (Table III) indicated that no correlation of the product yields and purity to the structure had been attempted, and certainly no investigation of the mechanism. Exploitation of any novel compound is dependent on obtaining the highest possible yields and these are obtained from ketoximes of aromatic substrates. The lowest yields have been observed for aldoximes and aliphatic ketoximes, with the added difficulty of separation of the carbonyl contaminents. For aralkyl ketoximes the yield is often influenced by the substituents attached to the carbon α to the oxime. There is not, however, a simple correlation with inductive and/or mesomeric effects; indeed substitution of the aromatic species also affects the yields obtained. This, to a certain extent, reduces the synthetic potential of the reaction unless it is possible in some way to influence the reaction mechanism - the ability to do so is very dependent on an understanding of the mechanism. Thus the authors undertook to both prepare novel compounds and investigate the mechanism using modern instrumental techniques.

Table III. Effect of Substituent on Yield of the Ponzio Reaction

No.	$_aR^1$	$_aR^2$	% gem-dinitro	% carbonyl	Ref
1	CH_3	OCH_3	0	-	38
2	[bicyclic dioxime structure with H₃C, HON, O-N, N-O, NOH, CH₃]		25	-	39
3	$C_2H_5O_2C$	Cl	48	-	40
4a	Ph	H	32	-	41
4b	Ph	H	38	-	42
5	Ph	OCH_3	38	30	38
6	p-NO_2Ph	OCH_3	41	29	38
7	m-NO_2Ph	OCH_3	47	26	38
8	m-NO_2Ph	H	27.5	-	42
9	Ph[b]	H	45-60	30-40	43
10	p-NO_2Ph	CH_3	70	30	44
11	Ph	Cl	69	-	40
12	Ph	Ph	77	15	37
13	Fluoren-9-one oxime		95+	1	37
14	2,4,5,7-Tetranitrofluoren-9-one oxime		95+	-	45

[a] for $R^1R^2C=NOH$

[b] 38 different aryl aldoximes were reacted with N_2O_5, meta and para derivatives resulted in 45-60%, ortho derivatives in 20-30%. Absolute yields for each reactant were not given.

n.b. Most of the novel polynitro compounds reported below are potentially high explosives and when handling them suitable precautions should be taken.

Polynitrofluorene Derivatives.

Recent work by Honey (45) on polynitrofluoren-9-one derivatives indicates that aromatic polynitropolycycles show interesting properties and it is this work that will be described in greater detail below. The availability to the author of pure N_2O_5 also permitted a comparison of the effect of the two different oxides of nitrogen in the Ponzio reaction. A difference in the respective behaviour of the two oxides of nitrogen was very noticeable for the reaction of fluoren-9-one oxime (46) **5** - see Schemes 1a (N_2O_4) and 1b (N_2O_5). With N_2O_4 preparation of 9,9-dinitrofluorene **9** occurs with no ring substitution, as observed previously (37). A number of variations in the reaction conditions were all found to produce **9** in near quantitative yields. Indeed it should be noted that **9** was also prepared from **5** using N_2O_4 in acetonitrile, rather than producing fluoren-9-one **1** as recently reported (47).

a - N_2O_4 in CH_2Cl_2, 99 % yield of **9** - R^1, R^4 = H, R^2, R^3 = NO_2

b - N_2O_5 in CH_2Cl_2, 76 % yield of **10** - R^1, R^2, R^3, R^4 = NO_2

(1)

However, when N_2O_5 in dichloromethane is utilised instead of N_2O_4, mono nitration of each of the rings occurs (Scheme 1b). The Ponzio reaction of oximes with N_2O_5 has, however, been reported (38),(40), and the ability of N_2O_5 to nitrate aromatic rings in a Ponzio reaction experiment is also documented (43).

This work led on to the preparation of a number of polynitrofluorene derivatives, the main routes to which are shown in Scheme 2. The preparation of 2,7,9,9-tetranitrofluorene **10** was achieved by a number of different methods which are indicated in Table IV.

Table IV. Preparative Routes to 2,7,9,9-tetranitrofluorene

Substrate	Reagent	Isolated yield	Overall yield[a]
9,9-Dinitrofluorene (**9**)	98+ % nitric acid	97 %	94 %
9,9-Dinitrofluorene (**9**)	16 % N_2O_5 enriched nitric acid	99 %	96 %
9,9-Dinitrofluorene (**9**)	23 % N_2O_5 in dichloromethane	78 %	75 %
Fluoren-9-one Oxime (**5**)	19 % N_2O_5 in dichloromethane	76 %	74 %
2,7-Dinitrofluoren-9-one oxime (**6**)	N_2O_4 in dichloromethane	~95 %	~86 %

[a] from fluoren-9-one **1**.

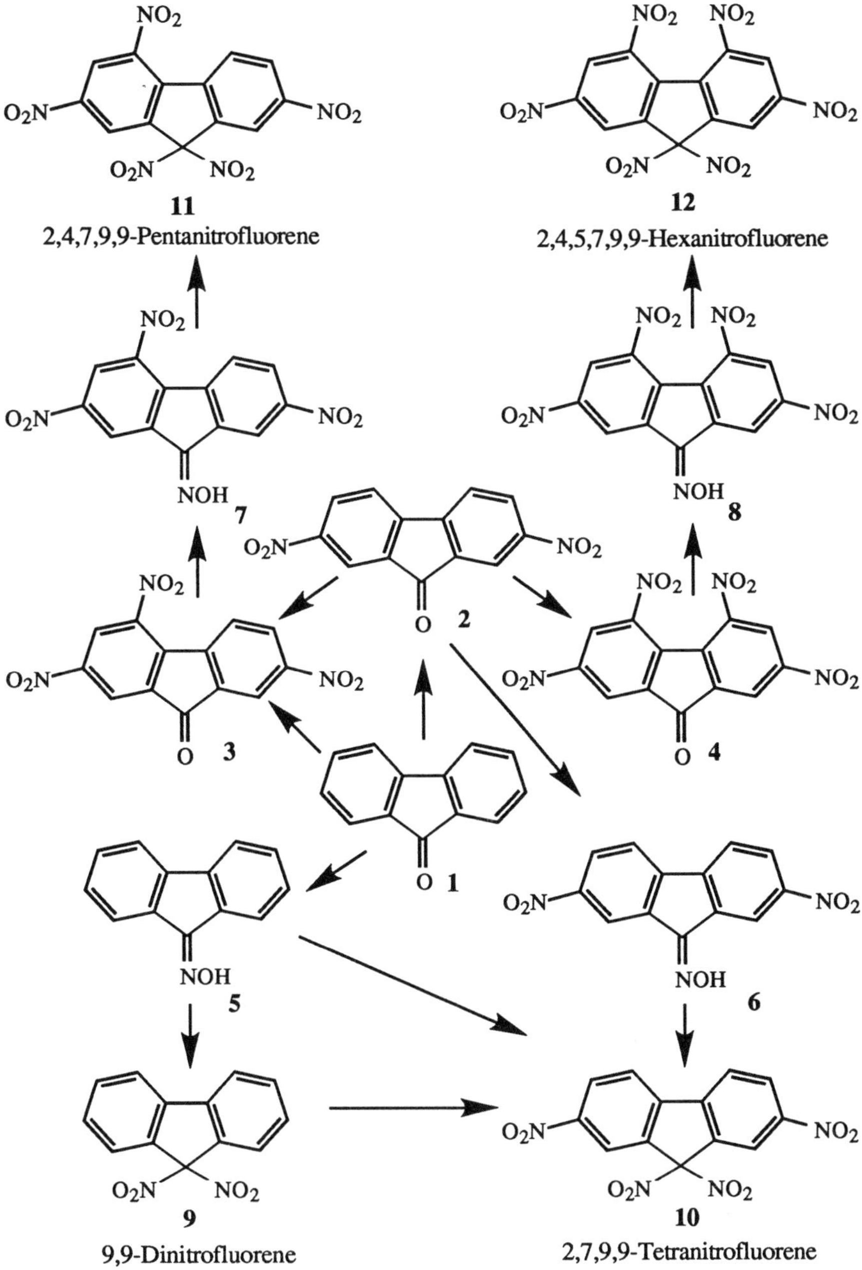

11
2,4,7,9,9-Pentanitrofluorene
12
2,4,5,7,9,9-Hexanitrofluorene
7
8
2
3
4
1
5
6
9
9,9-Dinitrofluorene
10
2,7,9,9-Tetranitrofluorene

(2)

The advantage obtained by using N_2O_5 in dichloromethane to prepare 2,7,9,9-tetranitrofluorene **10**, from fluoren-9-one oxime **5**, is a one-pot reaction. However the overall yield is much lower than that found in the two stage reaction using N_2O_4 in dichloromethane followed by reacting the 9,9-dinitrofluorene **9** with 98+% nitric acid. 2,7,9,9-Tetranitrofluorene **10** can also be prepared from 2,7-dinitrofluoren-9-one oxime **6** but this route does not confer any advantages because of the reduction in yield, from fluoren-9-one, and the need to purify the 2,7-dinitrofluoren-9-one **2** prior to preparation of the oxime **6**.

The high melting point of 2,7,9,9-tetranitrofluorene **10** is 294°C makes it very thermally stable but it does not have a particularly high oxygen balance (desirable for potential explosive use). Polynitrofluorene compounds with higher oxygen balance are 2,4,7,9,9-pentanitrofluorene **11** and 2,4,5,7,9,9-hexanitrofluorene **12**, and both compounds can be prepared via a three step process (using literature methods or variations thereof), these being ring nitration, oximation and Ponzio Reaction. Scheme 3 indicates the Ponzio reaction conditions used to achieve **11** and **12**.

(3)

7 ($R^1, R^2, R^4 = NO_2, R^3 = H$) - N_2O_4 in CH_2Cl_2, 30 mins, RT, 98% yield of **11**

8 ($R^1, R^2, R^3, R^4 = NO_2$) - N_2O_4 in CH_2Cl_2, 30 mins, RT, 98% yield of **12**

11 had a considerably lower melting point (114-5°C, from benzene) than that found for 2,7,9,9-tetranitrofluorene **10** owing to its unsymmetrical structure. It should be noted that 2,4,7,9,9-pentanitrofluorene **11** violently decomposed when attempts were made to recrystallise it from ethanol or methanol. The recrystallised material (38% overall yield from fluoren-9-one **1**) was also found to be unstable towards air and light, decomposing to form mainly 2,4,7-trinitrofluoren-9-one **3**.

The preparation of 2,4,5,7,9,9-hexanitrofluorene **12** resulted in an 81% overall yield from fluoren-9-one **1**. This material has an oxygen balance better than TNT and hence required care in its handling, especially with respect to electrostatic charge as indicated in Table V.

Table V. Properties of 2,4,5,7,9,9-Hexanitrofluorene

Property	Observation
Melting point	277-8°C.
Density	1.7 (approx., measured by helium pycnometry)
Figure of Insensitivity	59 cm (RDX = 80).
Mean gas volume	3.3 cm^3.
Temperature of ignition	smoke at 355°C.
Electrostatic test	Fired at 4.5 Joules 1st attempt, Fired at 0.45 Joules 1st attempt, no ignitions at 0.045 Joules.

Further explosiveness testing has not been performed, as yet, but the compound does show promise as a potential thermally stable explosive.

The various polynitrated fluorene products were best analysed by TLC, infra-red spectroscopy and melting point - various problems were experienced with other techniques. Thin layer chromatography (on Kieselgel 60F254 pre-coated plates using 4:1 dichloromethane:hexane) was found to separate the reactants and products easily due to the polarity of the substituent groups. For example :

2,7-dinitrofluoren-9-one Rf - 0.52,
2,7-dinitrofluoren-9-one oxime Rf - 0.08,
2,7,9,9-tetranitrofluorene Rf - 0.64.

The presence of impurities and/or variations of reaction products was best determined by infra-red spectroscopy. The carbonyl stretch frequency at ~1730 cm^{-1} was well separated from bands of other species to enable even small amounts of this, as an impurity, to be observed. The hydroxyl band frequency of the oxime could be observed in the 3030-3450 cm^{-1} region, although for 2,4,5,7-tetranitrofluoren-9-one oxime **8**, this appeared at 3600 cm^{-1} due to the presence of water of crystallisation. The gem-dinitro asymmetrical stretch frequency was observed at around 1578 cm^{-1}, which was well separated from the ring nitro asymmetrical stretch frequencies. Indeed differentiation of the three polynitrated compounds (**10, 11** and **12**) was possible from the ring nitro stretch pattern (Figure 1).

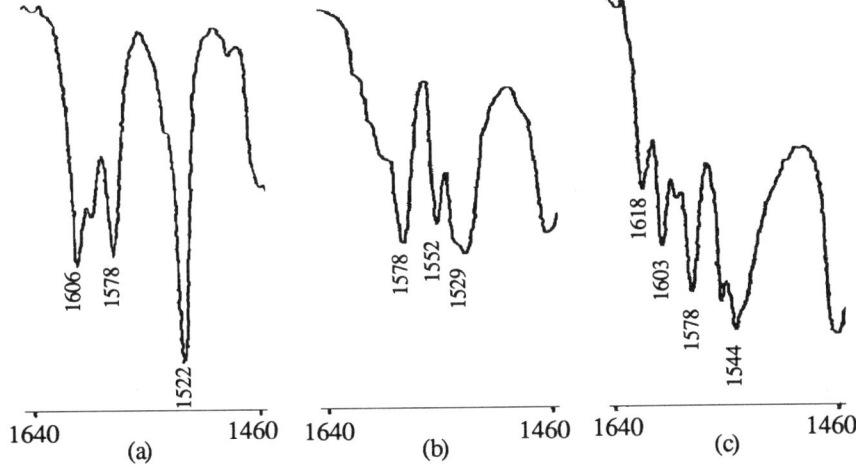

Figure 1. Infra-red spectra showing the ring nitration effect on the nitro asymmetrical stretch band of: (a) 2,7,9,9-tetranitrofluorene **10**, (b) 2,4,7,9,9-pentanitrofluorene **11** and (c) 2,4,5,7,9,9-hexanitrofluorene **12**.

Proton nmr using a 60Mz instrument could not differentiate between the carbonyl and gem-dinitro products which have similar proton substitution patterns. Also the low concentrations of the impurities in the gem-dinitro products made observation of the nmr signals difficult. Elemental analysis was performed on pure samples of **10** and **12** but gave low nitrogen and high carbon/hydrogen values due to decomposition of the sample in the solvent used. CI mass spectroscopy was found to be capable of confirming the structure but was poor at identification of impurities.

Polynitro(phenyl)ethane Derivatives.

After the synthetic success of preparing 2,4,5,7,9,9-hexanitrofluorene **12** other substrates were required which upon conversion, via the Ponzio reaction, would provide suitable polynitro compounds. The first substrates investigated were based on oximes of acetophenones which would produce dinitroethylbenzene derivatives, and similar compounds of this class, *viz* a 2:1 mixture of the 2,4-dinitro and 2,4,6-trinitro isomers are currently used as energetic plasticisers for binder applications. Here the nitro groups are positioned on the ring and the eutectic mixture containing the dinitro derivative is required to depress the melting point of the trinitro compound. This however reduces the available energy. Both the melting point and, by using solely trinitated species, the energy could be improved if some of the nitro groups were placed exocyclic to the phenyl ring, eg via the Ponzio reaction. In a similar fashion to fluoren-9-one oxime **5** the availability of N_2O_5 may provide a synthetic advantage over reaction with N_2O_4. Hence three substrates were chosen, with and without, a ring nitro group present (Scheme 4).

$$\begin{array}{c} \text{HON} \diagdown \text{CH}_2\text{R}^1 \\ \text{Ar}(\text{R}^2) \end{array} \xrightarrow{N_2O_4 \text{ (g)}} \begin{array}{c} \text{NO}_2 \\ \text{O}_2\text{N} - \overset{|}{\text{C}} - \text{CH}_2\text{R}^1 \\ \text{Ar}(\text{R}^2) \end{array} + \begin{array}{c} \text{O} \diagdown \text{CH}_2\text{R}^1 \\ \text{Ar}(\text{R}^2) \end{array} \quad (4)$$

13a - R^1 = H, R^2 = NO_2 **13b**   **13c**
14a - R^1 = NO_2, R^2 = H   **14b**   **14c**
15a - R^1, R^2 = H   **15b**   **15c**

4-Nitroacetophenone oxime **13a** had previously been reacted with N_2O_4 in dichloromethane and a mixture of 1,1-dinitro-1-(4-nitrophenyl)ethane **13b** and 4-nitroacetophenone **13c** had been reported (*44*). This was found to contain 70% of **13b** and 30% of **13c** which on chromatography yielded 61% of pure gem-dinitrated product. This reaction was repeated in the present work and a crude product weight, which equated to 91% of theoretical based on gem-dinitration, was achieved. Spectroscopic evidence confirmed the structures of the products which were not isolated pure. A similar percentage conversion was achieved for reaction in acetonitrile and, as with fluoren-9-one oxime **5**, the main product (gem-dinitro **13b**) differed from that reported previously (*47*). The ability of N_2O_4 to achieve conversion of the oxime in such diverse solvents as dichloromethane and acetonitrile indicates the versatility of the reaction.

The reaction of 4-nitroacetophenone oxime **13a** with N_2O_5 in dichloromethane was more rapid than the reaction with N_2O_4 but the crude product weight obtained equated to only 77% of theoretical based on gem-dinitration. Analysis of this product, by natural abundance ^{13}C nmr indicated a product ratio of 62% of **13b** and 38% **13c** which equates to conversion of 48% and 43% respectively. No other products were observed, in particular those which might have resulted from nitration of the aromatic ring (eg 1,1-dinitro-1-(2,4-dinitrophenyl)ethane), for the apparent reason that the introduction of the gem-dinitro substituents exert a strong deactivating influence on the aromatic ring. Hence N_2O_5 does not present any synthetic advantage over N_2O_4 in this instance.

In general in these reactions the oxime, although often not soluble in the reaction solvent, dissolves on addition of N_2O_4 which suggests that a heterogeneous reaction may be occurring. Hence the Ponzio reaction was attempted using solid oxime **13a** with gaseous N_2O_4. The oxime was observed to turn to a green liquid and then to yellow after about 12 minutes and when spectroscopically analysed a mixture of **13b** and **13c** was observed. The amount of **13c** was thought, based on consideration of the carbonyl to gem-dinitro band strength, to be greater than that found in the three solvent reactions. Although of no immediate synthetic interest this illustrates that the Ponzio reaction can be accomplished under an even wider diversity of reaction conditions than originally considered.

As with 4-nitroacetophenone oxime **13a**, benzoylnitromethane oxime **14a** was reacted in dichloromethane with both oxides of nitrogen (N_2O_4 and N_2O_5). In the first reaction (with N_2O_4) the product was found to be mixture of 1,1,2-trinitro-1-phenylethane **14b** and benzoylnitromethane **14c** (Scheme 4). **14a** reacted with N_2O_5 in dichloromethane virtually instantaneously and a temperature rise of 6°C in about 30 sec. was observed and with a crude product yield higher than for N_2O_4. Infra-red analysis indicated that some reaction of the aromatic ring with the N_2O_5 had occurred, but it was thought that this was nitration of the benzoylnitromethane after its formation, and not ring nitration of the gem-dinitro derivative.

The product **14c** from both the N_2O_4 and N_2O_5 reactions was found to be unstable to air/light and decomposition was observed after only 6 days. This form of decomposition had been seen previously only with the unsymmetrical 2,4,7,9,9-pentanitrofluorene **11** and certainly not with the 4-nitroacetophenone oxime reaction products (**13b**, **13c**). Hence, although N_2O_5 showed a synthetic advantage over N_2O_4, the reaction product could not be considered for any application because of the instability observed.

The reaction of acetophenone oxime **15a** with N_2O_4 in dichloromethane proceeded readily and completion was seen after about 5 minutes producing 1,1-dinitro-1-phenylethane **15b** and acetophenone **15c** (Scheme 4). The yield, and speed, of the reaction of acetophenone oxime **15a** with N_2O_4 in acetonitrile were, as with **13a**, higher than that in dichloromethane. The gas/solid phase reaction which had been performed on 4-nitroacetophenone oxime **13a** was repeated on acetophenone oxime **15a** and the flask was found to be hot to the touch, something which does not normally occur with the Ponzio reaction. The level of acetophenone **15c** present was found to be greater than that observed in the solvent reactions.

In brief there is a greater application potential for 1,1-dinitro-1-(4-nitrophenyl)ethane **13c** than **15c** because of the higher yields and greater number of nitro groups. However work on the oximes of other acetophenone derivatives is in abeyance pending an improvement in the yield of the respective gem-dinitro products, thereby reducing the level of the ketone impurities thus making separation easier. Increasing the number of ring nitro groups would give a significant increase in energy which could have application potential and this aspect is currently under investigation.

Polynitrophenol Derivatives.

A third category of compounds with potential as energetic materials are those based on nitrated phenol, the most utilised for explosives applications being picric acid (2,4,6-trinitrophenol). The presence of the hydroxyl group is sufficient to activate the ring to nitration, even when already nitro substituted, unlike the behaviour seen for the acetophenone oximes. Hence the compound chosen for investigation was salicylaldoxime (2-hydroxybenzaldehyde oxime) **16**. The stability of gem-dinitro compounds prepared from aldoximes, rather than ketoximes, has been questioned (*43*) but it is hoped that the presence of the hydroxyl (and nitro groups) will improve the stability.

The reaction of salicylaldoxime **16** in dichloromethane (Scheme 5) with N_2O_4 was found to cause the solvent to boil and slow addition was required. The oxime dissolved on mixing and a yellow product precipitated which was identified as 2-(dinitromethyl)-4-nitrophenol **17**. Infra-red spectroscopy confirmed the presence of the gem-dinitro species (1575 cm^{-1}) and the ring nitro group (1518 cm^{-1}). Proton nmr spectroscopy confirmed the 1,2,4-substitution pattern with natural abundance ^{13}C confirming the presence of 7 carbon atoms.

$$N_2O_4/CH_2Cl_2 = \mathbf{17} - R^1 = NO_2, R^2 = H, 13\%, \text{mp } 212\text{-}2.5°C$$
$$N_2O_4/CH_3CN = \mathbf{18} - R^1, R^2 = NO_2, 21\%, \text{mp } 244\text{-}5°C$$

(5)

There has usually been no effect on the substitution pattern of the Ponzio reaction product by variation of the solvent; merely on the ratio of the gem-dinitro species to the carbonyl impurity. However, when salicylaldoxime **16** was reacted in acetonitrile, as in dichloromethane, heat evolution was observed but no product precipitated. Isolation of the product by selective solvent extraction yielded a yellow material of high purity (Scheme 5).

The infra-red, nmr, TLC and melting point analysis proved this compound to be different from the dichloromethane reaction product, and the compound was identified as 4,6-dinitro-2-(dinitromethyl)phenol **18**. TLC indicated there to be only one compound but the initial interpretation of the natural abundance ^{13}C nmr spectrum suggested there to be two closely related isomers of a tetra-substituted compound. Apart from identifying the gem-dinitro (1551 cm^{-1}) and ring nitro (1533 cm^{-1}) bands analysis of the infra-red spectrum uncovered the origin of the two structures. The hydroxyl band had shifted from 3304 cm^{-1} in **17** to 3200 cm^{-1} in **18** (Figure 2), which was indicative of a change from intermolecular H-bonding to intramolecular H-bonding to the ortho substituent. Thus the two closely related isomers are derived from the hydroxyl being intramolecular H-bonded to both the 6-nitro and also one of the gem-dinitro groups in the 2-(dinitromethyl) group.

This is the first reported preparation of 4-nitro-2-(dinitromethyl)phenol **17** and 4,6-dinitro-2-(dinitromethyl)phenol **18** and **18** has a similar structure to picric acid, a known explosive. This however has advantages over picric acid in that its melting point is significantly higher (picric acid mp 122°C) and it is oxygen balanced to CO, which picric acid is not. There has been no significant reduction in the density through addition of the bulky gem-dinitromethyl group in **18** (ρ 1.712) compared to that of picric acid (ρ 1.767).

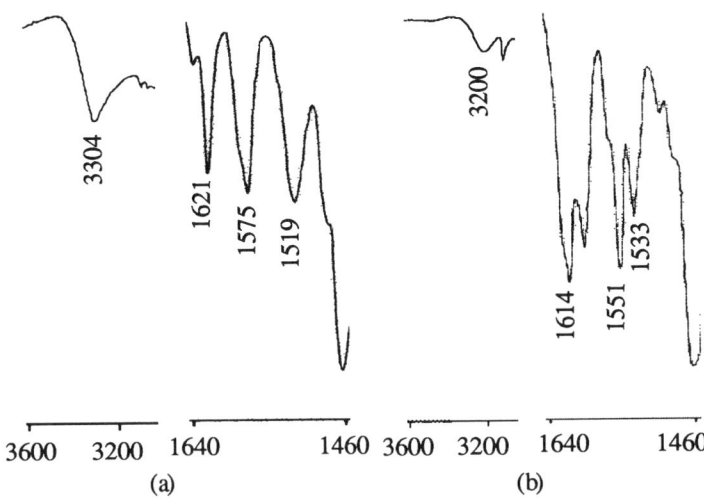

Figure 2. Infra-red spectra showing hydroxyl, gem-dinitro and ring nitro stretch bands of : (a) 4-nitro-2-(dinitromethyl)phenol **17** and (b) 4,6-dinitro-2-(dinitromethyl)phenol **18**.

The Ponzio reaction was observed to occur in a variety of different solvents using the salicylaldoxime substrate and the choice of solvent affected the product obtained in respect of the degree of ring nitration. This effect could be related to the solvent polarity, with the most polar solvents permitting dinitration, low polarity solvents permitting mono-nitration, and no nitration being observed with the least polar media. The yields obtained were, as expected, much lower than those obtained for the diaryl ketoxime (eg fluoren-9-one oxime **5**) and aralkyl ketoxime (eg acetophenone oxime **15a**) substrates. These are insufficient for exploitation but improved yields can be expected by ring nitration of the salicylaldehyde substrate prior to the Ponzio reaction.

Reaction Mechanism.

The Ponzio reaction, and the associated pseudonitrole reaction, have been known since the turn of the century (*48*) but a proposed mechanism (Scheme 6) was not published until 1968 (*37*). The identification of the mechanism may enable improvements to the yields to be achieved, especially for aralkyl ketoximes and, moreover, aldoximes.

The mechanism proposed in 1968 (scheme 6) indicated a series of radical reactions which at first sight appear feasible. The first step, abstraction of hydrogen, generates a viable initial intermediate, although there is, however, some doubt as to the validity, but not the type, of the subsequent series of reactions. The CIDNP nmr technique (Chemically Induced Dynamic Nuclear Polarisation) particularly with the ^{15}N nucleus probes the formation of specific products in radical reactions and not, as with esr, the presence of reactive radical intermediates. Ridd et al (*49*) have been using ^{15}NO$_2$ CIDNP studies to identify nitration pathways since the early 1980s (*50*). This would therefore provide a suitable method for investigation of the Ponzio reaction mechanism.

$$R_2C=N{-}OH + NO_2 \longrightarrow R_2C=N{-}O + HNO_2$$

$$R_2C=N{-}O + NO_2 \rightleftharpoons R_2C=\overset{+}{N}\begin{array}{c}O^-\\O-N=O\end{array} \longrightarrow R_2C\begin{array}{c}NO_2\\N=O\\NO_2\end{array} \longrightarrow R_2C\begin{array}{c}NO_2\\NO_2\end{array} + NO \quad (6)$$

In initial studies (51) on benzophenone oxime and fluoren-9-one oxime **5**, a CIDNP effect was observed in each reaction (Table VI). Difficulties were experienced with solubility of the oxime and the speed of the reaction, even at reduced temperatures, such that the CIDNP effect observed may not reflect the true level of enhancement for the reaction. 2,7-Dinitrofluoren-9-one oxime **6** was regarded as a better substrate because it would react more slowly than **5**, thus allowing room temperature investigation, but not as slowly as 2,4,5,7-tetranitrofluoren-9-one oxime **8**.

Table VI. ^{15}N CIDNP Effects Observed for Ponzio Reaction

Substrate[a]	reagent	temp (°C)	Effect observed during reaction on $^{15}NO_2$ signal from product
$Ph_2C=NOH$	$^{15}N_2O_4$	-25	enhanced emission of ca. 10x, reaction slow.
F=NOH **5**	$^{15}N_2O_4$	-15	enhanced emission, smaller than above but reaction faster. Mass Spec. analysis of the isolated product indicated only one ^{15}N present.
2,7-NO_2F=NOH **6**	$^{15}N_2O_4$	+21	enhanced emission, ca. 87x, reaction slow.
F=^{15}NOH	N_2O_4	-25	enhanced absorption of ca. 3x, short lived.

[a] F - fluoren-9-one.

The CIDNP nmr study (24 scans, FX200 FT-NMR, using ^{15}N nitrobenzene as standard) on 2,7-dinitrofluoren-9-one oxime **6** was started approximately 2.5 minutes after mixing the two reactants (0.1M). The reaction was then studied over the next three hours, this time being required due to the slower nature of the reaction of **6** compared to **5**. The first scan (2.5-6.5 minutes) and the last scan (140-170 minutes) are shown in Figure 3.

The enhanced emission, at the same frequency as the product, was about 87x that of the final product signal intensity - this suggesting that the reaction had occurred via the diffusion together of an alkyl radical and $^{15}NO_2$ to form, initially, a radical pair from which reaction and escape was possible. The results in Table VI, including the small and opposite enhancement observed with ^{15}N labelled fluoren-9-one oxime **5**, are consistent with reaction *via* an analogous radical pair. In the last case the ^{15}N label is in the alkyl group, one bond away from the radical centre, and has to be carried through to one of the nitro groups in the product for the CIDNP effect to be observed. The mechanism suggested in Scheme 7 was derived from these observations.

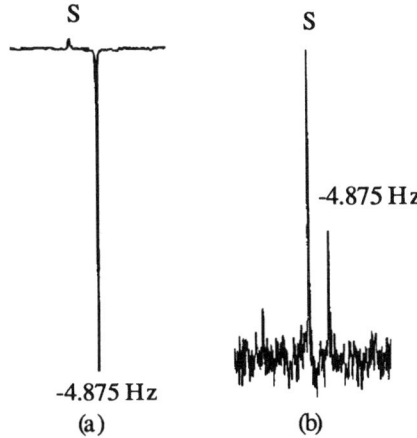

Figure 3. ^{15}N CIDNP experiment on 2,7-dinitrofluoren-9-one oxime **6**, (a) after 2.5-6.5 minutes and (b) after 140-170 minutes (S - ^{15}N nitrobenzene standard).

$$R_2C=N-OH + \cdot NO_2 \longrightarrow R_2C=N-O\cdot + HNO_2$$

$$R_2C=N-O \longleftrightarrow R_2\overset{\cdot}{C}-\overset{..}{N}=O \rightleftarrows \left[\begin{bmatrix} R_2\overset{\cdot}{C}-N=O \\ \updownarrow \\ R_2C=N-O\cdot \end{bmatrix} \cdot NO_2\right]$$

radical pair

$$\downarrow$$

$$R_2C\begin{matrix}N=O \\ NO_2\end{matrix} \quad (7)$$

$$R_2C\begin{matrix}N=O \\ NO_2\end{matrix} \xrightarrow{[O]} R_2C\begin{matrix}NO_2 \\ NO_2\end{matrix}$$

An experiment on the conversion of 2,4,7-trinitrofluoren-9-one oxime **7** to 2,4,7,9,9-pentanitrofluorene **11** was undertaken where the reactants/solvents were purged with O_2-free N_2 prior to, and during, the reaction. Despite this exclusion of oxygen **11** was obtained, thus supporting the hypothesis that N_2O_4 is involved in the final oxidation stage.

This work is still at the early stages and further experiments are envisaged with a view to confirming the proposed mechanism. A comparison of the Ponzio reaction products in Table 3 suggests a relationship between the stabilising influence of the groups attached to the oxime and the gem-dinitro yield obtained. Benzylic radicals are known to be more stable and less reactive than simple alkyl radicals (26), thus, in the case of fluoren-9-one oxime, the carbon radical formed in the proposed mechanism would be stabilised through the conjugated ring system. A reduction in the conjugation and/or the number of electron withdrawing groups (ie decreased radical stability) results in a reduction in the yield of the Ponzio reaction product. Improvements in product yields can therefore be achieved by addition of groups which would stabilise the proposed carbon radical, but such additions would not improve the yields for substrates which could not accommodate such incorporation of stabilising groups.

The mechanism indicated in Scheme 7 does not indicate the route by which the carbonyl impurity is produced and, because the yield of carbonyl products is not commonly quoted for Ponzio reaction studies, any relationship with radical stability cannot be considered (if indeed the reaction is radical). However, radicals of alkyl compounds are known to be less stable and more reactive than aryl compounds, and this may have some influence on the type of product.

Conclusions.

The preparation of 2,4,5,7,9,9-hexanitrofluorene **12** can be accomplished easily by reaction of 2,4,5,7-tetranitrofluoren-9-one oxime **8** with either N_2O_4 or N_2O_5 in dichloromethane. N_2O_5 does not confer any advantage in this reaction over N_2O_4 except in speed (about 5 minutes rather than 25 minutes). The properties measured so far indicate that this compound has a potential use as a thermally stable explosive. 2,7,9,9-Tetranitrofluorene **10** has a higher melting point but the lower oxygen balance makes it less attractive for this purpose.

The polynitro(phenyl)ethane derivatives **13b-15b** are potential binder plasticisers, similar to eutectic mixtures of 2,4,6-trinitroethylbenzene and 2,4-dinitroethylbenzene. It is hoped that, because some of the nitro groups are exocyclic to the aromatic ring, the plasticisation effect can be achieved without the need for a mixture.

4,6-Dinitro-2-(dinitromethyl)-phenol **18** could be regarded as a potential picric acid replacement with the added advantages of higher oxygen balance and melting point. Improvements in yields are expected for the Ponzio reaction via the 4,6-dinitrosalicylaldoxime derivative.

$^{15}NO_2$ CIDNP nmr experiments have shown enhanced emission/absorption signals indicating radical nitration of the substrate. The results have allowed a mechanism to be proposed (Scheme 7) based on the formation of a stabilised carbon radical. Further $^{15}NO_2$ CIDNP experiments will be undertaken to corroborate this proposal and to confirm that both oxides of nitrogen undergo similar Ponzio reaction routes.

References.

1   Noble, P. Jr.; Borgardt, F. G. and Reed, W. L., *Chem. Rev.*, **1964**, *64*, 19.
2   Olah, G. A.; Ramaiah, P.; Prakash, G. K. S. and Gilardi, R., *J. Org. Chem.*, **1993**, *58*(3), 763.

3 Marchand, A. P.; Sharma, R.; Zope, U. R.; Watson, W. H. and Kashyap, R. P., *J. Org. Chem.*, **1993**, *58*(3), 759.
4 Larson, H. O., in *Chemistry of Nitro and Nitroso Groups Pt 1;* Feuer, H. Ed, Interscience, New York, 1969, Chpt 6.
5 Padeken, H. G.; von Schickh, u O. and Segnitz, A., in *Houben-Weyl, Methoden der Organischen Chemie, Band 10/1*, Muller, E. Ed, Georg Thieme Verlag, Stuttgart, 1971, Chpt 1.
6 Meyer and Stuber, *Ber. dtsch. Chem. Ges.*, **1872**, *5*, 203.
7 Okhlobystina, L. V.; Fainzilberg, A. A. and Novikov, S. S., *Izv. Akad. Nauk. SSSR. Otd. Khim. Nauk.*, **1962**, 517. CA 57:14920h (1962).
8 Kornblum, N.; Larson, H. O.; Blackwood, R. K.; Mooberry, D. D.; Oliveto, E. P. and Graham, G. E., *J. Am. Chem. Soc.*, **1956**, *78*, 1497.
9 ter Meer, E., *Liebig's Ann. Chem.*, **1876**, *181*, 1.
10 Hawthorne, M. F., *J. Am. Chem. Soc.*, **1956**, *78*, 4980.
11 Kaplan, R. B. and Shechter. H., *J. Am. Chem. Soc.*, **1961**, *83*, 3535.
12 Hamel, E. E.; Dehn, J. S.; Love, J. A.; Scigliano, J. J. and Swift, A. H., *Ind. Eng. Chem. Prod. Res. and Devel.*, **1962**, *1*, 108.
13 Matacz, Z.; Piotrowska, H. and Urbanski, T., *Pol. J. Chem.*, **1979**, *53*(1), 187.
14 Kornblum, N.; Singh, H. K. and Kelly, W. J., *J. Org. Chem.*, **1983**, *48*(3), 332.
15 Garver, L. C.; Grakauskas, V. and Baum, K., *J. Org. Chem.*, **1985**, *50*, 1699.
16 Marchand, A. P., *Tetrahedron*, **1988**, *44*(9), 2377.
17 Archibald, T. G.; Garver, L. C.; Baum, K. and Cohen, M. C., *J. Org. Chem.*, **1989**, *54*, 2869.
18 Baum, K.; Bigelow, S. S.; Nguyen, N. V. and Archibald, T., *J. Org. Chem.*, **1992**, *57*, 235.
19 Titov, A. I., *J. Gen. Chem. (USSR)*, **1948**, *18*, 1312. CA 43:4217 (1949).
20 Titov, A. I., *Tetrahedron*, **1963**, *19*, 557.
21 Archibald, T. G. and Baum, K., *J. Org. Chem.*, **1988**, *53*(20), 4645.
22 Ponzio, G., *Gazz. Chim. Ital., Rome*, **1906**, *36* II, 98.
23 Nielsen, A. T., *J. Org. Chem.*, **1962**, 27, 1993.
24 Waykole, L. M.; Shen, C-C. and Paquette, L. A., *J. Org. Chem.*, **1988**, *53*(21), 4969.
25 Emmons, W. D. and Pagano, A. S., *J. Am. Chem. Soc.*, **1955**, *77*, 4557.
26 Sykes, P., in *A Guidebook to Mechanism in Organic Chemistry*, 4th Ed, Longman Group Ltd, London, 1975.
27 Iffland, D. C. and Criner, G. X., *J. Am. Chem. Soc.*, **1953**, *75*(19), 4047.
28 Iffland, D. C. and Yen, T-F., *J. Am. Chem. Soc.*, **1954**, *76*, 4083.
29 Patchett, A. A.; Hoffman, F.; Giarrusso, F. F.; Schwam, H. and Arth, G. E., *J. Org. Chem.*, **1962**, 27, 3822.
30 Scholl, R., *Ber. dtsch. Chem. Ges.*, **1888**, *21*, 506.
31 Ungnade, H. E. and Kissinger, L. W., *J. Org. Chem.*, **1959**, *24*, 666.
32 Grakauskas, V., *J. Org. Chem.*, **1973**, *38*(17), 2999.
33 Luk'yanov, O. A. and Pokhvisneva, G. V., *Izv. Akad. Nauk. SSSR., Ser. Khim.*, **1991**, 2797.
34 Dave, P. R.; Ferraro, M.; Ammon, H. and Choi, C. S., *J. Org. Chem.*, **1990**, *55*(14), 4459.
35 Krauch, H. and Kunz, W., in *Organic Named Reactions*, J Wiley & Sons Inc., New York, **1964**, 363.

36 Riebsomer, J. L., *Chem. Rev. (ACS)*, **1945**, *36*, 157.
37 Frojmovic, M. M. and Just, G., *Can. J. Chem.*, **1968**, *46*, 3719.
38 Luk'yanov, O. A. and Pokhvisneva, G. V., *Izvestiya Akademii Nauk., SSSR*, **1991**, 2148.
39 Fruttero, R.; Ferrarotti, B.; Gasco, A. and Calestani, G., *Liebig's Ann. Chem.*, **1988**, 1017.
40 Luk'yanov, O. A. and Zhiguleva, T. I., *Izvestiya Akademii Nauk., SSSR*, **1982**, 1423.
41 Crombie, L. and Roughley, B. S., *Tetrahedron*, **1986**, *42*(12), 3147.
42 Fieser, L. F. and Doering, W. v. E., *J. Am. Chem. Soc.*, **1946**, *68*, 2252.
43 Kolesetskaya, G. I.; Tselinskii, I. V. and Bagal, L. I., *Zh. Org. Khim.*, **1970**, *6*(2), 334.
44 Norris, R. K. and Randles, D., *Aust. J. Chem.*, **1979**, *32*, 1487.
45 Honey, P. J., M. Phil. Thesis, Hatfield Polytechnic, UK, **1991**.
46 Moore, F. J. and Huntress, E. H., *J. Am. Chem. Soc.*, **1927**, *49*, 2618. The author observed yields in excess of 98% and the reaction was complete after 10 minutes at reflux denoted by precipitation of the oxime.
47 Shim, S. B.; Kim, K. and Kim, Y. H., *Tetrahedron Letters*, **1987**, *28*(6), 645.
48 Ponzio, G., *J. Prakt. Chem.*, **1906**, *73*, 494.
49 Ridd, J. H;, Trevellick, S. and Sandall, J. P. B., *J. Chem. Soc., Perkin Trans. 2*, **1993**, 1073 and refs contained therein.
50 Ridd, J. H. and Sandall, J. P. B., J. Chem. Soc., *Chem. Commun.*, **1981**, 402.
51 Millar, R W, unpublished work, **1991**.

© **British Crown Copyright 1995, DRA. Published with the permission of the Controller of Her Britannic Majesty's Stationery Office.**

RECEIVED January 16, 1996

Chapter 14

New Nitration and Nitrolysis Procedures in the Synthesis of Energetic Materials

Philip F. Pagoria, Alexander R. Mitchell, Robert D. Schmidt, Clifford L. Coon, and Edward S. Jessop

Energetic Materials Center, Lawrence Livermore National Laboratory, P.O. Box 808, L-282, Livermore, CA 94550

The development of new nitrolysis and nitration reagents are essential to the research and development of energetic materials. In this paper we describe the development of two new nitrolysis reagents, trifluoromethanesulfonic anhydride/ HNO_3/ N_2O_5 and trifluoroacetic anhydride/ HNO_3/ N_2O_5, which were uniquely successful in the nitrolysis of secondary amide groups to yield several new, bicyclic nitramines. An alternative method for the synthesis of secondary nitramines by the mild nitrolysis of N-*tert*-butoxycarbonyl (N-BOC) derivatives and the attempted synthesis of ^{18}O-labeled 2,4,6-trinitrotoluene using trifluoromethanesulfonic acid and $NaN^{18}O_3$ at elevated temperatures are also described.

There are a variety of known reagents for the nitrolysis of secondary amide or *tert*-butyl amines to nitramines including 20% N_2O_5/ HNO_3[1], Ac_2O/ HNO_3[2], Ac_2O/ NH_4NO_3[3], HNO_3/ H_2SO_4[4], trifluoroacetic anhydride (TFAA)/ 100% HNO_3[5], 100% HNO_3/ P_2O_5[2], or TFAA/ NH_4NO_3[6]. These reagents have been used extensively in the synthesis of energetic materials. In this paper we describe the development of two new nitrolyzing mixtures used for the synthesis of 2,5,7,9-tetranitro-2,5,7,9-tetraazabicyclo[4.3.0]nonan-8-one (**K-56**) (**1**) and 2,4,6,8-tetranitro-2,4,6,8-tetraazabicyclo[3.3.0]octane (**Bicyclo-HMX**) (**2**). We also describe the use of other known nitrolysis reagents in the synthesis of new energetic materials, and the effect of the strength of the nitrating medium on yields and product distribution is discussed.

Recently we investigated the synthesis of a series of energetic materials possessing the nitrourea moiety (Figure 1). Previously reported properties of nitrourea compounds[7] suggested they would make excellent candidates as both insensitive and highly energetic materials. Tetranitroglycouril (**TNGU**)[8], first synthesized by the nitration of glycouril using 100% HNO_3 and P_2O_5 at 50°C, has one of the highest densities of known C,H,N,O- based energetic materials (2.04 g/cc). 1,4-Dinitroglycouril (**DNGU**), an analog of TNGU, also has an attractive crystal density of 1.992 g/cc[9] and was considered for some time as an insensitive, highly energetic material. The insensitivity may be attributed to the intermolecular hydrogen bonding of the mononitrourea.

Synthesis of 2,6,8-trinitro-2,4,6,8-tetraazabicyclo[3.3.0]octan-3-one (3) and 2,4,6,8-tetranitro-2,4,6,8-tetraazabicyclo[3.3.0]octan-3-one (4). The preparation and subsequent nitration of 2,4,6,8-tetraazabicyclo[3.3.0]octan-3-one dihydrochloride salt (7) to yield 2,4,6,8-tetranitro-2,4,6,8-tetraazabicyclo[3.3.0]octan-3-one (4) was first described by Wenjie and coworkers[10]. Compound 4 was considered interesting as an intermediate in a more efficient route to 2,4,6,8-tetranitro-2,4,6,8-tetraazabicyclo[3.3.0]octane (2), first synthesized in our laboratories and described

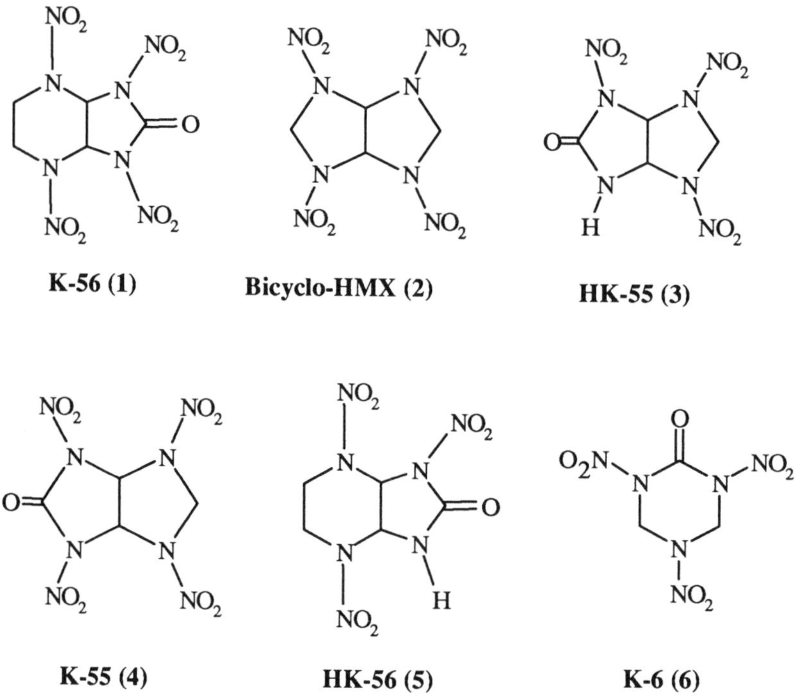

K-56 (1) **Bicyclo-HMX (2)** **HK-55 (3)**

K-55 (4) **HK-56 (5)** **K-6 (6)**

Figure 1. New energetic materials described in this paper.

later in this paper. In an attempt to reproduce the synthesis of compound 4, we found that if the reaction temperature was maintained at <15 °C the nitration does not go to completion and 2,6,8-trinitro-2,4,6,8-tetraazabicyclo[3.3.0]octan-3-one (**HK-55**) (3) precipitates from the reaction mixture in 53% yield (Fig. 2). When the reaction temperature is raised to 20-50 °C, the nitration is complete, and compound 4 is obtained exclusively in 40% yield (Fig. 2). These results initiated a study of the effects of temperature and nitration reagent on the yields of 3 and 4 (Table 1). It was found that the use of 90% HNO_3/Ac_2O as the nitrating agent at 0-40 °C gave exclusively 3 in 72% yield while the use of 100% HNO_3/ trifluoroacetic anhydride (**TFAA**), a powerful nitrating agent, gave exclusively 4 over the same temperature range. Compound 3 proved to be interesting because of its high density (ρ= 1.905 g/cc)[11] and relative insensitivity to shock (61 cm on the drop hammer machine vs. 32 cm for HMX). TIGER code calculations suggest it has approximately the same energy as HMX.[12]

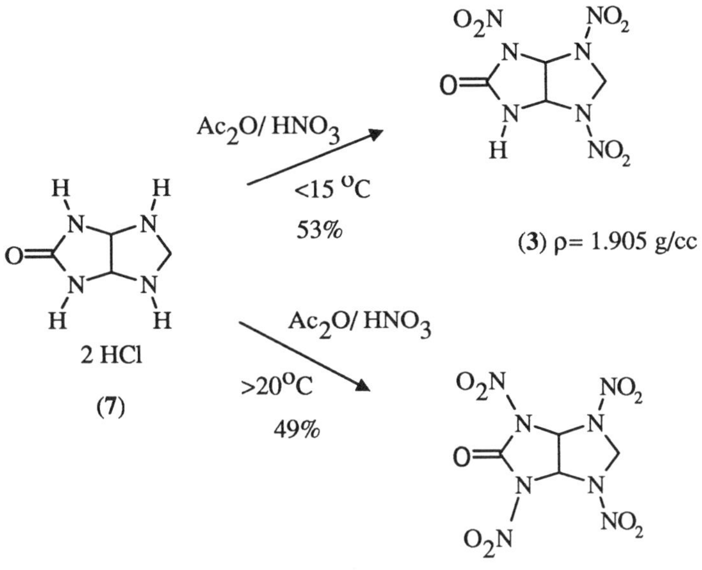

Figure 2. Effect of temperature on the nitration products of **7**.

Table 1. Effect of the nitrating medium on the yields of **3** and **4**

Compound HNO$_3$/	90% HNO$_3$/Ac$_2$O	100% HNO$_3$/Ac$_2$O	100% TFAA
(3)	72%	53% (<15°C)	0
(4)	0	49% (20-50 °C)	51%

Synthesis of 2,5,7,9-tetranitro-2,5,7,9-tetraazabicyclo[4.3.0]nonan-8-one (1)[13] and 2,5,7-trinitro-2,5,7,9-tetraazabicyclo[4.3.0]nonan-8-one (5). We have also synthesized the homologues of **3** and **4**, 2,5,7-trinitro-2,5,7,9-tetraazabicyclo[4.3.0]nonan-8-one (**5**) and 2,5,7,9-tetranitro-2,5,7,9-tetraazabicyclo[4.3.0]nonan-8-one (**1**).[13] Compound **1** is interesting because of its high molecular density of 1.969 g/cc.[13] The synthesis of **1** involved the first use of the new nitrolysis mixture, trifluoromethanesulfonic acid anhydride (**TFMSAA**)/20% N$_2$O$_5$/HNO$_3$ (Figure 3).

The syntheses of **1** and **5** involved a multi-step reaction sequence in which 1,3-diacetyl-2-imidazolone (**8**)[14] is brominated and reacted with ethylenedinitramine (**EDNA**)[4] in the presence of triethylamine to yield a mixture of 7,9-diacetyl-2,5-dinitro-2,5,7,9-tetraazabicyclo[4.3.0]nonan-8-one (**9**) and 7-acetyl-2,5-dinitro-2,5,7,9-tetraazabicyclo[4.3.0]nonan-8-one (**10**).[15] The diacetyl compound is easily converted to the mono-acetyl derivative by treatment with ammonium hydroxide in CH$_3$CN with an overall yield of **10** of 32%. Nitration of

10 with a 20% N_2O_5/ HNO_3[16] mixture at room temperature yielded 7-acetyl-2,5,9-trinitro-2,5,7,9-tetraazabicyclo[4.3.0]nonan-8-one (**11**) in 80% yield. The nitrolysis of the final acetyl group on **11** proved to be very difficult. Several nitrolysis reagents

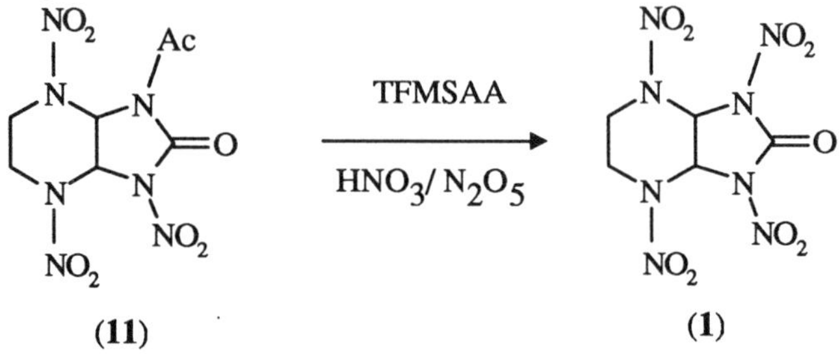

Figure 3. Use of TFMSAA/ 20% N_2O_5 /HNO_3 for the synthesis of **1**.

were tried including 20% N_2O_5/ HNO_3 at 25-50 °C, TFAA/20% N_2O_5/ HNO_3 at 20-40 °C, Ac_2O/20% N_2O_5/HNO_3 at 25-50 °C and TFAA/100% HNO_3, but all resulted in recovery of starting material. The development of the new, powerful nitrolysis mixture, trifluoromethanesulfonic anhydride (TFMSAA) and a 20% N_2O_5/ HNO_3 solution, allowed the nitrolysis of the final acetyl group to give **1** in 69 % yield. This reagent was chosen on the premise that the strength of the nitrolysis reagent was related to the strength of the conjugate acid of the anhydride used. It was also of interest because the reagent may be removed with methylene chloride, thus avoiding an aqueous work-up, This was based on a concern that aqueous work-up may lead to at least partial hydrolysis of the dinitrourea moiety. Later studies indicated that the dinitrourea was relatively stable to acidic hydrolysis procedures but was readily hydrolyzed under basic conditions.

The hydrolysis of **11** with aqueous HCl in refluxing CH_3CN gave **5** in 85% yield (Figure 4). Compound **5** was found to have a density of 1.84 g/cc[17]. Nitration of **5** with a 20% N_2O_5/ HNO_3 mixture at 40 °C gave **1** in 90% yield, providing an alternative method for the synthesis of **1**.

Graindorge and Lescop[18] reported an improved synthesis of K-56, analogous to the synthesis of K-55 reported by Wenjie and coworkers[10]. The synthesis involved the reaction of ethylene diformamide with glyoxal to yield the 1,4-diformyl-2,3-dihydroxypiperidine followed by condensation with urea in the presence of acid catalyst to give 2,4,7,9-tetraazabicyclo[4.3.0]nonane dihydrochloride salt which upon nitration gave K-56 in good yields.

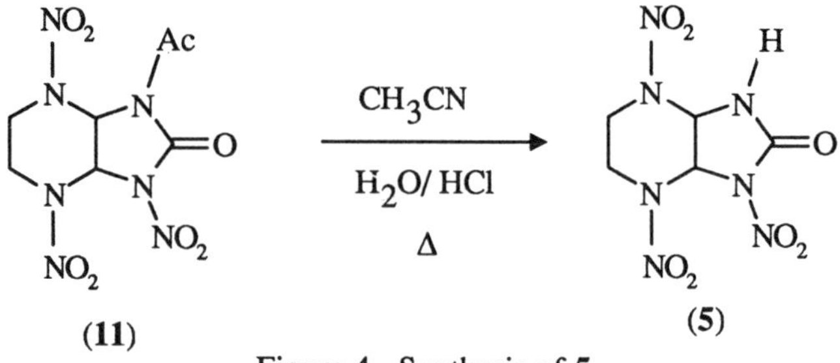

Figure 4. Synthesis of **5**.

Synthesis of 2-oxo-1,3,5-trinitro-1,3,5-triazacyclohexane (K-6). The synthesis of 2-oxo-1,3,5-trinitro-1,3,5-triazacyclohexane (K-6)[19] (**6**) involved the preparation and subsequent nitration of 2-oxo-5-*tert*-butyl-1,3,5-triazacyclohexane (5-*tert*-butyltriazone, TBT) (**12**) (Figure 5). Our choice of TBT as the precursor to K-6 was based on the work of Adolph and Cichra[5] which showed that N-*tert*-butyl derivatives of secondary amines are readily nitrolyzed to the corresponding secondary nitramines using either HNO_3/H_2SO_4 or trifluoroacetic anhydride (TFAA)/100% HNO_3.

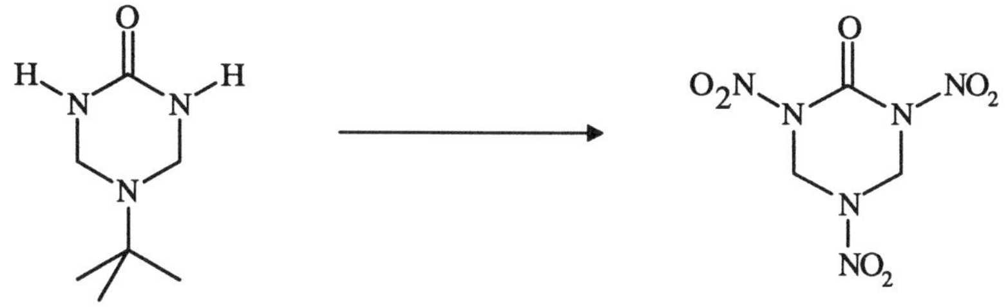

Figure 5. Synthesis of **6**.

Our synthesis of the previously unknown 5-*tert*-butyltriazone uses a variation of a procedure reported by Petersen.[20] *tert*-Butylamine, 37% formalin, and urea at 50-55 °C react for 12 hours to yield TBT in 52% yield.

Nitrolysis of **12** utilizes a mixture of TFAA and HNO_3 at 0 °C with subsequent warming to 40-45 °C. Caution must be exercised at this stage as rapid warming of the reaction mixture can cause a vigorous exotherm which greatly decreases the yield. Cooling and removal of the nitration components under reduced pressure gives **6** as a white powder. Recrystallization from acetonitrile or ethyl acetate gives **6** as white plates in 57% yield. The use of a relatively volatile nitration reagent such as HNO_3/TFAA allows removal under vacuum, thereby eliminating the need for aqueous work up. This is desirable in cases where dinitrourea compounds are susceptible to hydrolysis. Subsequent studies showed that **6** was relatively stable in aqueous nitric acid solutions.

Table II. Effect of nitrolysis reagent on the yield of **6**.

Nitrating Agent	Yield
TFAA/ N_2O_5/ HNO_3	57%
TFAA/ 100% HNO_3	43%
Ac_2O/ 100% HNO_3	57%
NO_2BF_4	40%
H_2SO_4/ HNO_3	0
Ac_2O/ 90% HNO_3	21%
Ac_2O/ 70% HNO_3	0

TFAA= Trifluoroacetic acid anhydride
N_2O_5/ HNO_3= 20% by weight of N_2O_5 in 100% Nitric acid

A study of the nitrolysis of **12** was performed to determine the best method for scale-up of the synthesis of **6** and is summarized in Table II. All of the methods investigated followed the same general procedure in which **12** is slowly added to a cooled solution of the nitrolyzing reaction medium and then allowed to warm to a desired reaction temperature. The work-up consisted of pouring the reaction mixture into ice water and collecting the precipitated product. In each case the crude reaction product gave an ^{1}H-nmr spectrum identical to that of authentic **6**. The nitrolysis reagent chosen for scale-up was $Ac_2O/$ 90% HNO_3, a decision based on safety and cost considerations instead of yield.

Synthesis of 2,4,6,8-tetranitro-2,4,6,8-tetraazabicyclo[3.3.0]octane (Bicyclo-HMX) (2). Extensive research on numerous routes to Bicyclo-HMX (**2**) was carried out at both Department of Defense[21] and Department of Energy laboratories. Compound **2** generated considerable interest as a target material because it was predicted to be more energetic than HMX through higher oxygen balance and a predicted higher heat of formation due to the strained bicyclic ring system. These included attempts at the synthesis of **2** from 2,4,6,8-tetraacetyl-2,4,6,8-tetraazabicyclo[3.3.0]octane by nitrolysis of the acetyl groups which yielded only decomposition products.[22] Koppes and coworkers[21] reported the synthesis of the only known analog of **2** possessing 2,4,6,8-tetranitro substitution, 3,3,6,6-tetrakis(trifluoromethyl)-2,4,6,8-tetranitro-2,4,6,8-tetraazabicyclo[3.3.0]octane, obtained from the nitration of 3,3,6,6-tetrakis(trifluoromethyl)-2,4,6,8-tetraazabicyclo[3.3.0]octane.

Our approach to the synthesis of **2** involved the nitrolysis of the intermediate, 2,4-dipropionyl-6,8-dinitrobicyclo[3.3.0]octane (**13**) (Figure 6). Compound **13** was synthesized by the condensation of methylenedinitramine (**MEDINA**) with 1,3-dipropionyl-4,5-dibromoimidazolidine in acetonitrile in the presence of triethylamine.[15] This condensation yields a mixture of bicyclic products from which **13** may be isolated by silica gel chromatography using a Chromatotron (Harrison Research, 840 Moana Ct, Palo Alto, CA). Compound **13** cannot be nitrolyzed directly to **2** but decomposes in the presence of almost all nitrating agents, such as 90% nitric acid, 100% nitric acid, mixtures of nitric and sulfuric acids, and nitronium tetrafluoroborate.

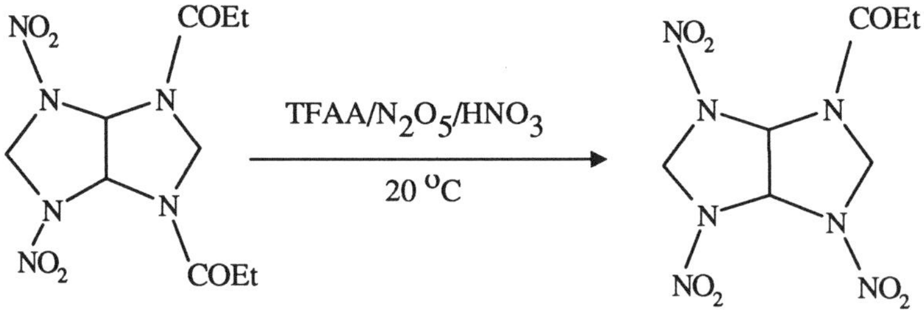

Figure 6. Nitrolysis of **13**.

The unique combination of a 20% solution of dinitrogen pentoxide in 100% nitric acid and trifluoroacetic anhydride converted **13** to its trinitro derivative, **14**, in greater than 90% yield. It is interesting to note that the diacetyl derivative of **13** gave only a 10% yield of **14** under identical conditions. The ease of nitrolysis of a propionyl group over an acetyl group has been noted previously,[2,22] but not to the extent demonstrated in this reaction. This new nitrolyzing mixture must be handled remotely and in small quantities. During this work several explosions occurred causing serious damage.

The efficient conversion of **13** to **14** with N_2O_5, 100% nitric acid and trifluoroacetic anhydride led us to investigate the effect of higher reaction temperatures and longer reaction times in the hope of nitrolyzing the final acyl group. It was found that nitrolysis at 47°C for 4h led to the surprising formation of the fully nitrated 2-trifluoroacetoxy derivative (**16**) in greater than 90% yield (Figure 7.). This compound may have resulted from the oxidation of **13** to an iminium ion intermediate which was trapped by adventitious trifluoroacetate anion followed by subsequent nitrolysis of the propionyl group. The structure of **16** was confirmed by x-ray crystallography.[11] Compound **16** is a moisture sensitive material which was subject to ring opening and decomposition in the presence of both mild acid and base. There was no evidence that this reaction proceeded through **2** as an intermediate, suggesting alternative procedures would be needed to convert **14** to **2**.

The conversion of **14** to the final product **2** was achieved by stirring a solution of **14** in a 20% N_2O_5 /100% nitric acid mixture at ambient temperature for 18 hours (Figure 8). Other nitrating agents, including 100% nitric acid, nitric acid/sulfuric acid, nitric acid/oleum, and nitronium tetrafluoroborate, were studied, but these gave either no reaction or complete decomposition of **14**. Bicyclo-HMX (**2**) is a white, crystalline solid with a density of 1.87 g/cc[11] that is stable at ambient conditions. The density of **2** is lower than 1,3,5,7-tetranitro-1,3,5,7-tetraazacyclooctane (HMX) (ρ=1.91 g/cc), allowing the prediction that **2** and HMX

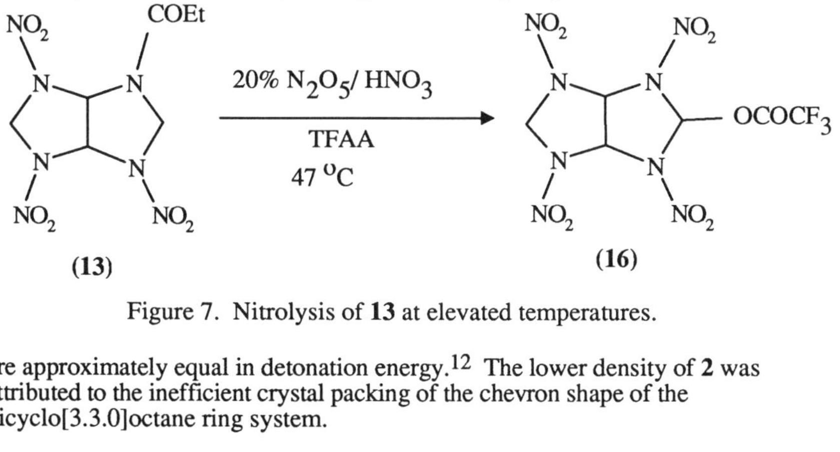

Figure 7. Nitrolysis of **13** at elevated temperatures.

are approximately equal in detonation energy.[12] The lower density of **2** was attributed to the inefficient crystal packing of the chevron shape of the bicyclo[3.3.0]octane ring system.

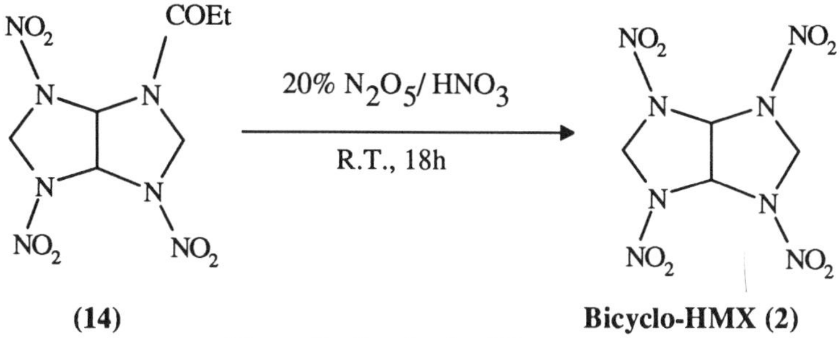

Figure 8. Synthesis of **2**.

Synthesis of 1,4-dideoxy-1,4-dinitro-*neo*-inositol-2,5-dinitrate (LLM-101) (17). *Neo*-inositol explosives were investigated to provide a new, inexpensive class of explosives with energy similar to that of RDX and HMX and as ingredients of new energetic polymer formulations. *Neo*-inositols were first prepared by Dinwoodie and Fort [23] who reported their synthesis and properties in 1966. The starting materials for neo-inositol explosives are glyoxal and nitromethane, both of which are potentially inexpensive. These reagents combine in the presence of mild base to yield 1,4-dideoxy-1,4-dinitro-*neo*-inositol (1,2,4,5-tetrahydroxy-3,6-dinitrocyclohexane) (**DNNI**) (**18**) in 40% yield.[23] Nitration of **18** with a variety of reagents, including Ac$_2$O/ HNO$_3$, HNO$_3$/ H$_2$SO$_4$, and 100% HNO$_3$ in CHCl$_3$, all yielded the highly oxidized and very shock sensitive, 1,4-dideoxy-1,4-dinitro-*neo*-inositol-2,3,5,6-tetranitrate. This was converted to LLM-101 in 80% yield by trituration with EtOH, in what may be considered a transesterification reaction (Figure 9). This method was undesirable because it goes through the sensitive tetranitrate which should not be isolated in anhydrous form. By using the less powerful nitrating agent, 90% HNO$_3$/ Ac$_2$O, LLM-101 may be

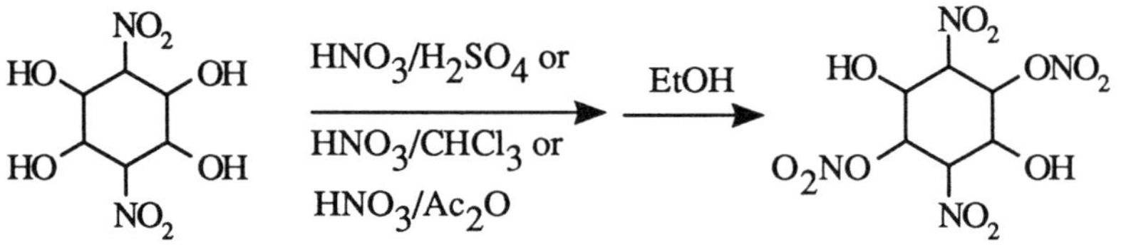

Figure 9. Synthesis of **17**.

synthesized directly in 70% yield (Figure 10). Its density of 1.87 g/cc[11] is significantly higher than predicted (1.80-1.82 g/cc), probably because **LLM-101** possesses a point symmetry. The thermal stability of **LLM-101** is also very good. In DSC runs no indication of decomposition appears prior to the exotherm that peaks at 243°C.

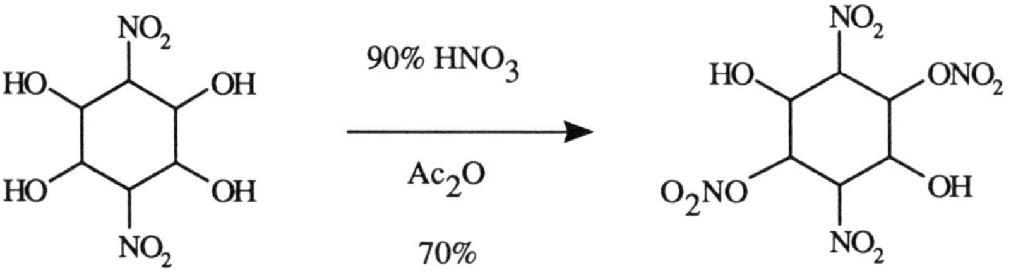

Figure 10. One-step synthesis of **17**.

Attempted Synthesis of ^{18}O-labeled TNT. The utility of TFMSA as a solvent for the nitration of aromatics was demonstrated by Coon and coworkers[24] who found that mono- and di-nitrotoluenes could be prepared in almost quantitative yield using HNO$_3$ in CF$_3$SO$_3$H at temperatures at or below 20°C. We extended this procedure using NaNO$_3$ or KNO$_3$ as the source of nitronium ion and elevated reaction temperatures in the synthesis of 2,4,6-trinitrotoluene (TNT). This procedure was investigated as a method to make ^{18}O-labeled TNT from NaN^{18}O$_3$. Our initial study concentrated on the use of KNO$_3$. A series of experiments were performed in an attempt to optimize the conditions before the procedure was applied to making the ^{18}O-labeled material. The treatment of toluene with a 3.18/1 ratio of KNO$_3$ to toluene yielded a 40:60 mixture of di- and trinitrotoluene even at reflux temperatures. When the amount of KNO$_3$ was increased to a 4/1 molar ratio at 125-30 °C for 10h an 87% yield of TNT is obtained. The progress of the reaction was followed by TLC analysis and the purity of the product was determined by ^{1}H-nmr. When the amount of KNO$_3$ was decreased to 3.5 /1 molar excess an 80% yield of TNT is obtained at a reaction temperature of 125-35 °C for 8h. The addition of an equal molar amount of H$_2$SO$_4$ to the reaction mixture and allowing the reaction to run under the conditions described above gave a 83% yield of TNT. The sulfuric acid was added in an attempt to increase the acidity of the mixture to allow the possible use of lower reaction temperatures. The addition did increase the yield slightly and gave a lighter colored crude product but high reaction temperatures were still required to drive the nitration to completion. We also investigated the use of NaNO$_3$ using the best conditions developed in the KNO$_3$ study. The NaNO$_3$ nitrations consistently gave lower yields of TNT and required slightly longer reaction times to drive the reaction to completion. Unfortunately NaN^{18}O$_3$ was the only reagent available for the ^{18}O-labeling experiment. Nitration of toluene with NaN^{18}O$_3$ (95% ^{18}O) in TFMSA in the presence of 3 molar equivalents of H$_2$SO$_4$ at 125-35 °C gave a 65% yield of TNT. A mass spectrum indicated the major product (87%) consisted of only two ^{18}O atoms per TNT molecule indicating substantial exchange of the oxygen, although there was small amounts of the tri- and tetra-^{18}O-substituted product. Rajendran and coworkers[25] have reported exchange of oxygens occurs readily for inorganic nitrites and nitrates but that nitro groups show no exchange of oxygens. These results suggest the first nitration was facile but the subsequent nitrations more difficult and the long reaction times allowed exchange of the oxygens of NaN^{18}O$_3$.

Use of N-tert-Butoxycarbonyl (BOC) Derivatives for the Synthesis of Nitramines. The synthesis of secondary nitramines has been studied extensively with respect to substrate and conditions of nitrolysis.[2] *N,N*-Dialkyl derivatives of acetamide and propionamide are readily nitrolyzed, whereas *N,N*-dialkylurethanes derived from methanol and ethanol are resistant to nitrolysis by nitric acid/trifluoroacetic anhydride at 50° C (Figure 11).

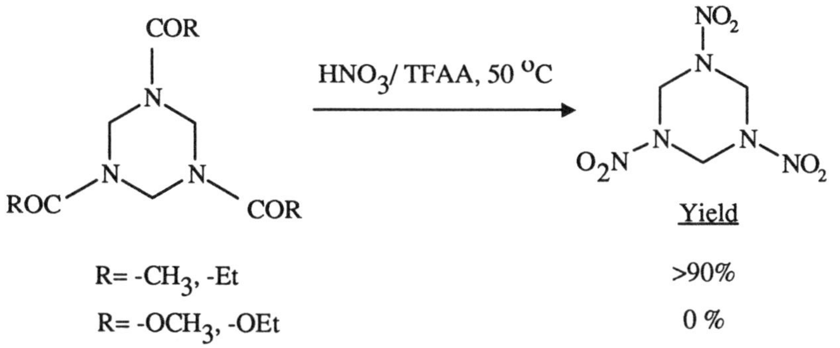

	Yield
R= -CH$_3$, -Et	>90%
R= -OCH$_3$, -OEt	0 %

Figure 11. Nitrolysis of amides and urethanes.

Our interest in developing mild and safe procedures for the preparation of nitramines led us to examine nitrolysis reactions employing urethane substrates derived from *tert*-butanol (*N-tert*-butoxycarbonyl derivatives). The *tert*-

butoxycarbonyl (BOC) group is extensively used in peptide synthesis as an acid-labile amine protecting group.[26-28] BOC amino acid derivatives undergo a rapid acidolytic cleavage via a S_N1 process under relatively mild conditions (trifluoroacetic acid at 0° C; HCl in acetic acid, 1,4-dioxane, ethyl ether or nitromethane at 25° C) with the release of isobutene, CO_2 and deprotected amino acid derivative[29].

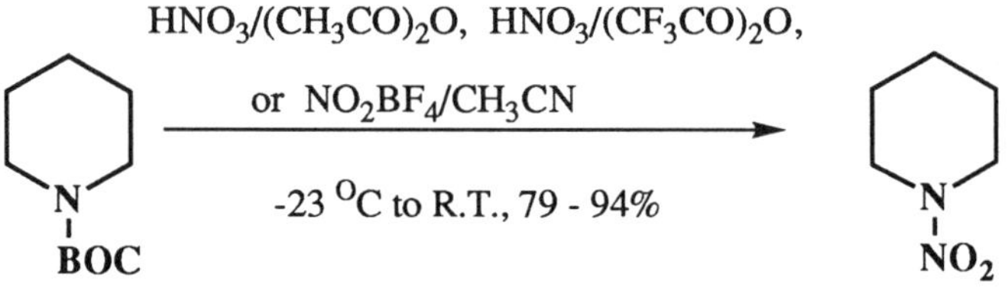

Figure 12. Nitrolysis of N-BOC-piperidine.

The conversion of N-BOC derivatives to the corresponding nitramines requires mild nitration conditions. The avoidance of excess acid and high temperature favors nitrolysis over the competing cleavage reaction. Thus, treatment of N-BOC-piperidine with 100% nitric acid gave only traces (<5%) of 1-

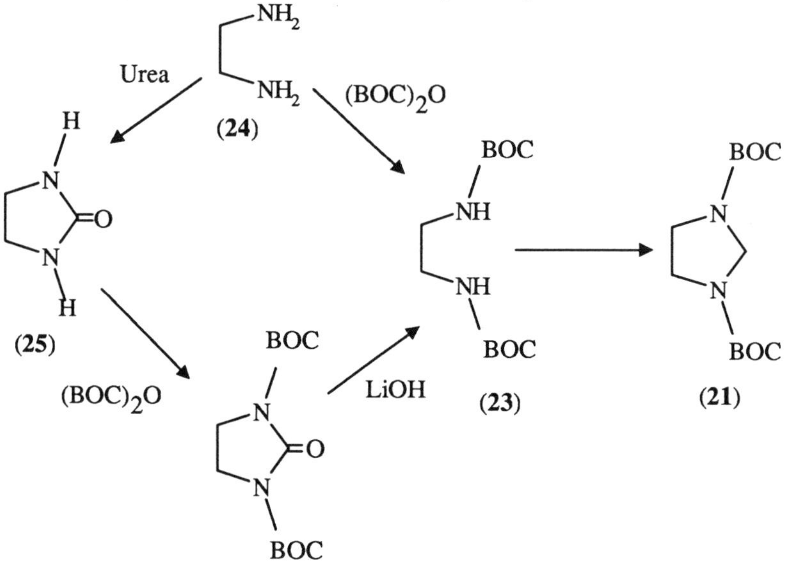

Figure 13. Synthesis of **21**.

nitropiperidine on thin layer chromatography. The use of nitric acid in combination with acetic or trifluoroacetic anhydrides converted N-BOC-piperidine (**19**) to 1-nitro-piperidine (**20**) in 87-94% yields while nitronium tetrafluoroborate[30] in acetonitrile effected the same conversion in 79% yield (Figure 12).
Nitrogen pentoxide in dichloromethane proved to be an exceptionally mild nitration reagent in the nitrolysis of N-BOC-piperidine (**19**) which gave a 1:1 mixture of unreacted N-BOC-piperidine (**19**) and 1-nitropiperidine (**20**).

We used the nitrolysis of 1,3-bis-BOC-1,3-diazacyclopentane (**21**) to 1,3-dinitro-1,3-diazacyclopentane (**22**)[31,32] as a model for the synthesis of larger cyclic methylenenitramines. The preparation of 1,3-bis-BOC-1,3-diazacyclopentane (**21**) from *N,N*-bis-BOC-ethylenediamine (**23**), derived from either ethylenediamine (**24**) or imidazolidone (**25**), is illustrated in Figure 13. Di-*tert*-butyl dicarbonate [(BOC)$_2$O] reacted with **24** to yield **23** under standard conditions.[33] The reaction of **25** with (BOC)$_2$O, however, requires catalysis by 4-dimethylaminopyridine[34] to produce 1,3-bis-BOC-imidazolidone (**26**). Treatment of **26** with aqueous lithium hydroxide to provide **23** furnishes an example of a regioselective hydrolysis promoted by the BOC group on a *N,N'*-disubstituted urea. *N*-BOC derivatives of lactams have been similarly hydrolyzed to cleanly provide the corresponding acyclic BOC- amino acid derivatives.[34] The nearly quantitative formation of 1,3-bis-BOC-1,3-diazacyclopentane (**21**) from *N,N*-bis-BOC-ethylenediamine (**23**) and formaldehyde demonstrates that the bulky *tert*-butoxycarbonyl groups do not provide a steric barrier to ring formation. Treatment of **21** with nitric acid/acetic

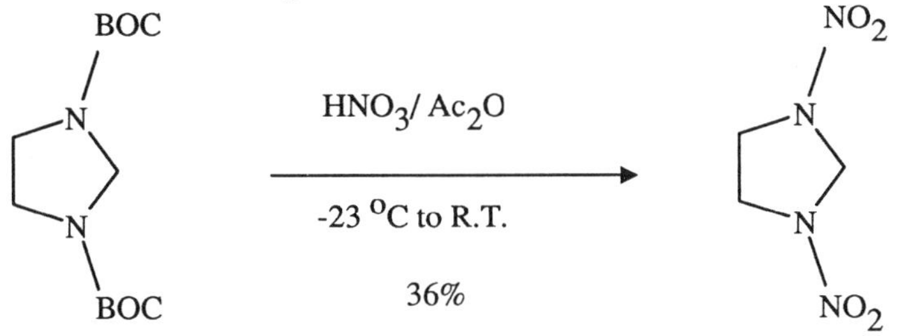

Figure 14. Nitrolysis of **21**.

anhydride or nitronium tetrafluoroborate gives **22** in moderate yield (Figure 14).

The nitrolysis of a *N*-BOC derivative of a primary amine was studied using *N,N'*-bis-BOC-ethylenediamine (**23**). When this substrate was treated with nitric acid/acetic anhydride, the BOC groups were surprisingly neither nitrolyzed nor cleaved. *N*-Nitration did occur, however, to give *N,N'*-bis-BOC-

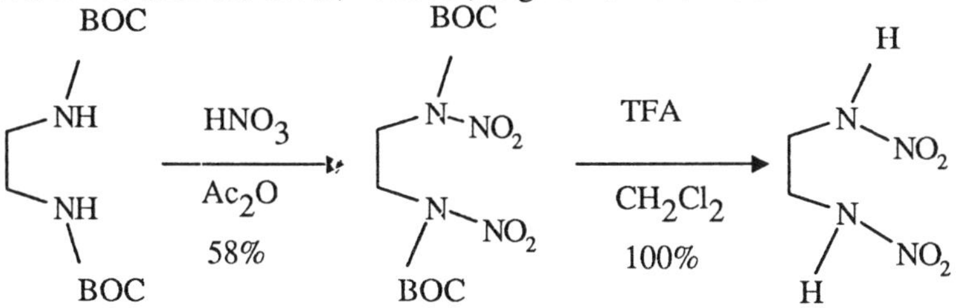

Figure 15. Synthesis of EDNA.

ethylenedinitramine (**27**) in 58% yield. Treatment of this product with trifluoroacetic acid-dichloromethane-1:1 effected rapid removal of the BOC groups to yield ethylenedinitramine (**28**, **EDNA**) in quantitative yield (Figure 15).

The reaction of hexamethylenetetramine (**29**) with (BOC)$_2$O provides 1,3,5-tris-BOC-hexahydro-s-triazine (**30**) as illustrated in Figure 16. The direct nitrolysis of **30** to RDX (**31**) proceeded in low yield (5%) using nitric acid/trifluoroacetic anhydride. In contrast, conversion of **30** to 1,3,5-triacetylhexahydro-s-triazine (**32**) (refluxing acetic anhydride-acetic acid-1:1, 5h.) followed by nitrolysis (nitric acid/trifluoroacetic anhydride) gave RDX in overall yields of greater than 70%. The direct conversion of **30** and related *N*-BOC derivatives to RDX will be examined using nitration reagents such as NO$_2$BF$_4$/CH$_3$CN and N$_2$O$_5$/CH$_2$Cl$_2$ at low temperature.

In conclusion, we have developed a mild nitrolysis reaction employing urethane substrates derived from *tert*-butanol (*N-tert*-butoxycarbonyl derivatives). The use of this newly developed chemistry to prepare new energetic materials is under investigation.

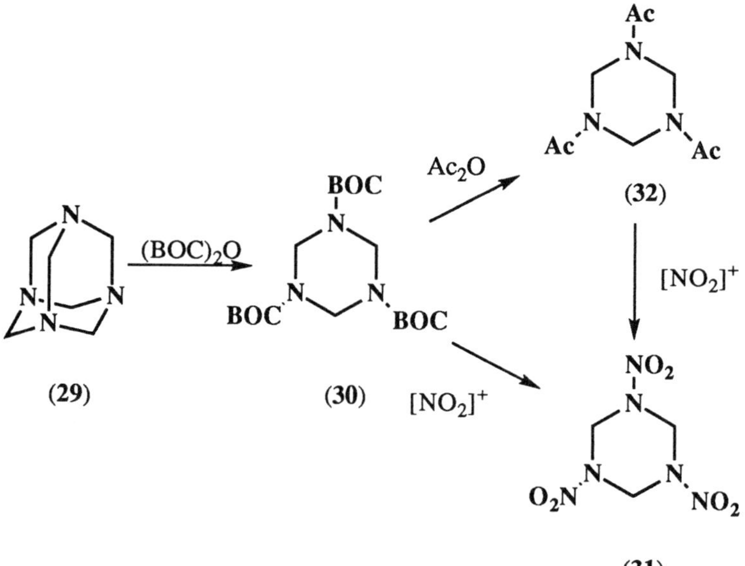

Figure 16. Synthesis of RDX from **30**.

Work performed under the auspices of the U.S. Department of Energy by the Lawrence Livermore National Laboratory under contract No. W-7405-ENG-48.

Literature Cited

1. *Energy and Technology Review*, Lear, R.D., McGuire, R., Eds.; Lawrence Livermore National Laboratory,;: Livermore, CA, January, 1988. Report No. UCRL-52000-88-1.
2. Gilbert, E.E., .Leccacorvi, J.R, Warmann, M. In *Industrial and Laboratory Nitrations,* Albright, L.F. , Hanson C., ED.; American Chemical Soc., Washington, D.C. 1976, 327-340.
3. Dampawan, P., Zajac, Jr., W. W., *Synthesis,* **1983**, 545

4. Bachmann,W.E. Horton, W.J., Jenner, E.L., MacNaughton, N.W., Maxwell, III, C.E. *J. Am. Chem. Soc.*, **1950**, 72, 3132,
5. Cichra, D.A. and Adolph, H., *J. Org. Chem.*, **1982**, 47, 2474.
6. Crivello, J. V. *J. Org. Chem.*, **1981**, 46, 3056).
7. a) McKay, A.F., Manchester, D.F. *J. Am. Chem. Soc.*, **1949**, 71, 1970; b) McKay, A.F., Wright, G.F. *J. Am. Chem. Soc.*, **1948**, 70, 3990; c) McKay, A.F., Bryce, J.R.G., Rivington, D.E. *Can. J. Chem.*, 29, **1951**, 382 d) McKay, A.F., Park, W.R.R., Viron, S.J. *J. Am. Chem. Soc.*, 72, **1950**, 3659 e) Brian, R.C., Lamberton, A.H. *J. Chem. Soc.*, **1949**, 1633 ; f) Desseigne, C. *Mem. Poudres*, **1948**, 30, 111 ; g) Caesar, G.V., Goldfrank, M. U.S. Pat. 2,400,288. *Chem. Abstr.*, B, 40, 4525.
8. Boileau, J., Emeury, J.M., Kehren, J.P.A.*Ger. Offen.*, 2,435,651, Feb. 6, 1975. *Encyclopedia of Explosives and Related Items*. Federoff, B.T., Ed., Picatinny Arsenal, Dover, NJ, 1960, Vol. 1, p. A65,.
9. Boileau, J., Wimmer, E., Gilardi, R., Stinecipher, M.M., Gallo, R., Pierrot, M. *Acta. Cryst.*, **1988**, C44, 696.
10. Wenjie, L., Guifen, H., Miaohong, C. In *Proc. Int. Symp. on Pyrotechn. and Expl.*, Jing D., Ed.; China Academic Publishers: Beijing, China, , 1987; pg 187.
11. X-ray crystallography performed by Richard Gilardi, Naval Research Laboratory, Washington D.C.
12. Cowperthwaite, M., and Zwisler, W.H. *TIGER Computer Program Documentation* SRI Publication No. Z106, Stanford Research Institute, Menlo Park, CA, 1973.
13. Flippen-Anderson, J.L., George, C., Gilardi, R. C. *Acta Cryst.*, **1990**, C46, 1122.
14. Duschinsky, R. Dolan, L.A. *J. Am. Chem. Soc.*, **1946**, 68, 2350. (and references cited within)
15. Luk'yanov, O.A.. Onishchenko, A.A Kosygin, V.S. Tartakovskii, V.A. *Izv. Akad. Nauk. SSSR, Ser. Khim (engl)*, **1979**, 1495.
16. A 20% N_2O_5/ HNO_3 solution is made by adding an excess amount of N_2O_4 to 100% HNO_3 (enough for ~40 % solution), cooling to 0 °C, and ozonizing until the solution is colorless. We have found that this consistently gives an 18-22 % solution of dinitrogen pentoxide.
17. George, C., Gilardi, R., Flippen-Anderson, J.L. *Acta Cryst.*, **1992**, C48, 1527.
18. Graindorge, H.R., Lescop, P.A., 209th American Chemical Society Meeting, Anaheim, CA, April, 1995.
19 Mitchell, A.R., Pagoria, P.F., Coon, C.L., Jessop, E.S., Poco, J.F., Tarver, C.M., Breithaupt, R.D., Moody, G.L., *Prop. Explos. Pyrotechn.*, **1995**, "Synthesis, Scale-up, and Characterization of K-6", Report No. UCRL-LR-109404 (1992), Lawrence Livermore National Laboratory, Livermore, CA, USA.
b) Gilardi, R., Flippen-Anderson, J.L., George, C. *Acta. Cryst.*, **1990**, C46, 706. c).
20. Petersen, H. *Synthesis*, **1973**, 267.
21. Koppes, W.M., Chaykovsky, M., Adolph, H. G., Gilardi, R., George C. *J. Org. Chem.*, **1987**, 52, 1113 and references cited therein.
22. Smart; G.N.R., Wright, G.F. *Can. Jour. Res.*, **1984**, 26B , 284.
23. A. H. Dinwoddie and G. Fort, Brit. Pat. 1,107,907, "Explosive Compound", 1966.
24. Coon, C.L., Blucher, W.G., Hill, M.E., *J.Org. Chem.*, **1973**, 38, 4243-48.
25. Rajendran, G., Santini, R.E., Van Etten, R.L., *J. Am. Chem. Soc.*, **1987**, 109, 4357-62.
26. Carpino, L. A. *Acc. Chem. Res.* **1973**, 6, 191-198.
27. Wünsch, E. *Methoden der Organischen Chemie (Houben-Weyl);* Georg Thieme Verlag: Stuttgart, 1974; Vol. 15/1, pp 117-129.

28. Green, T. W.; Wuts, P. G. M. *Protective Groups in Organic Synthesis,* 2nd ed.; Wiley: New York, NY, 1991; pp 327-330.
29 Losse, G.; Zeidler, D; Grieshaber, T. *Liebigs Ann. Chem.* **1968,** *715,* 196-203.
30. Olah, G. A.; Malhotra, R; Narang, S. C. *Nitration: Methods and Mechanisms,* VCH Publishers, Inc.: New York, NY, 1989; pp 278-286.
31. Goodman, L. *J. Am. Chem. Soc.* **1953,** *75,* 3019-3020.
32. Willer, R. *J. Org. Chem.* **1984,** *49,* 5147-5150.
33. Moroder, L.; Hallett, A.; Wünsch, E.; Keller, O.; Wersin, G. *Hoppe-Seyler's Z. Physiol. Chem.* **1976,** *357,* 1651-1653.
34. Flynn, D. L.; Zelle, R. E.; Grieco, P. A. *J. Org. Chem.*; **1983,** *48,* 2424-26.

RECEIVED September 26, 1995

Chapter 15

Flow Reactor Nitrations Using Dinitrogen Pentoxide

N. C. Paul

Defence Research Agency, Fort Halstead, Sevenoaks, Kent TN14 7BP, United Kingdom

The production of energetic compounds by nitration processes is potentially very hazardous. The use of a flow reactor not only minimises the amount of reacting species and product at any one time but also allows better reaction control. Because of the high reactivity of dinitrogen pentoxide such nitrations lend themselves very readily to flow reactor processes. Dinitrogen pentoxide solutions in an inert solvent can also show greater specificity than conventional nitrating agents where there are two or more potential reactive sites on the substrate molecule which may react at different rates. The scale-up of processes is facilitated by a flow reactor technique where slower competing reactions can be eliminated by immediate quenching of the reacted stream. The application of this technique to the production of nitrato oxetanes and nitrato oxiranes is presented.

Flow reactors are used in some chemical production processes (*1-3*) and possess many advantages over batch procedures. For a batch procedure the reactants are charged to the reaction vessel, well mixed and the reaction allowed to proceed for a specific time. Batch reactions are unsteady state operations where the composition changes with time although, at any one instant the composition in the reactor is uniform. In contrast, steady-state flow reactors are charged continuously and a continuous product stream is produced. There are two ideal types of flow reactor; Plug Flow and constant flow stirred tank reactors (CFSTR). Plug Flow is characterised by an orderly flow of the mixed reactants through the reactor with no back mixing and the residence time in the reactor is the same for all elements. In a CFSTR there is a constant feed, the tank contents are stirred and uniform and the exit stream has the same composition as the the liquid in the reactor. A production process employing a flow reactor allows a continuous production and the plant is generally much more compact.

Processes using Plug Flow reactors are, in general, intrinsically safer for two reasons. Firstly, there is a reduced inventory of potentially hazardous materials within the reactor and secondly, there is a greater degree of control of the reaction since the rate of reactants feed can be precisely regulated. This control, together with the low volume reaction zone means that exothermic reactions can be more easily contained. These factors are of special significance when hazardous materials, such as explosives, are concerned. In addition to these advantages, control of both reaction time and temperature can be employed to suppress slower secondary reactions leading to purer products.

Not all reaction systems are suitable for a flow reactor process and certain criteria should be met if such a process is to be considered. For example, if only a small quantity of a product is required then a flow system would often be inappropriate and where the necessary reaction times are long a flow system may not offer a significant advantage.

An ideal system for a flow reactor process is therefore one where large quantities are required and where the principal reaction is fast and any secondary reactions relatively slow. Two such ideal reaction systems are the nitration of 3-hydroxypropylene oxide (glycidol) and 3-hydroxymethyl-3-methyloxetane (HMMO) with dinitrogen pentoxide in an inert organic solvent. The required products from these substrates; glycidyl nitrate (GLYN) and 3-nitratomethyl-3-methyloxetane (NMMO) are essential precursors for the production of a new generation of energetic binders for explosive and propellant formulations (4,5). Initial investigations were carried out on a custom built laboratory scale tubular flow reactor.

Experimental Details

Flow Reactor Design. A simple, all glass jacketed tubular reactor was constructed as shown in Figure 1. To promote efficient mixing of the two reactant streams the reactor was packed with 2-3mm diameter glass beads and the total free volume was about 5cm^3. A stainless steel temperature probe was placed at the exit point to monitor the reaction exotherm.

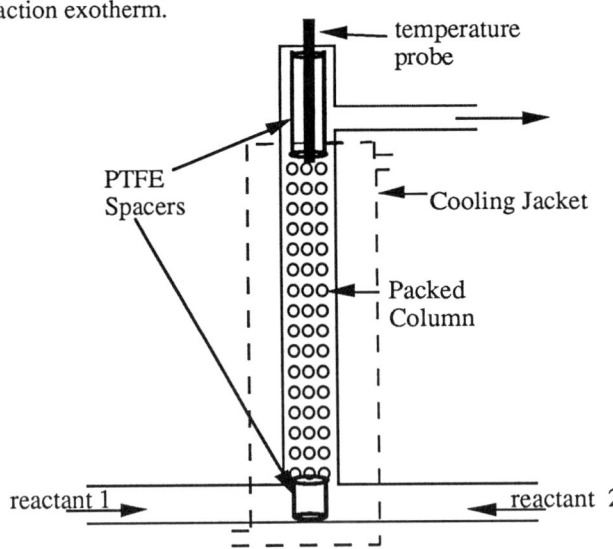

Figure 1. Co-current Laboratory Flow Reactor

Flow Nitration System. A schematic of the system is shown in Figure 2 and consists of two jacketed glass reservoir vessels holding the two reactants, held at sub-ambient temperature by a flow of cooling liquid from a refrigerated circulating bath. This same cooling system also supplied the flow reactor cooling jacket.

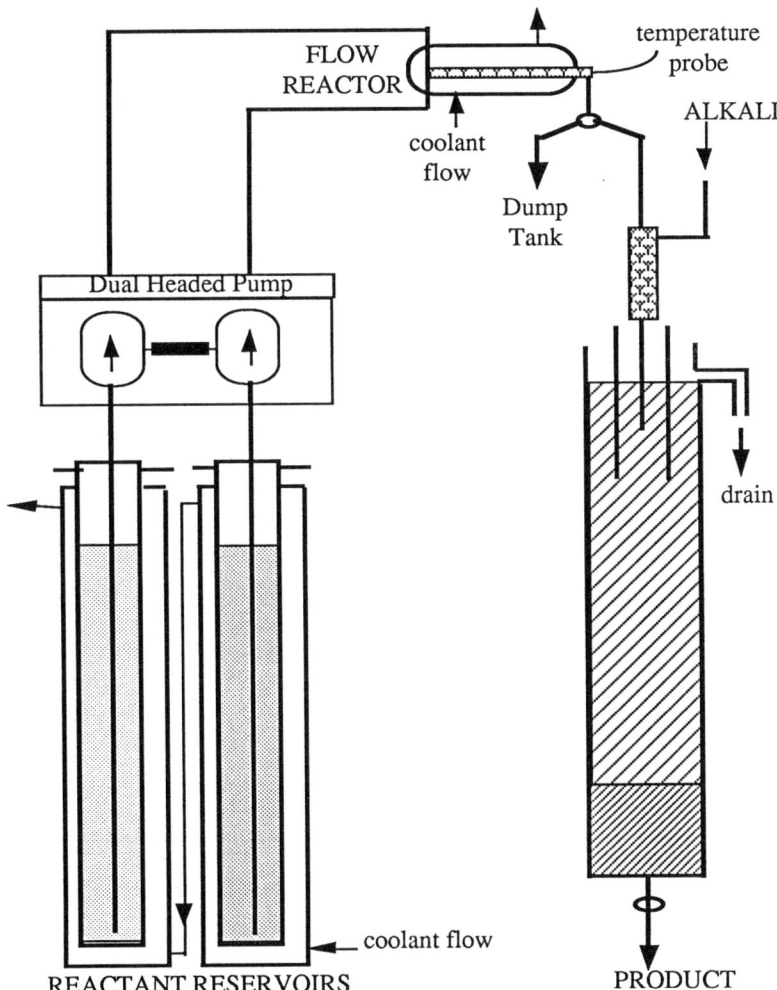

Figure 2. Schematic of Flow Nitration System.

The reactant reservoirs had additional ports to provide a dry nitrogen bleed to exclude the ingress of moisture. Reactants were pumped from the reservoirs using a dual barrel piston pump. For short production runs the product solution was run into a stirred tank of aqueous alkali solution to neutralise any excess dinitrogen pentoxide and the nitric acid formed in the reaction. The product remained in the lower organic solvent layer which was separated by decantation.

Dinitrogen Pentoxide Production. Initial production was by the gas phase oxidation of dinitrogen tetroxide using ozone with cold-trapping of the pure solid (6, 7). In this process a stream of gaseous dinitrogen tetroxide is mixed with a stream of ozone in oxygen generated by a commercial ozoniser. An excess of ozone is used to ensure complete oxidation to dinitrogen pentoxide which is then isolated as a white crystalline solid in a series of cold traps at -60 to -80°C. Dinitrogen pentoxide can be stored for long periods at -60°C without decomposition and solutions are prepared immediately before use. This operation is time consuming and limited in scale and later work was carried out using a 'solvent' process where the ozone oxidation was carried out in within a dichloromethane solvent. This process design has been reported elsewhere (8).

Dinitrogen pentoxide produced by the gas phase method is a very pure material and no analysis is necessary. A known weight is dissolved in a known volume of dry solvent just prior to nitration. The product obtained from the larger scale 'solvent' process can contain small amounts of both nitric acid and dinitrogen tetroxide and the composition of this medium was determined by laser raman spectroscopy using calibration standards constructed from synthetic mixtures of the three components.

Continuous Quench and Neutralisation. For extended runs the output from the flow reactor passed into a flow neutraliser, mixed with aqueous alkali and then passed to a continuous separator. The denser organic solvent, containing the product, was removed from the bottom of the separator whilst the aqueous wash solution overflowed at the top.

Operating Parameters and Procedure for NMMO and GLYN. A continuous dry air or nitrogen blanket was maintained over the reactant reservoirs and the cooling circulator switched on. The coolant temperature was typically set at -10°C. One reservoir was filled with a solution of dinitrogen pentoxide in dichloromethane and the other with a dichloromethane solution of either HMMO or Glycidol. The concentration of dinitrogen pentoxide was generally in the range of 10 to 15% w/v and the concentration of the substrate to be nitrated was adjusted such that, for equal volumes, a 5 to 10% excess of dinitrogen pentoxide was maintained. The aqueous alkali feed to the neutraliser was started and adjusted so that a twofold excess over the dinitrogen pentoxide rate was fed to the neutraliser. On starting the nitration the output from the reactor was fed via a three way valve to a dump tank and the reactor exit temperature monitored. With a feed rate for each reactant of about 50ml per minute the exit temperature rose to 15 to 20°C and when this temperature stabilised the product stream was fed to the neutraliser. After neutralisation with aqueous alkali the dichloromethane solution was separated in a continuous separator.

Work-up Procedure and Analysis. The dichloromethane solution was dried over anhydrous calcium chloride and filtered. The materials produced required no further purification or isolation and after concentration were used directly in polymerisation steps to produce energetic binders. For this next step in the production of energetic binders it was only necessary to concentrate the solutions to the required level by removal of the bulk of the solvent on a rotary evaporator. The concentrated solutions were then used directly in the polymerisation reaction. Samples of the product concentrates were analysed for purity and assayed by gas liquid chromatography.

Results and Discussion

Initial laboratory synthesis of NMMO used a batch procedure (9) in which the dinitrogen pentoxide was added to the HMMO; both in dichloromethane solvent. This procedure could not be scaled above about 1 mole due to the increasing level of secondary reactions. The nitration of the hydroxyl group was found to very exothermic and required slow addition of the nitrating agent to contain it. With larger scale operations the addition times were necessarily prolonged and secondary reactions took place. A study of the nitration of HMMO with dinitrogen pentoxide showed that the initial nitration of the hydroxyl group (Reaction [1]) is extremely fast, reaction taking place within seconds.

$$HMMO \xrightarrow{N_2O_5} NMMO + HNO_3 \quad [1]$$

With prolonged reaction times slower ring opening reactions can occur with both nitric acid produced and any excess dinitrogen pentoxide (Reactions [2], [3]).

$$NMMO \xrightarrow{N_2O_5} \text{Metriol trinitrate} \quad [2]$$

$$NMMO \xrightarrow{HNO_3} \text{Metriol dinitrate} \quad [3]$$

The main concern, in the batch reaction, is the ring opening by nitric acid, which is formed in stoichometric amounts from the hydroxyl nitration [1], to produce metriol dinitrate [2]. This may then further react to form metriol trinitrate [3]; which could also arise through a ring opening reaction with excess dinitrogen pentoxide. In addition to these secondary reactions there is also the possibility of oligomerisation by an acid catalysed ring opening mechanism.

The nitration of glycidol shows a similar behaviour but the secondary reactions become more prominent since the ring opening of oxiranes is more facile than with oxetanes. Two major by-products have been identified as glycerol 1,2-dinitrate, from the ring opening with nitric acid, and nitroglycerine, by ring opening with dinitrogen pentoxide (Reactions [4],[5],[6]). Nitroglycerine could also be formed by further nitration of the dinitrate.

$$\text{Glycidol} \xrightarrow{N_2O_5} \text{Glycidyl nitrate} + HNO_3 \quad [4]$$

$$\text{(Glycidyl nitrate)} \xrightarrow{HNO_3} \text{Glycerol 1,2 dinitrate} \quad [5]$$

$$\text{(Glycidyl nitrate)} \xrightarrow{N_2O_5} \text{Nitroglycerine} \quad [6]$$

Small production runs were first carried out, for each substrate, to ascertain the optimum conditions and subsequently, the scale of operations was increased to the capacity of the dinitrogen pentoxide production. A range of reactant flow rates and concentrations were employed and, provided the exotherm was sufficiently controlled, almost quantitative yields of high purity materials were obtained.

NMMO Production. The results for NMMO are summarised in Table I and show that the technique is capable of producing very high purity material on a routine basis. In the NMMO runs an impure product was obtained (Run 5 Table I) when a fault occurred in the cooling circulator. There was insufficient control of the reaction exotherm resulting in an 'overnitration' and some metriol trinitrate was produced through the secondary reactions described. Metriol trinitrate could arise from the ring opening of the oxetane ring in NMMO with the small excess of dinitrogen pentoxide or by ring opening with nitric acid followed by further nitration of of hydroxyl group formed. It is thought that the latter mechanism is the more likely.

This highlights the importance of monitoring and controlling the reaction exotherm.

Table I. Flow Reactor Nitration of HMMO with Dinitrogen Pentoxide
(§) 'overnitration' product formed ~ 3% metriol trinitrate

Run No.	Exotherm Temp. (°C)	REACTANT			PRODUCT		
		Scale (moles)	Concn (%w/v)	Flow rate (cm³/min)	Yield (g)	Theor. (%)	Purity (%)
1	12	0.95	4.9	50	139	99.4	99.6
2	14	4.9	10.0	40	719	99.7	99.5
3	16	4.2	9.4	50	610	99.6	99.6
4	14	4.0	9.9	60	560	99.6	99.8
5	29	15.1	4.0	50	2590	97	96 (§)
6	21	19.6	10.0	70	2876	99.8	99.9
7	22	36.1	14.0	60	5290	99.7	99.6
8	24	37.7	14.0	100	5500	99.1	99.6
9	24	37.7	14.0	150	5500	99.1	99.6
10	23	43.2	14.0	100	6100	96.3	99.6

GLYN Production. The results of a series of runs were similarly carried out using glycidol as the substrate and the results shown in Table II. Once again an incomplete control of the exotherm in one experiment (Run 14 Table II) resulted in the formation of some glycerol 1,2-dinitrate through ring opening of the GLYN product with nitric acid. No nitroglycerine was observed indicating that the ring opening reaction with nitric acid proceeds faster than that with dinitrogen pentoxide.

Table II. Flow Reactor Nitration of Glycidol with Dinitrogen Pentoxide
(§) Product contaminated with glycerol 1,2-dinitrate

Run No.	Exotherm Temp. (°C)	REACTANT			PRODUCT		
		Scale (moles)	Concn (%w/v)	Flow rate (cm³/min)	Yield (g)	Theor. (%)	Purity (%)
11	17	0.85	7.9	50	94	97	99.6
12	17	1.9	7.0	50	221	98	99.6
13	15	10.1	10.3	45	1162	98	99.6
14	26	24.6	10.0	60	2490	85	98 (§)
15	15	30.8	8.3	60	3500	95	99.9
16	19	31.8	12.0	60	3780	99	99.5
17	17	33.8	10.0	65	4010	99.8	99.6
18	18	36.8	10.0	60	4040	92	99.9
19	16	39.0	10.5	40	4640	99.8	99.9
20	16	40.5	10.0	40	4460	93	99.9

Purity of Dinitrogen Pentoxide. Dinitrogen pentoxide produced by the large scale solvent process was less pure than that obtained by the gas phase oxidation and trapping technique. Material from the solvent process contained a small amount of nitric acid and, in some cases, traces of dinitrogen tetroxide. The purity of the dinitrogen pentoxide nitrating agent was first thought to be of paramount importance but a comparison of results with material from the two production methods showed that the flow process described can tolerate moderate amounts of nitric acid and small traces of dinitrogen tetroxide. For NMMO, Runs 4-10 and for GLYN, Runs 13-20 (Table III) were carried out with dinitrogen pentoxide from the solvent process.

Table III. Effect of Purity of Dinitrogen Pentoxide on Yield and Purity of Products

	Run	Composition of nitrating agent in Dichloromethane Solvent (% w/w)			PRODUCT	
		N_2O_5	HNO_3	N_2O_4	Yield % theor.	Purity %
N M M O	1	12.2	-	-	99.4	99.6
	2	12.6	-	-	99.7	99.5
	3	14.7	-	-	99.6	99.6
	4	12.8	1.1	0.5	99.6	99.8
	5	13.7	0.8	0.6	97	96
	6	11.2	0.6	< 0.5	99.8	99.9
	7	12.8	1.3	< 0.5	99.7	99.6
	8	13.9	0.3	0.6	99.1	99.6
	9	15.7	1.3	0.6	99.1	99.6
	10	10.4	0.9	0.5	96.3	99.6
G L Y N	11	12.6	-	-	97	99.6
	12	13.5	-	-	98	99.6
	13	12.2	1.3	< 0.5	98	99.6
	14	13.6	1.9	< 0.5	85	98
	15	12.2	0.7	0.6	95	99.9
	16	15.2	0.9	< 0.5	99	99.5
	17	10.6	3.4	0.8	99.8	99.6
	18	10.9	0.4	0.5	92	99.9
	19	15.7	0.9	0.6	99.8	99.9
	20	14.8	1.2	0.7	93	99.9

In the presence of a small excess of dinitrogen pentoxide and, within the time scale of the flow reactor residence time, secondary reactions did not occur provided the reaction exotherm was contained. Even in the worst case noted (Run 17), where the dinitrogen pentoxide/nitric acid/dinitrogen tetroxide weight ratio was 3/1/0.1, high purity GLYN (99.6%) was obtained in an almost quantitative yield (99.8% of theoretical).

Conclusions

A flow reactor nitration system for the production of energetic compounds has been demonstrated to confer considerable advantages over a batch process. In the field of energetic materials the main attraction is the significant decrease in the operating hazards normally associated with this class of compounds.

The combination of a flow reactor and an automatic quench and neutralisation step allowed the suppression of secondary reactions that are evident in a batch process. The process is very amenable to scale-up and gives products in high yield and of exceptional purity. This latter aspect is of prime importance in the production of the target compounds reported here since the subsequent polymerisation step to produce an energetic polymer is very susceptible to impurities. The materials produced by this process required no further purification.

The process has been successfully applied to the production of NMMO and GLYN. These are necessary precursors for a range of novel energetic polymeric binders which have applications in the formulation of high energy propellant and explosive compositions that are less vulnerable to accidental stimuli.

Flow reactor nitrations are of general applicability to the safer production of energetic compounds.

Literature Cited

1. Raczynski, S.; Polish Patent 21648 (**1935**).
2. Simmons, W.H.; Forster, A., Bowden, R C., *Ind. Chemist.* **1948**, *24* , 530.
3. Nilssen, A., US Patent 2737522 (**1956**).
4. Colclough, M. E., *et al*; "Energetic Polymers as Binders in Composite Propellants and Explosives" *Polymers for Advanced Technologies* **1994**, *5*, 554-560.
5. Arber, A., *et al*; "Novel Energetic Monomers & Polymers prepared using Dinitrogen Pentoxide Chemistry" *Proc. 21st Annual Conference of I.C.T. on Technology of Polymer Compounds & Energetic Materials, 3-6 July 1990.* Fraunhofer Institut für Chemische Technologie, Karlsruhe, FRG.
6. Golding, P., Millar, R.W., Paul, N. C., Richards, D. H.; *Tetrahed. Letts.* **1988**, *29*, 2731-2734.
7. Harris, A. D., Trebellas, J. C., Jonassen, H. B.; *Inorg. Synth.* McGraw-Hill: NY, **1967**, Vol. 9; 83-88.
8. Bagg, G.E.G., Arber, A.; UK Patent 2252309 (**1994**).
9. Golding, P., Millar, R.W., Paul, N. C.; UK Patent 2239018 (**1992**).

© British Crown Copyright. DRA 1995

RECEIVED January 10, 1996

Chapter 16

Modeling To Gain Insight into Thermal Decomposition of Dinitrotoluene

J. L. Case, Richard V. C. Carr, and M. S. Simpson

Air Products and Chemicals, Inc., 7201 Hamilton Boulevard, Allentown, PA 18195

Modeling of the results obtained in a low thermal mass adiabatic bomb calorimeter provides insight into a multiple reaction thermal decomposition of dinitrotoluene (DNT). Comparisons of predicted self-heat rates and rates of pressure rise between a single and multiple reaction model with experimental data obtained in a bomb calorimeter were performed. The multiple reaction model predictions provided a significantly improved correspondence to the observed experimental data. In the proposed multiple reaction model DNT reacts with itself to form a nonvolatile intermediate at temperatures below 300°C. At higher temperatures the intermediate product undergoes rapid decomposition characterized by a high heat of reaction, large activation energy and a massive release of noncondensable gases. The proposed multiple reaction model is consistent with recent isothermal experiments, the analytical characterization of isolated intermediates and with the body of data reported in the literature.

Reaction and Engineering Models

A fundamental reaction model, which describes thermochemistry, stoichiometry and kinetics, is a powerful concept because the model can be validated in an experimental device and then applied to a commercial piece of equipment. For decomposition of DNT, calorimetry results were modeled to develop a fundamental reaction model. In contrast to a fundamental reaction model, an engineering model characterizes the environment in which the reaction is occurring and includes relationships describing the conservation of mass and energy, the influence of vapor-liquid equilibrium and the equilibration of pressure (for a closed system) or vent rate (open system). The strength of the engineering model is that the fundamental principals on which it is based can be applied to any environment provided reasonable assumptions are made in its development.

The combination of reaction and engineering models is essential not only to the validation of the fundamental reaction model, but also in applying that validated model to the design of commercial equipment. In the former case, the engineering model describes the results obtained from the experimental apparatus, while in the latter it is predicting behavior. The simplest forms for both models assumes that the reaction model may be described with a single reaction and the engineering model considers only an energy balance.

Standard Techniques for Adiabatic and Isothermal Calorimetry

Calorimetric data can frequently be used to determine kinetic constants which are used to test various reaction models (1). Often a reaction model can be validated using such an approach. Using data from isothermal experiments performed at different temperatures is one traditional approach. However, in many commercial applications, a decomposition reaction which leads to a thermal runaway is more accurately represented by an adiabatic process. Both approaches will yield the same kinetic constants if only a single reaction is responsible for heat generation and if other phenomena which influence the energy balance, such as vaporization, are suppressed.

Kinetic Constants from Isothermal Data. Kinetic constants may be obtained directly from experimental explosive critical temperatures determined at two different sample sizes and/or shapes. If an exothermically decomposing material is exposed to a constant boundary temperature, the sample may either experience a thermal runaway or achieve a steady-state temperature gradient which transfers the heat of decomposition to the surrounding constant temperature bath. Frank-Kamentski (2) developed an energy balance for this problem and termed the minimum surface temperature which results in a thermal runaway as the explosive critical temperature. Using a first order Arrhenius kinetic expression (rate of decomposition is directly proportional to the Arrhenius expression times the amount of unreacted material) he obtained an algebraic expression which relates the activation energy and frequency factor to the critical temperature, thermal conductivity, density, heat of reaction, geometrical shape factor and significant dimension (slab half thickness or inner radius). The experimental determination of two explosive critical temperatures combined with the simultaneous solution of the Frank-Kamentski algebraic expressions yields the following relationships for the activation energy and frequency factor:

$$E = \frac{R\ Tc1\ Tc2}{(Tc2-Tc1)} \ Ln \left\{ \frac{r1^2\ Tc2^2}{r2^2\ Tc1^2} \ \frac{Sh2}{Sh1} \right\} \quad (1)$$

$$A = \left\{ \frac{Lam\ Sh1\ R}{Ro\ DHr\ E} \right\} \left(\frac{Tc1^2}{r1^2} \right) EXP\left(E/(R\ Tc1)\right) \quad (2)$$

where:

A	=	frequency factor (inverse seconds)
E	=	activation energy (kcal/mole)
R	=	gas constant (kcal/mole °K)
Lam	=	thermal conductivity (cal/cm sec °K)
Ro	=	density (g/cc)
DHr	=	heat of decomposition (cal/g)
Tc	=	experimentally determined critical temperature (°K)
Sh	=	geometrical shape factor (0.88 for slab, 1.5 for cylinder)
r	=	significant dimension (slab half thickness or inner radius)

Experimental Explosive Critical Temperatures Determined for Two Sample Sizes. The standard Henkin-McGill (3) test procedure was applied to small samples of DNT shaped into a thin slab to determine the explosive critical temperature. The thin layer of sample is placed in a blasting cap which is then placed in an oil bath maintained at different temperatures. A thermal runaway is assumed to have occurred if the blasting cap ruptures. The lowest temperature which results in a blasting cap rupture is selected as the critical temperature. Because the sample is small rupture, should it occur, is nearly instantaneous. This standard test is quick and relatively inexpensive. The small sample size also results in relatively high explosive critical temperatures. A second group of experiments using larger samples of DNT involved placing one liter of material in a stainless steel cylinder which was then immersed in an oil bath maintained at constant temperature. This larger sample size results in lower explosive critical temperatures and requires a longer time for a thermal runaway to develop. Monitoring the sample's radial temperature distribution can provide some warning that a thermal runaway is beginning to develop. Zinn and Mader (4) have also provided some generic graphical solutions to the unsteady-state energy balance based on generalized numerical solutions. Pressure was monitored for the one liter cook-off tests and a nitrogen pad was applied to the ullage volume to suppress the vaporization of the DNT. Both the Henkin-McGill and, especially, the one liter cook-off tests should be performed by researchers who specialize in the testing of explosive materials. The isothermal DNT work was performed by the Center for Explosives Technology and Research in Socorro, New Mexico. The one liter cook-off tests were performed in a firing chamber which safely contained one of the experiments that developed into a thermal runaway and destroyed the testing apparatus.

Table I Presents the Results of the Isothermal Calorimetry. The experimentally determined explosive critical temperatures for the Henkin-McGill and the one liter cook-off experiments are presented in Table I. Shape factors, half-slab thickness or inner radius, transport and physical properties, heat of reaction and the values for the derived activation energy and frequency factor are also included. There is a large difference between the critical temperatures for the two sample sizes. This is desirable, if a single mechanism is capable of describing the decomposition over this broad range.

Table I. Results of Dinitrotoluene Isothermal Calorimetry

Property	Henkin-McGill	One Liter Cook-off
Tc, (°K)	370	198
Sh	0.88	2.5
r, (cm)	0.031	4.42
Lam, (cal/cm sec °K)	0.00029	0.00029
Ro, (g/cc)	1.2	1.2
DHr, (cal/g)	300	300
E, (kcal/mole)	33.25	
A, (inverse sec)	3.35E9	

However, with more complex chemistry, it is unlikely that one mechanism would dominate over a 200°C range of temperature. Clearly, this inherent difficulty with the isothermal technique can be overcome by selecting intermediate sample sizes which yield explosive critical temperatures that are closer to each other.

Kinetic Constants from Adiabatic Data. Kinetic constants may be obtained directly from adiabatic self-heat rates. For a single reaction mechanism, conversion is assumed to be complete and directly proportional to measured temperature. The following development assumes a first order Arrhenius kinetic expression for a comparison with the expressions developed for the isothermal treatment (the approach is applicable to any reaction order). The moles of DNT (N) at any given time is directly proportional to the measured temperature (T):

$$N = m T + b \ ; \ N=No \ @ \ T=To \text{ and } N=0 \ @ \ T=Tf \quad (3)$$

$$N' = - A \ EXP \ (-E/R/T) \ N \quad (4)$$

$$m = - No / (Tf - To) \ ; \ b = - m \ Tf \quad (5)$$

$$N = m (T-Tf) = - No (T-Tf)/(Tf-To) = No (Tf-T)/(Tf-To) \quad (6)$$

Equation (6) relates the decomposition conversion to the measured thermal data. The derivative of equation (6) with respect to time, N', may be equated with the first order Arrhenius rate expression as represented by equation (4):

$$N' = -No \ T'/(Tf-To) = -A \ EXP \ (-E/R/T) \ No \ (Tf-T)/(Tf-To) \quad (7)$$

$$T' = A \ EXP \ (-E/R/T) \ (Tf-T) \quad (8)$$

$$Ln \ [T'/(Tf-T)] = Ln \ (A) - E/R/T \quad (9)$$

(T') divided by the difference between the final and measured temperatures against inverse temperature will result in a straight line for a first order reaction. The frequency factor is equal to the inverse natural logarithm of the intercept and the activation energy is equal to the negative slope of the data multiplied by the gas constant. The left-hand side of equation (9) is equal to the natural logarithm of the rate constant and the data is typically reported in this fashion. The above determination of kinetic constants from measured self-heat rate data is the standard approach that is applied to adiabatic calorimeters (1). Equation (3) requires the instrument to be capable of measuring the final temperature corresponding to complete conversion. The phi factor reflects the extent to which the thermal mass of the test cell tempers the heat given off in the decomposition reaction. It is numerically equal to the sum of the test cell and sample thermal mass divided by the sample thermal mass. A high phi factor reduces the final temperature that the sample experiences, and may limit higher temperature chemistry. Modern adiabatic calorimeters surround the test cell with a radiant heater whose temperature is controlled to be equal to that being measured in the actual sample. The instruments ability to track the sample temperature and maintain adiabatic operation may also limit the temperature range where the above equations are applicable for analyzing experimental data.

Single Reaction Decomposition Model

A single step reaction for the thermal decomposition of DNT was proposed by C. Grelecki (5) and is shown below:

$$CH_3\text{-}C_6H_3\text{-}(NO_2)_2 \rightarrow 3\,C + 4\,CO + N_2 + 3\,H_2 \qquad (10)$$

Kinetic parameters were manipulated to provide the best fit between predicted and self-heat rates measured in the accelerated rate calorimeter (ARC). An approach, similar to the above standard technique, but consistent with the heat of reaction prescribed by the above reaction was used. This reaction model was then combined with various engineering models of commercial equipment and used in safety analyses. The engineering models predicted that DNT vaporization significantly tempers the decomposition reaction provided pressure is maintained at a low value. Recent advances in low thermal inertia adiabatic bomb calorimeters offered an opportunity to validate this fundamental DNT decomposition model.

New Advances in Adiabatic Calorimeters. The family of latest devices, originally developed by Fauske Associates (6) in cooperation with AICHE's Design Institute for Emergency Relief Design, offer good sensitivity at exotherm detection, excellent tracking capability to maintain adiabatic operation and can be operated in a semi-open configuration. The PHI-TEC II as manufactured by Hazard Evaluation Laboratory was selected to validate the single reaction DNT decomposition model. The PHI-TEC II as configured for this work is comprised of a thin wall test cell with a volume of 120 cubic centimeters, equipped with a large hole in its top (~ 1.5 cm in diameter) which is placed in a larger pressure vessel having a volume of four liters. Figure 1 presents a simplified schematic of the device. The test cell is completely surrounded by a radiant heater maintained at a temperature equal to that measured in the test cell liquid. The heater is capable of tracking liquid temperature, in other words

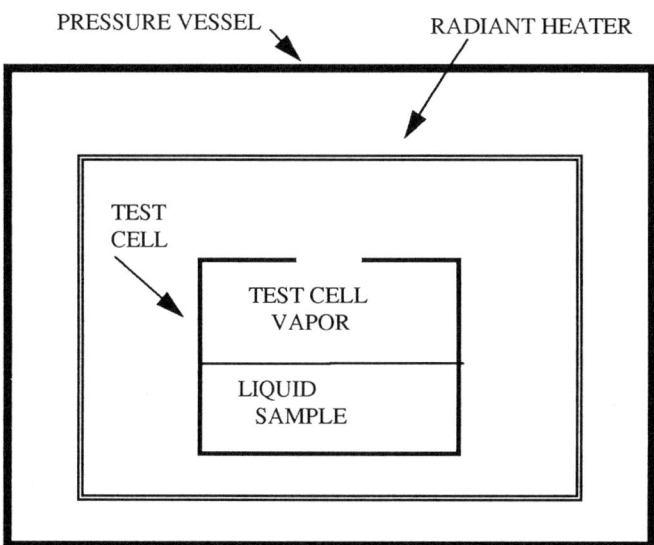

Figure 1. Simple Sketch of PHI-TEC II Adiabatic Calorimeter

keeping the test cell adiabatic from radiant heat transfer, up to a rate of ~ 1000 °C per minute. A phi factor of unity corresponds to an adiabatic path for the sample. The phi factor for the PHI TEC II device in the DNT work was 1.07, as opposed to a value of seven used in ARC experiments. This near unity value is possible because of the thin wall of the test cell and that pressure equilibration between the inside and outside of the test cell prevents rupture as the decomposition reaction progresses. The PHI-TEC II test cell is open to a four liter pressure vessel, which provides a 40 fold increase over the ullage volume of the test cell. This large volume reduces the increase in system pressure as noncondensable gas is generated by the reactions. This lower pressure allows for the possibility of vaporization to influence the self-heat rate and provides an important insight into the process chemistry. A magnetic stirrer provides homogeneity to the test cell liquid. The PHI-TEC II offers direct temperature measurement of the test cell liquid, essentially adiabatic operation that minimizes pressure increase and allows for vaporization to influence the self-heat rate. For the experiments reported in this work, exotherm detection occurred between 220 and 230 °C with the maximum self-heat rate being observed within ninety minutes.

Engineering Model of PHI-TEC II Device. An engineering model of the PHI-TEC II device describes its behavior using a set of differential equations with time as the independent variable. Huff (7) has presented a model to predict system behavior in an ARC. Typically, ten to twenty differential equations were used to model the PHI-TEC II adiabatic calorimeter and they include rate expressions for each reaction, mole balances for each chemical specie, a mathematical statement of vapor-liquid equilibrium for each volatile specie, pressure equilibration between the test cell and pressure vessel and a global energy balance. The key assumption in the device engineering model is that while the pressure in the test cell and pressure vessel are equal, only the vapor in the test cell is in equilibrium with the test cell liquid. The device engineering models used in this work are obviously more complicated than the standard approaches described previously which extracted kinetic constants from experimental data using a only a global energy balance.

Comparison of Model Prediction and Experimental Data for the Single Decomposition Reaction. Figure 2 presents actual experimental self-heat rate (pure 2,4-DNT) and pressure versus inverse temperature for the PHI-TEC II experimental data compared to that predicted using the single reaction decomposition model in combination with an engineering model of the PHI-TEC II device. Self-heat rates are represented with the logarithmic scale on the left-handed axis, and range six orders of magnitude from 0.01 to 10000 C/min. The experimental self-heat rate data exhibit nearly linear behavior during the initial phase of the exotherm. The slope of the line is related to the global activation energy. In the range of -1.65 to -1.6 °K^{-1} (~330 °C) the experimental self-heat rate exhibits a double inflection, actually dropping off before steeply rising to values in excess of the device's capability of keeping its test cell adiabatic. The model predicted self-heat rate is in excellent agreement with experimental data during the initial phase of the exotherm. However, in the region where the experimental data exhibits a double inflection, the model predicts a single maximum point. At this maximum point, the model is predicting that the heat required to boil DNT is matched by the heat generated by the decomposition reaction, and the temperature stabilizes at 335 °C, until the total mass of dinitrotoluene is depleted by both decomposition and vaporization. This phenomena is commonly referred to, within the specialized field of over pressure protection, as exotherm self-tempering due to reactant vaporization. Comparing model predicted and experimental self-heat rates suggest that while vaporization does provide significant tempering, some of the DNT has been converted to a relatively nonvolatile material which rapidly decomposes before it has the opportunity to vaporize.

Comparison of model versus experimental pressure provides additional insight. Pressure is represented with the use of the linear right-handed axis. There is only a minimal increase in pressure during the initial phase of the exotherm. In fact, the experimental data reveals that the real increase in pressure occurs after the double inflection, when the experimental apparatus is no longer adiabatic. A comparison of model prediction and experimental data indicates significant differences even in the early phase of the exotherm. The evidence is compelling in that the reaction or reactions which are responsible for the self-heat rates in the early, or low temperature, induction phase of the exotherm cannot be the direct single-step decomposition previously shown, because significantly less than the predicted amount of noncondensable gas is observed. This conclusion is very consistent with the isothermal one liter cook-off experiments to determine explosive critical temperatures (described previously). These experiments, which had very small ullage volumes, did not generate noncondensable gas at a rate consistent with that predicted by the single reaction mechanism.

Actual and Representative Chemistry

If the single decomposition reaction is inconsistent with experimental results, what is the real chemistry? T. Brill (8) has published an extensive review of the kinetics and mechanisms for the thermal decomposition of nitroaromatic compounds. He proposes three reaction regimes: (1) an induction phase, (2) an autocatalyzed acceleratory phase and (3) a decay phase involving ring break-up and the generation of stable noncondensable gas. In his own words:

> " ... many products are possible because all of the oxidized forms of $-CH_3$ and reduced forms of $-NO_2$ are present and each displays its own intermolcular and intramolecular chemistry ... anthranil, alcohol, aldehyde, carboxylic acid, nitroso, oxime, nitrile, nitrone, nitroxide, azo, and azoxy groups form. Disruption of the aromatic ring also occurs, leading to gas products including NO, CO, CO_2, N_2 and H_2O. An insoluble high molecular weight 'explosive coke' having the elemental composition of $C_6H_3N_2O_3$ forms from TNT."

DNT thermal decomposition is overwhelmingly complex and not particularly amenable to simple reaction modeling. For a model system, representative of the actual chemistry, one might propose that DNT reacts with itself to form nitroanthranil and water. Another molecule of DNT reacts very quickly with the nitroanthranil to form an azo-dimer and a second molecule of water. The dimerization is assumed to be the induction reaction which produces a nonvolatile, yet still potentially reactive heavy material. The model is now comprised of three reactions which include a single induction reaction, dimerization of DNT, and two decomposition reactions involving a volatile (DNT) and a nonvolatile (azo-dimer). Of course in the real chemistry, many other intermediates are formed, which upon further reaction with themselves and DNT, form the 'explosive coke' that Brill refers in the above quotation and is based on the findings of Dacons (9) who isolated intermediates during the thermal decomposition of TNT. Dacons did find small amounts of dinitroanthranil and Tsang (10) isolated anthranil during the decomposition of 2-nitrotoluene using the techniques of shock tube analysis. The following two reactions are combined with equation (10) to form the multiple reaction model:

$$2\ CH_3\text{-}C_6H_3\text{-}(NO_2)_2 \rightarrow C_{14}H_8N_4O_6 + 2\ H_2O \qquad (11)$$

$$C_{14}H_8N_4O_6 \rightarrow 8\ C + 6\ CO + 4\ H_2 + 2\ N_2 \qquad (12)$$

The model assumes that the azo-dimer decomposes to elemental carbon and noncondensable gas.

Isolating Intermediates During the Decomposition of DNT. The multiple reaction model represents the nonvolatile induction reaction product as an azo-dimer. It is an oligomer of nitroanthranil and can be considered as a dehydrated form of DNT. To compare azo-dimer with an analysis of real material requires interrupting the adiabatic exotherm in order to obtain a sample. To isolate a sufficient amount of material for analysis, the DNT must be maintained at a fixed temperature for extended periods of time, as was practiced by Dacons in his work with TNT. Of course, the longer the material is held at this fixed temperature, the more likely the composition will differ from material that is truly on an adiabatic temperature-time path. Two samples of concentrated explosive coke were collected by interrupting a 1.2 gram sample of DNT in the ARC apparatus maintained at 245°C after 30 and 120 minutes respectively. The 30 minute exposure resulted in a 15% conversion by weight to explosive coke. The 120 minute exposure resulted in an 85% conversion. Material which was not converted to explosive coke, was assumed to be either unreacted DNT or noncondensable gas. Numerous analytical tests were performed on the explosive coke and a portion of the material was returned to the ARC to complete the decomposition process. This fully decomposed explosive coke was termed exhausted coke.

Analysis of Isolated Explosive Coke. Table II presents the elemental analysis, on a mole basis, of DNT, nitroanthranil, azo-dimer, the laboratory explosive cokes, exhausted coke, solid residue collected form the PHI-TEC II experiment and the modeled reaction product, elemental carbon. As stated earlier, nitroanthranil and the

Table II. Elemental Analysis of Collected Samples

	C	H	O	N
DNT	7	6	4	2
NA and Azo-dimer	7	4	3	2
30 minute explosive coke	7	3.7	3.3	1.6
120 minute explosive coke	7	4.4	2.5	1.7
exhausted coke	7	4.3	1.5	1.5
PHI-TEC residue	7	2.0	0.7	0.7
model residue	7	0	0	0

azo-dimer are dehydrated forms of DNT and show the stoichiometric loss of two hydrogen and one oxygen in the table. The explosive cokes are similar to the model dimer, but do reflect the loss of some nitrogen as well. The exhausted coke reflects a further loss of nitrogen and oxygen. The PHI-TEC residue reflects the greatest loss in hydrogen, oxygen and nitrogen. The trend in the elemental analysis indicates higher carbon content in the PHI-TEC residue. This material has experienced the most adiabatic path, at least up to the tracking limitation of the device. The model assumes that the final form of the nonvolatile residue is elemental carbon and is, at least, consistent with the trend in real material.

FT-IR analysis indicates that the functional groups NO_2, $ArCO_2H$, $ArOH$ and $CO_3^=$ are present in the 120 minute explosive coke. The exhausted coke displays all of them but NO_2. The results of the solids probe mass spectroscopy indicate that many fragments disassociated from the 120 minute explosive coke have molecular weights higher than DNT. TGA indicates no significant weight loss of the 120 minute explosive coke until 300 C. DSC indicates that, above this temperature, an exothermic reaction occurs yielding 625 cal/gram. In all of the analyses of concentrated explosive coke, the results support an aromatic material with a molecular weight greater than DNT which retains some nitro group functionality.

Thermochemistry of DNT Decomposition. Table III provides the heats of formation (in kcal/mole) for DNT, nitroanthranil, azo-dimer and a collection of different possible final products grouped to reflect their bonding to oxygen. An exothermic reaction requires a negative heat of reaction (DHR) which is equal to the difference between the sum of the heats of formation of the products and reactants. Products with larger negative values for their heats of formation (relative to the reactants) result in larger exothermic heats of reaction. Given the relative values

Table III Thermochemistry of DNT Decomposition

$DHR = DH_f^{prod} - DH_f^{reac}$		exothermic:	DH_f^{prod}: large negative DH_f^{reac}: large positive	
	comp:		DH_f:	
	DNT		-9.96	
	Nitroanthranil		-8.3	
	Azo Dimer		-98.6	
comp:	DH_f:		comp:	DH_f:
Nitrogen:			Carbon/Hydrogen:	
NH_3	-11		CH_3COOH	-105
N_2	0		HCOOH	-87
NO_2	8		CH_3OH	-48
N_2O	19.6		CH_2O	-29
NO	21.6		CH_4	-18
Carbon:			Hydrogen:	
CO_2	-94.5		H_2O	-57.8
CO	-26.4		H_2	0
C	0			

between DNT and the nitrogen group of products, it is apparent that an exothermic decomposition of DNT requires the product oxygen to be bonded to either hydrogen or carbon, and not to nitrogen. A similar argument applies for an exothermic

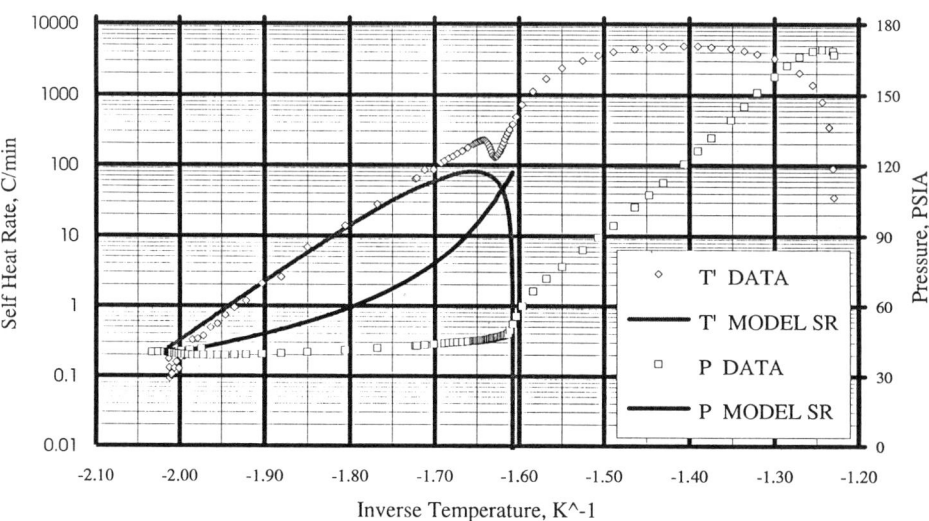

Figure 2. Comparison of PHI-TEC II Experimental Data (Pure 2,4-DNT) and Single Reaction Model Prediction

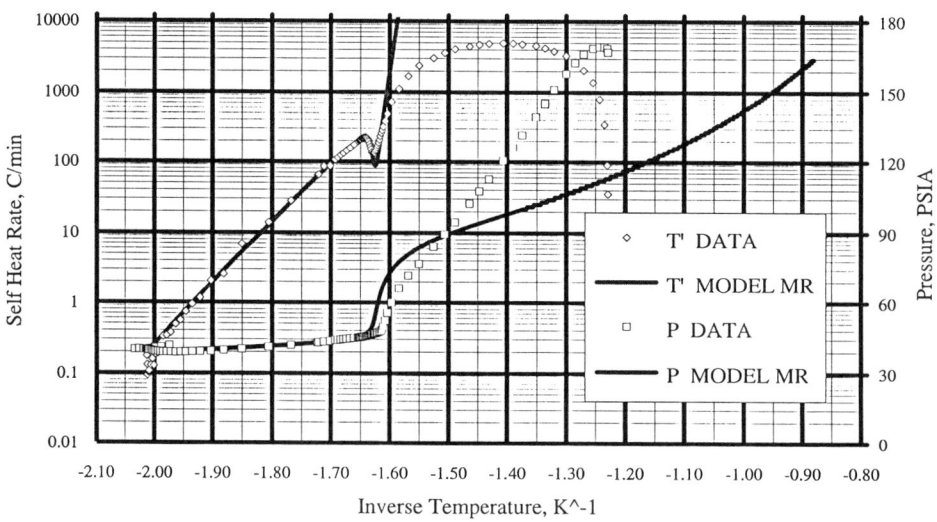

Figure 3. Comparison of PHI-TEC II Experimental Data (Pure 2,4-DNT) and Multiple Reaction Model Prediction

induction reaction. The formation of nitroanthranil and water from DNT is exothermic because of the hydrogen-oxygen bond in water. The loss of nitro functionality (oxygen bonded to nitrogen) in the exhausted coke is very consistent with the corresponding loss of exothermic reactivity.

Comparison of Prediction and Experimental Data for the Multiple Decomposition Reaction Model. The three reaction model is, obviously, a simplification of the real chemistry. The use of a simple model to represent such a complex system has merit only if it can reproduce the experimental behavior in terms of what is actually measured, pressure and temperature. Figure 3 is a replot the PHI-TEC II experimental pressure and self-heat rate and compares this data with the predicted behavior of the multiple decomposition reaction model. The simple three reaction model, comprised of one induction and two decomposition reactions, provides a remarkable improvement between model predicted and experimentally measured self-heat rate. Deviation between model prediction and experimental data at the inverse temperature of -1.60 °K^{-1} and greater is, again, explained by the inability of the calorimeter to keep the test cell adiabatic with its radiant heaters. The model is perfectly adiabatic through out the exotherm. Figure 3 reflects a similar improvement between model predicted and experimental pressure. Once the reaction model parameters have been adjusted to obtain a suitable fit, the engineering model can be used to analyze the predicted behavior of those things which are not experimentally measured. The model predicted behavior indicates that the tempering of the induction exotherm is a result of DNT vaporization. However, DNT vaporization is strongly linked to the generation of noncondensable gas produced by the decomposition reactions. Table IV provides reaction model parameters for the predicted behavior of pure 2,4-DNT as shown in Figure 3. The mixed DNT

Table IV DNT Multiple Decomposition Reaction Model Parameters

	Pure 2,4 DNT	Mixed DNT
Frequency Factor (1/sec):		
Azo-dimer Formation	0.25E14	0.24E15
Azo-dimer Decomposition	0.30E70	0.70E69
DNT Decomposition	0.30E70	0.70E69
Activation Energy (cal/mole):		
Azo-dimer Formation	42000	45000
Azo-dimer Decomposition	200000	200000
DNT Decomposition	200000	200000
Heat of Reaction (-):		
Azo-dimer Formation	0.5	0.425
Azo-dimer Decomposition	1.5	1.5
DNT Decomposition	1.0	1.0
Mole Carbon and Noncondensable Gas per Mole Reactant (-):		
Azo-dimer Formation	0/0.5	0/1
Azo-dimer Decomposition	6/14	6/10
DNT Decomposition	3/8	3/8

parameters correspond to composition of 80% 2,4 and 20% 2,6 DNT (based on comparisons with additional experiment data). The heats of reaction are represented as multipliers based on the heat of reaction for the single decomposition reaction (equation(10)). Deviation of the stoichiometric and thermochemical model parameters from those values corresponding to the representative chemistry provided a more precise fit between model prediction and experimental data. These deviations seemed appropriate, since in the real system there are many more reactions taking place than the three included in the representative model.

Summary

A three reaction model is proposed for the characterization of the thermal decomposition of DNT. This multiple reaction model is comprised of a single exothermic induction reaction which generates a nonvolatile intermediate with minimal noncondensable gas generation. This intermediate and DNT can both undergo direct exothermic decomposition reactions yielding noncondensable gas and elemental carbon. For modeling purposes an azo-dimer of nitroanthranil or dehydrated DNT is used to represent the intermediate product which results from the induction reaction. A sample of actual intermediate material has been isolated. Its analytical characterization, in general, supports the premise that it is nonvolatile and capable of exothermic decomposition from the condensed phase. While the three reaction model is a simple representation of the real chemistry, it provides a closer correspondence to experimental data than that which is obtained with a single decomposition reaction model. This experimental data is time dependent pressure and temperature, as obtained in a low thermal inertia adiabatic bomb calorimeter.

Acknowledgments:

The authors gratefully acknowledge the experimental work performed by the Center for Explosives and Technology Research and Hazard Evaluation Laboratory, Ltd.

Literature Cited:

1. Townsend, D. I. *Thermochim Acta* 1980, 37, 1.
2. Frank-Kamenetskii, D. A. *Acta Physicochem. USSR* 1939, 10, 365.
3. Henkin, H.; McGill, R. *Ind. Engr. Chem.* 1952, 44, 1391.
4. Zinn, J.; Mader, C. L. *J. Appl. Phys.* 1960, 31,323.
5. Grelecki, C;*Fundamentals of Fire and Explosion Hazards Evaluation*; AICHE Today Series; AICHE:New York, NY, 1976, pp. A10-A11.
6. Leung, J. C.; Fauske H. K.; Fisher, H. G. *Thermochim Acta* 1986, 104, 13.
7. Huff, J. E. *Plant/Operations Progress 1984, 3, 50.*
8. Brill, T. B.; James, K. J. *Chem. Rev.* 1993, 93, 2667.
9. Dacons, J. C.; Adolph, H. G.; Kamlet, M. J. *J. Phys. Chem.* 1970,74,3035.
10. Tsang, W.;*Thermal Stability Characteristics of Nitroaromatic Compounds*; U.S. Army Research Office:Research Triangle Park, NC, 1986,pp 23-31.

RECEIVED October 24, 1995

Chapter 17

Removal and Destruction of Tetranitromethane from Nitric Acid

T. L. Guggenheim[1], C. M. Evans[2], R. R. Odle[1], S. M. Fukuyama[3], and G. L. Warner[4]

[1]GE Plastics, 1 Lexan Lane, Mount Vernon, IN 47620
[2]Chemetics International Company Ltd., 1818 Cornwall Avenue, Vancouver, British Columbia V6J 1C7, Canada
[3]GE Plastics, Parkersburg, WV 26102
[4]GE Corporate Development and Research Center, Schenectady, NY 12301

Tetranitromethane (**TNM**) is a common by-product resulting from the nitration of organic substrates. Understanding the genesis and fate of TNM in nitration processes is paramount to safe plant operation since mixtures of TNM and certain organic materials can be powerful explosives. Nitration processes involving recycling nitric acid and/or sulfuric acid can be particularly challenging since TNM can build up in these recycle acid streams to the point of becoming a significant hazard. Methods to remove TNM from nitric acid and sulfuric acid are not well known.

This paper discusses methods to concentrate and destroy TNM in both nitric only and mixed acid nitrations (nitric acid/sulfuric acid). A generic practical process to remove TNM from a nitric acid recycle stream is delineated. The process can be applied to mixed acid systems as well.

Tetranitromethane (**TNM**, structure 1), bp 126°C, presents a myriad of challenges to chemists and engineers when present in a manufacturing process. It is very soluble in

$$O_2N-C(NO_2)_3$$

1

strong nitric acid (>30% by weight in 99% nitric acid) but not in 40% nitric acid (~0.5% by weight), or in 85% sulfuric acid (~0.8% by weight). It is toxic material (*1*). TNM is also an oxidizer that forms powerful explosives when mixed with combustible organic materials such as benzene, nitrobenzene, or toluene (*2*). TNM is a common by-product of nitrations. For example, TNM is formed in the mixed acid nitration of toluene to produce trinitrotoluene (*3*). The presence of TNM has been held responsible for explosions in trinitrotoluene manufacturing facilities. It has been postulated that mixtures of toluene and TNM, which are as sensitive as nitroglycerin, have condensed in off gas lines and detonated (4). TNM by itself was not detonated even when using 10 grams of a high explosive as a detonator (2).

It is crucial to know if TNM is present in any process stream when designing a nitration facility. The genesis and fate of TNM must be well understood within the system in order to design and operate a safe process. Analysis of TNM in nitric acid or sulfuric acid can be performed by high pressure liquid chromatographic techniques.

Most nitration facilities require recycling of acids (sulfuric and nitric acid) to provide for an economical and environmentally sound process (5). The possibility of TNM concentrating in an acid recycle stream, should it be present in a nitration process, is real if a purge point or a destruction point for TNM does not exist. This paper reviews processes that involve tetranitromethane, reviews known methods for destroying tetranitromethane, and finally details an economically safe method to remove TNM from recycle nitric acid streams using principles practiced in mixed acid nitrations and nitric acid concentration processes.

Cases Involving Tetranitromethane

As already noted, the mixed acid nitration of toluene to produce trinitrotoluene affords a small amount of TNM (3). The TNM is probably removed from reaction mixtures by sparging with air or nitrogen, or it simply volatilizes at elevated temperatures. The resulting gas stream may be scrubbed with water or dilute nitric acid to recover oxides of nitrogen (NOx) as nitric acid, the TNM being destroyed in the weak nitric acid (*vide infra*). The gas stream may also be scrubbed with excess aqueous sodium hydroxide in an appropriate unit operation, the caustic converting the TNM to the sodium salt of nitroform (2) and subsequently to sodium carbonate and sodium nitrite, equation 1 (6,7,8).

$$C(NO_2)_4 \xrightarrow[H_2O]{ex.\ NaOH} (NO_2)_3CNa \longrightarrow Na_2CO_3 + NaNO_2 \quad (equ.\ 1)$$
$$\underline{1} \qquad\qquad\qquad\qquad \underline{2}$$

Two Japanese patents describe the removal of TNM from the mixed acid nitration of anthraquinones (9,10). Basically, the patents teach that a mixed acid reaction containing nitrated anthraquinones can be sparged with air to remove a gas stream composed of nitric acid, oxides of nitrogen and TNM. The gas stream is then scrubbed with dilute sodium hydroxide. For example, a continuous process was shown wherein a spent sulfuric acid stream, composed of 409 kg/h of nitric acid, 372 kg/h of sulfuric acid, 37.2 kg/h of water, was combined with 110.4 kg/h of recovered nitric acid, 16.1 kg/h of 99% nitric acid , and 20.8 kg/h of anthraquinone. The resulting slurry was heated at 80°C for 0.5 h and 60°C for 3h. The slurry was sparged with 1.3 m3/h of air at 60-80°C to give a mixed gas stream at 3.3 m3/h containing nitric acid, TNM, oxides of nitrogen, and air. The reaction mixture was filtered to recover 1,5-dinitroanthraquinone. The waste gas stream was scrubbed with caustic to destroy the TNM. *The removal of TNM from a mixed acid system via gas sparging hinges on the high vapor pressure of TNM in mixed acid systems.*

The nitration of benzene or nitrobenzene to form *m*-dinitrobenzene is typically carried out in a mixed acid system (*11*). We have studied the nitration of benzene in an all nitric acid process. For example the nitration of benzene (one part by weight) in 99% nitric acid (10 parts by weight) at 0-60°C afforded 82% *m*-dinitrobenzene, 8.9% o-dinitrobenzene, 1.7% p-dinitrobenzene, and traces of tetranitromethane. The isolation of m-dinitrobenzene from strong nitric acid by the distillative removal of nitric acid followed by crystallization of the product from the still bottoms has been described (*12*). If such a process were practiced, then the TNM entering a distillation column would necessarily split between the nitric acid overheads stream and the column bottoms stream. Our studies in related systems have shown that most of the TNM goes overhead. *The process described above for the removal of TNM from mixed acid nitrations is not applicable to an all nitric nitration as we have shown that sparging nitric acid containing TNM does not effectively remove it (vide infra).*

An all nitric acid process to make 4-nitro-N-methylphthalimide (4) from N-methylphthalimide (3) has been described, equation 2 (*13*). The reaction is typically run

<chemical structure: N-methylphthalimide (3) + 99% HNO₃ at 50°C → 4-nitro-N-methylphthalimide (4)> (equ. 2)

by slowly dissolving N-methylphthalimide (1 part by weight) in 99% nitric acid (12 parts by weight) and heating at 50°C for 3-6h. TNM is a byproduct of this nitration. The product (4) can be isolated in several ways. The reaction mixture can be diluted with water to precipitate the product, which can then be collected and washed via filtration technology. The filtrate (weak nitric acid) contains the TNM generated in the process. For such a process to be feasible the nitric acid must be recovered, the TNM removed from the recovered nitric acid, and the acid reconcentrated to 99%, not necessarily in that order.

The above nitration reaction mixture can also be fed to an evaporative unit wherein most of the strong nitric acid is removed as a vapor stream and the bottoms stream contacted with water to precipitate the product thus producing a weak nitric acid stream, similar to the process described in reference 12. The nitric acid streams must be continuously reconcentrated and recycled to provide for an economical process. If the TNM is not removed/destroyed in one of these streams, then the TNM can build up to levels where it can exceed the solubility limit and separate as a discrete phase.

Phase separation of TNM can be very hazardous depending on the design of the process. Undetected pools of TNM can collect in low points in the process. Phase separated TNM can extract organics from a process stream. Conversely, phase separated organics present in the process can extract TNM from a process stream. Other scenarios can lead to detonatable mixtures of TNM and oxidizable organics concentrating in a process. The energy content of mixtures of oxidizer (TNM, nitric acid), fuel (organic material), and diluents can be estimated (*14*). Impurities such as TNM or NOx, can sensitize detonatable mixtures. All potentially dangerous plant streams must be tested for explosion and fire hazards under the direction of qualified investigators. Mixtures of nitric acid, containing dissolved TNM, and organics can be safe as long as the oxygen balanced scenario has not been compromised (*14*). It is essential that the TNM concentration in any stream in a nitration process be controlled to well below its solubility limit (*vide infra*).

TNM present in nitric acid only nitrations is more difficult to deal with than in mixed acid systems since simple sparging does not remove TNM from nitric acid that is 40-99% in strength. The rest of this paper will deal with the removal of TNM from nitric acid streams, as it applies to nitric only nitrations that involve acid recycle streams.

Known Methods to Remove TNM from Nitric Acid Streams

Consider in more detail a system similar to that described for the nitric acid only nitration of N-methylphthalimide or benzene to illustrate methods to remove TNM from nitric acid streams. The nitration of these substrates requires the use of 10 parts >90% nitric acid to one part substrate to effect rapid conversion to the desired product. The substrate and nitric acid are fed to a nitrator, either in a batch mode or continuous mode, and then to an evaporative unit where the bulk of the strong nitric acid is flashed off, strong nitric acid being >85%, Figure 1.

The effluent stream from the evaporator is brought into contact with water to precipitate (or phase separate) the product, the resulting weak nitric acid (weak by virtue of being diluted with water) and product are separated, the weak nitric acid purified/recovered in an appropriate unit operation (organic destruction unit or concentrator), and the product purified by one of any number of conventional means. The TNM generated in the nitration splits between the two nitric acid recycle streams, the majority going with the strong nitric acid recycle stream.

The strong nitric stream and the weak nitric stream are combined and reconcentrated via distillation to provide >90% nitric acid. Alternatively, the reconcentration can be done by distilling the nitric acid in the presence of strong sulfuric acid (~85%), in turn producing weak sulfuric acid. Distillation of the water from the weak sulfuric acid provides 85% sulfuric acid that can be recycled to the nitric acid concentrator. Such a system is known as a NAC SAC, nitric acid concentration, sulfuric acid concentration (*5*). Any TNM entering a NAC SAC codistills with the strong nitric acid overheads product, none being destroyed in the system. It has been shown that TNM is stable in 70% nitric acid at 80°C, and in mixtures of 70% nitric/85% sulfuric acid at 100°C for days. TNM is also stable in 99% nitric acid at 80°C (*15*). The NAC SAC as described is able to produce 99% nitric acid when using the correct ratio of 85% sulfuric acid and dilute nitric acid of appropriate strength.

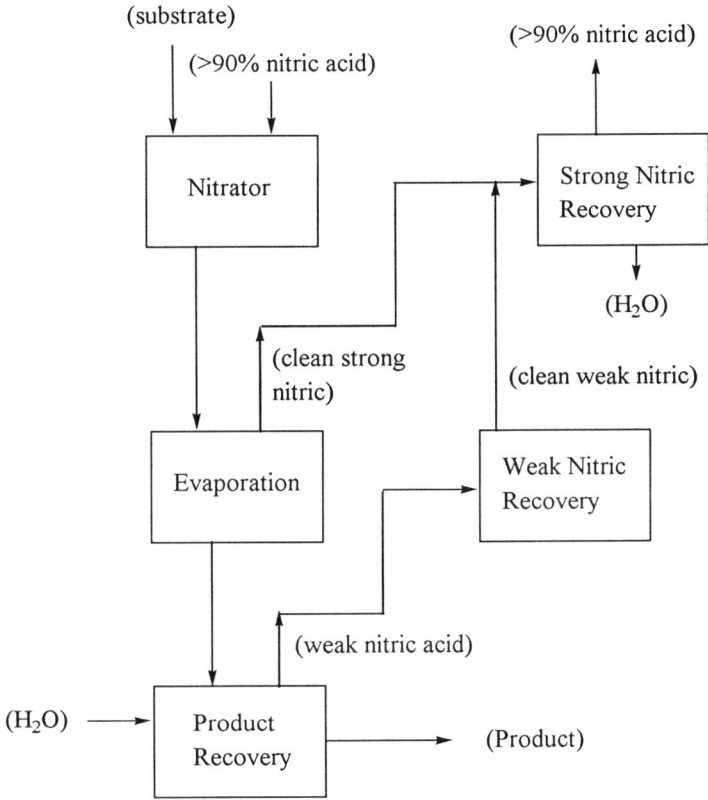

Figure 1. Simplified All Nitric Acid Nitration and Acid Recovery Scheme.

There are few practical methods known to remove TNM from nitric acid and still recover the nitric acid (*16*). Probably the best method is described by Fossan who has patented a process to remove azeotropically TNM from 99% nitric acid (*17*). He has shown that 99% nitric acid (960 kg) containing 2% by weight TNM can be introduced to a distillation column to afford an azeotropic overheads product (66.7 kg) that is 30% by weight TNM. Such a process could be used in tandem with a NAC SAC to remove TNM from recycle nitric acid streams. The TNM azeotrope can be treated with caustic to destroy the TNM (*8*) or diluted with water to phase separate the TNM, the TNM recovered and then used for a particular application (*17*). The main problem with these methods is that a considerable amount of nitric acid is also destroyed.

Work at C.I.L. Explosives coupled the distillative removal method of Fossan, a pyrolysis unit, and a scrubbing system to destroy TNM in strong nitric acid, Figure 2 (*18*). The 30% TNM azeotrope is sent to a pyrolysis unit operated at 250°C wherein the TNM is converted to carbon dioxide and about 40% of the nitric acid is decomposed to NOx (oxides of nitrogen). The residence time in the pyrolysis unit is on the order of 1 second. The vapor stream from the pyrolysis unit is sent to an appropriately designed scrubber where the nitric acid and NOx is recovered as weak nitric acid (*5*). The weak nitric acid can then be reconcentrated in a NAC SAC.

Taken together, the whole process can remove TNM from nitric acid recycle streams. Only a small slip stream of TNM contaminated strong nitric acid may have to be fed to a TNM concentration unit as shown if Figure 2 as long as the TNM removal rate is greater than the TNM production rate in any particular nitration process.

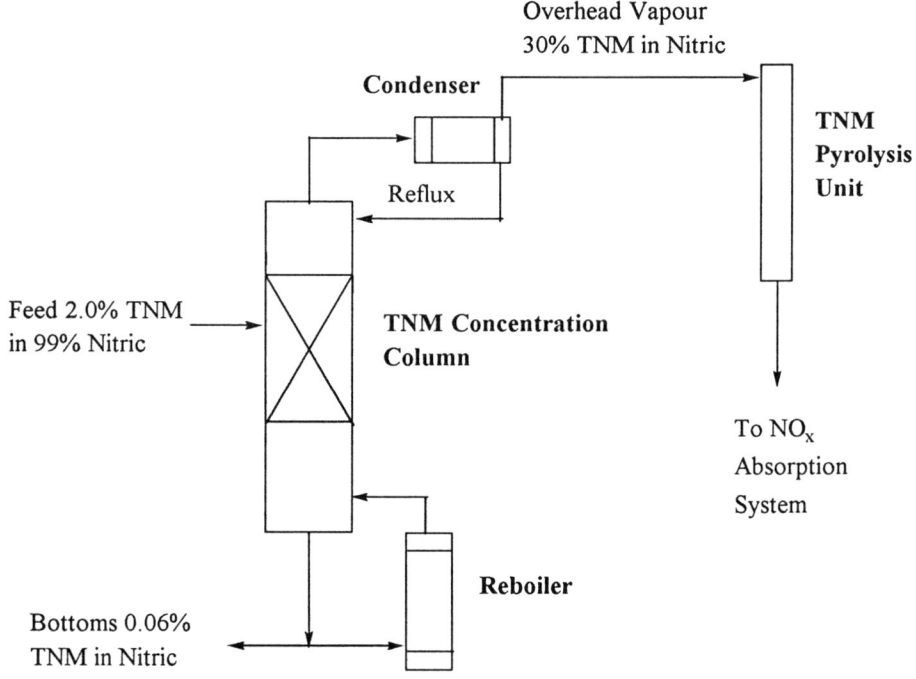

Figure 2. Isolation and Destruction of the Nitric Acid/TNM Azeotrope.

Nitration processes that involve <50% nitric acid recycle streams may be able to rely on the solvolysis of TNM to nitroform (trinitromethane) as a mode of controlling the TNM to acceptable levels. Henglein and Frank have studied the solvolytic conversion of TNM to nitroform in aqueous solutions of pH 5 to 7 (*19*). If TNM were solvolyzed in dilute nitric acid to nitroform, which in turn was solvolyzed to nitric acid (or NOx) and carbonic acid, then one might have a process to control TNM concentrations in a process.

For illustrative purposes, reconsider the process described in Figure 1. A nitration reaction mixture is fed to an evaporative unit to remove the bulk of the nitric acid and most of the TNM. The product stream will contain traces of TNM if removal in the evaporative unit is not complete. Phase separation of the product via dilution with water provides a weak nitric acid of desired strength to facilitate isolation of the product. However, the amount of water used must be controlled to insure the continued solubility of TNM; additionally the amount of water added impacts the capacity of the unit operation used to recover/reconcentrate the weak nitric acid after product removal. Assume the TNM solvolyzes in the weak nitric acid to innocuous products prior to reconcentration of the acid in a NAC SAC. The residence time in the weak nitric loop, the strength of the acid and the process temperature must be controlled to assure TNM destruction. If the amount of TNM destroyed in a weak nitric acid loop exceeds the rate of formation of TNM in the nitration reaction, then the TNM concentrations in the plant can be controlled safely as long as the solubility limit of TNM in any process stream is not approached.

The authors are unaware of any published studies pertaining to the stability of TNM or nitroform in dilute nitric acid or nitric-sulfuric acid mixtures. We decided to study the solvolysis of these species in acids to determine if the scheme outlined above had any merit. Along with these studies we determined the solubilities of TNM in certain acids.

Solubility and Stability of TNM and Nitroform in Acid Systems.

The solubility of TNM in various acids was determined experimentally at 25°C and 60°C, see Table I. Considerable scatter (± 0.2% by weight) in the data was occasionally observed and was believed to be the result of dissolved NOx in the nitric acid employed. The solubility of TNM in nitric acid did not appear to be a function of temperature. The prudent investigator assessing the safety of a process should determine the solubility of TNM in the acids that are actually present at possible process temperatures. A safety factor should be designed to assure that TNM will not phase separate from any process stream (i.e., a concentration level at 10, 20, or 50% of the solubility limit might be considered an acceptable safety factor). Nitroform is fairly soluble in water, 16.7g per 100 mL at 0°C, 193.8g per 100 mL at 60°C (20). Nitroform can be prepared from the reaction of TNM with hydrogen peroxide and aqueous potassium hydroxide, followed by acidification and extraction, equation 3 (21).

$$C(NO_2)_4 \xrightarrow[\text{2. HCl}]{\text{1. KOH, H}_2\text{O}_2\text{, H}_2\text{O}} (NO_2)_3CH \quad \text{(equ. 3)}$$

(TNM) (Nitroform)

Table I. Solubilities of TNM in Nitric Acid and Sulfuric Acid

Acid Composition	25C (wt. %)	60C (wt.%)
Water	0.1	0.08
20% Nitric Acid	0.25	0.23
40% Nitric Acid	0.54	0.53
60% Nitric Acid	0.72	0.7
70% Nitric Acid	1.0-1.5	1.0-1.5
80% Nitric Acid	5.4	5.4
90% Nitric Acid	15.3	15.3
99% Nitric Acid	>30.0	>30.0
98% Sulfuric Acid	0.8-1.0	2.5-3.1
86% Sulfuric Acid	0.7-0.8	not determined
86%Sulfuric/90%Nitric (2/1 by weight)	0.7-0.8	not determined

The solvolysis of TNM in 40% nitric acid was carried out in sealed tubes at 60 and 90°C. The kinetics were not straightforward: reproducible rates of TNM disappearance were not obtained nor was complete TNM destruction in an expected timeframe observed. Therefore, no attempt was made to calculate rates of TNM disappearance or derive rate equations. Periodically a brown gas, oxides of nitrogen, formed during the solvolysis experiments. Experiments that generated the brown gas usually showed a faster rate of disappearance of TNM. Rates of TNM disappearance at 60°C were more reproducible than at 90°C, Table II shows typical solvolysis data at 60°C.

Table II. Solvolysis of TNM in 40% Nitric Acid and 60°C

Time (h)	TNM (ppm by weight)
0	4687
2.3	1957
7.2	1680
27.3	1705
96	1571
120	1384

The solvolysis of TNM in 40% nitric acid at 90°C was repeated several times in sealed tubes and it was observed that in approximately half the experiments the level of TNM dropped to 600 ppm from an initial value of 4600 ppm within 6h, while in others the TNM level fell to ~1000 ppm within 6h. Table III shows the slowest rate data observed at 90°C. The amount of nitroform generated in the experiment was also measured. Solvolysis experiments at 90°C that generated NOx (brown gas) consistently displayed the faster TNM disappearance behavior (Table IV).

Table III. Solvolysis of TNM in 40% Nitric Acid at 90°C

Time (h)	TNM (ppm by weight)	Nitroform (ppm by weight)
0	4600	0
2.5	2600	270
21.5	700	800
43.5	275	1050

Table IV. Effect of NOx on the Solvolysis of TNM in 40% Nitric Acid and 90°C

Time (h)	TNM (ppm by weight) NOx Observed	TNM (ppm by weight) Colorless Reaction
0	4600	4600
2	850	1600
4	780	1400
6	600	1000

The moles of nitroform produced did not equal the moles of TNM destroyed, indicating that the TNM was destroyed by another mechanism than solvolysis to nitroform or that the nitroform was not stable to the reaction conditions. In an effort to better understand the decomposition process nitroform was solvolyzed in 40% nitric acid at 110°C in sealed tubes. It was found that the nitroform was destroyed within five hours. Analysis of the liquid phase failed to detect any measurable quantities of TNM or nitroform, and GCMS analysis of the head space showed the presence of carbon dioxide, carbon monoxide and oxides of nitrogen.

Nazin, Manelis, and Dubovitskii have observed the accelerated rate of destruction of nitroform in the presence of nitric oxide and nitrogen dioxide in the gas phase (22). Perhaps the same phenomenon was operative when we observed brown gas (NOx) accompanying the accelerated rate of disappearance of TNM or nitroform in sealed tubes at elevated temperatures. Sodium nitrite dissolved in nitric acid produces nitrous acid, which in turn decomposes to nitric acid, nitric oxide and water (23). Nitric oxide is oxidized to nitrogen dioxide in the presence of oxygen (24). A tube was charged with 40% nitric acid containing 4600 ppm (by weight) TNM and 5 mole equivalents of sodium nitrite. The tube was sealed and heated at 90°C. The level of TNM decreased to less than 300 ppm within 5 hours reproducibly. A 40% nitric acid solution of nitroform (5900 ppm by weight) was subjected to the same experimental conditions: the level fell to 76 ppm within 5.5 hours.

Nitric oxide/oxygen compressed gas was also found to accelerate the decomposition of TNM and nitroform in 40% nitric acid at 90°C. Nitrosyl chloride, sulfur dioxide, iron nitrate, and hydroxylamine sulfate did not significantly enhance the rate of disappearance of TNM or nitroform in 40% nitric acid at 90°C.

Related studies showed that TNM is very stable in 99% nitric acid at 60°C, and in 70% nitric acid/85% sulfuric acid (1:2 by weight mixture) at 90°C for 4h; at 140°C in a sealed tube TNM was not stable in this acid mixture, 65% of the TNM being destroyed within 3h.

Removal of TNM from Nitric Acid via a Mixed Acid Approach

A curious observation was made when preparing a TNM spiked mixture of 85% sulfuric acid:70% nitric acid (2:1 by weight) for solvolysis studies. Difficulty was encountered preparing a 5000 ppm (by weight) TNM mixture even though that was below the solubility limit for TNM: the amount of TNM actually in solution was considerably less than expected! Apparently the vapor pressure of TNM is greater in sulfuric than in nitric acid, and the TNM was volatilizing from the mixed acid system. Two patents cited earlier showed that TNM could be sparged from sulfuric acid/nitric acid nitration reaction mixtures (9,10), leading to speculation that TNM could be sparged from clean acids mixtures.

A mixture of acids (110g of 85% sulfuric acid, 55g of 70% nitric acid) containing TNM (0.33g) was charged to a 250 mL vessel, fitted with means for an air sparge, a reflux condenser topped with a dry ice trap, and means for agitation. The mixture was heated at 70-90°C and sparged with 20 mL/min. of air, and sampled periodically to determine the amount of TNM remaining in the mixed acid system. It was found that 50% of the TNM had been removed after 30 minutes, 90% within 2 hours, and that within 3 hours <1% of the TNM remained in solution. The dry ice trap was found to hold the TNM that had been removed from the system. *Sparging 70% nitric acid containing TNM under the same conditions for 2 hours only removed 35% of the TNM within 2 hours, and 50% after 3 hours.*

This sparging process coupled with a NAC SAC and a pyrolysis unit/scrubber can remove TNM from nitric acid. Consider the system in Figure 3. Recycled nitric acid containing TNM is fed to a packed column with 85% sulfuric acid in a ratio such that 99% nitric acid can be subsequently recovered. The mixed acid travels down the column while air or nitrogen is sparged through the bed removing a portion of the TNM and some nitric acid.

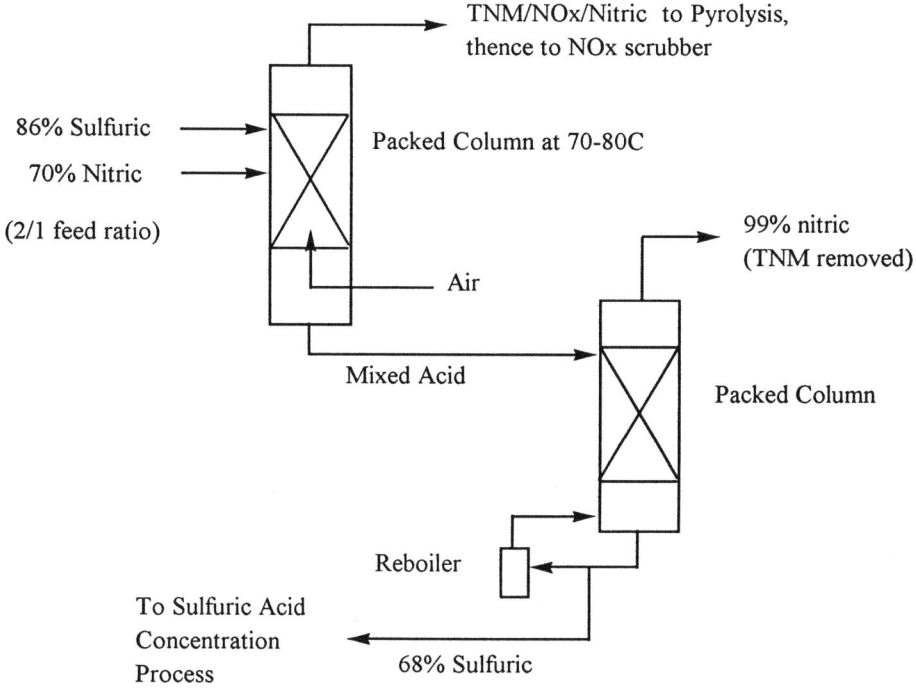

Figure 3. Process to Remove TNM from Nitric Acid Using a NAC SAC.

The vapor stream off the sparged column is sent to a pyrolysis unit to destroy the TNM, and the vapor from the pyrolysis unit is sent to an appropriate scrubber to reclaim the NOx as weak nitric acid (*vide ante*). The mixed acid stream from the sparging column is then sent to a conventional NAC SAC where 99% nitric acid (containing less TNM than if the mixed acid had not been sparged) is distilled overhead and weak sulfuric acid (68%) is a bottoms product. Such a TNM removal system can be employed in the all nitric nitration processes described for benzene or N-methylphthalimide described earlier. The TNM removal efficiency of the sparging column need only be that required to remove slightly more TNM than is produced in the nitration process and to insure that the TNM solubility limit in any process stream is not approached. In this way TNM will not build up in recycle acid loops or phase separate from solution. Additional engineering is required for the implementation of such a process.

The sparging column has been piloted on a laboratory scale. A mixture of 85% sulfuric acid/70% nitric acid (2/1 by weight) containing 6000 ppm TNM was fed to a jacketed glass column (12"x1.5") packed with glass helices. Air was metered into the bottom of the packing. The amount of TNM removed from the mixed acid was measured as a function of the jacket temperature, the acid flow rate, and the air flow rate. It was shown that with an acid flow rate of 2 mL/min, an air flow rate of 1 liter/min, and with a jacket temperature of 90°C, that 90% of the TNM was removed from the acid mixture along with 3% of the nitric acid. The conditions can be balanced to minimize nitric acid loss and TNM removal efficiency. In theory, the process can use nitric acid strength of 30-100% and sulfuric acid of 50-100% strength.

Recall the observation that TNM codistills with 99% nitric acid when 70% nitric acid, contaminated with TNM, and 85% sulfuric acid are fed to a NAC SAC. This phenomenon could be the exaggerated behavior noted when mixing TNM contaminated 70% nitric acid with 85% sulfuric acid at room temperature and finding that TNM had volatilized from the system.

Conclusions

The presence of TNM in nitration processes presents many challenges, safety chief among them. It is imperative that TNM not phase separate from process streams and that the solubility limit of TNM in any process stream not be approached. We have shown that TNM solvolyzes in 40% nitric acid at an appreciable rate. TNM can be removed from mixed acid systems via sparging with air and subsequently destroyed in an appropriately designed pyrolyzer/scrubbing process. A process to remove TNM from recycled nitric acid by first combining it with sulfuric acid, then sparging the mixture with air, and finally distillation of the nitric acid in a NAC SAC has been shown to be effective on the laboratory scale. The TNM removal rate from a nitration process must keep pace with the amount of TNM formed in the system so that TNM does not build up in any process recycle stream (25).

The presence of TNM in a nitration process is far from insurmountable. A safe plant can be designed by first understanding its genesis, its concentration levels in process streams, its solubility characteristics in process streams, and its rate of destruction (if any) within the process. With this data in hand, a TNM concentration/destruction method can be properly designed to control TNM within the nitration operation.

Experimental

The experiments described in the text need little clarification. Standard laboratory equipment was used to conduct the solubility and solvolysis studies, and the sparging experiments. TNM was analyzed by high pressure liquid chromatography (HPLC) using a Waters Resolve C18 5 micron Radial-Pak column, an isocratic solvent gradient using 50/50 acetonitrile/water 2.0 mL/min, and UV detection at 280 nm. Quantification of trinitromethane and TNM in clean acid mixtures can also be determined by HPLC using a DuPont ODS C18 analytical column using 65% water (0.1M in tetramethylammonium chloride plus 0.1% methanol)/35% methanol in acetonitrile (1:9) as eluent at 1.75 mL/min and UV detection at 280 nm. Acid samples were diluted with acetonitrile prior to analysis.

Literature Cited

1. The Sigma-Aldrich Corporation material safety data sheet for TNM shows that the oral LD50 (rat) is 130 mg/kg, inhalation LC50 (rat) is 18 ppm/4h, and it can cause methemoglobinemia.
2. Urbanski, T., *Chemistry and Technology of Explosives*, Pergamon Press, Oxford, England, 1985, Vol. 1.
3. Holahan, F.S.; Castorina, T.C; Autera, J.R.; Helf, S.; *J. Am. Chem. Soc.* **1962**, *Vol. 84*, 756.
4. Holahan, F.S.; PATR 2695, AD 241129, July, 1960.
5. Evans, C.M.; 'Practical Considerations in the Recovery and Concentration of Nitration Spent Acids', presentation at the Anaheim American Chemical Society meeting, Nitration Symposium, 4/4/95
6. Mundy, R.A.; Gilbert, E.E.; US Army Armament Research and Development Command, Large Caliber Weapon Systems Laboratory, Dover, New Jersey; Technical Report ARLCD-TR-78027, *'Nitroform Recovery and Isolation Studies'*, May 1978.
7. A similar method of TNM removal from trinitrotoluene waste gas streams has been patented: Gilligan, W.H.; Hall, T.N.; *US Patent 4,003,977*, January 18, 1977.
8. Glover has studied the destruction of TNM in water with sodium hydroxide and sodium nitrite, Glover, D.J.; *The Journal of Physical Chemistry*, **1970**, *74*, 21. Glover, D.J.; *US Patent*, 3,125,606, March 17, 1964.
9. Miki, T.; Obata, K.; *Japanese Patent*, Kokai Patent No. SHO 57(1982)-156445, September 27, 1982.
10. *Japanese Patent*, Patent No. 57-302, January, 6, 1982.
11. Cook, N.C.; Davis, G.C., *US Patent 3,981,933*, September 21, 1976.
12. Schroeder, B.; Thelen, B; Auge, W.; Thiem, K.-W., *US Patent 4,261,908*, April 14, 1981.
13. Odle, R.R., Groeneweg, P.G.; *US Patent 4,902,809*, February 20, 1990.
14. There are many parameters to consider when evaluating the explosive hazards of materials. An introduction to the processes used in these evaluations can be found here: C. Grelecki, in his course notes, 'Fundamentals of Fire and Explosion Hazards Evaluation', AIChE Today Series, American Institute of Chemical Engineers, 345 East 47 Street, New York, N.Y. 10017. Experts need to be consulted to aid in these types of evaluations.
15. Hager, K.F.; *Industrial and Engineering Chemistry*, **1949**, 2168. This paper details the production of TNM via the reaction of acetylene with strong nitric acid in the presence of mercury nitrate. A pilot plant employing this process was built in Newark, NJ and was destroyed by an explosion in 1953 (see ref. 2, Urbanski, Vol. 3, page 299). Hager showed that trinitromethane is stable in 85% nitric acid. Trinitromethane is rapidly converted to TNM in excess 99% nitric acid.
16. The destruction of TNM with alcohols and inorganic ions has been studied, S.L. Walters; T.C. Bruice; *J. Am. Chem. Soc.*, **1971**, *93*, 2269.
17. Fossan, K.R., *US Patent 3,781,374*, Dec. 25, 1973.
18. Chin, C.-H.; Edmonds, A.C.F.; Evans, C.M., *US Patent 4,713,232*, December 15, 1987.

19. Frank, A.J.; Henglein, A.; *Berichte der Bunsen Gesellschaft*, **1976**, *7*, 590.
20. *The Merck Index, Ninth Edition*, Edited by M. Windholz, Published by Merck and Co., Inc., **1976**, p. 1248.
21. Brown, L.H.; Geckler, R.D.; 'Research in Nitropolymers and their application to Solid Smokeless Propellants', Aerojet Engrg. Corp., Quarterly Summary Report 371, April, 15, **1949**. Trinitromethane explodes when heated rapidly. Mixtures of organic material and pure trinitromethane should be avoided as they may be potentially unstable.
22. Nazin, G.N.; Manelis, G.B.; Dubovitskii, F.I.; *Izv. Akad. Nauk SSSR, Seriya Khim.* **1969**, *5*, 1035.
23. Mellor, J.W.; "Mellor's Comprehensive Treatise on Inorganic and Theoretical Chemistry", volume VIII Supplement II, N (Part II), John Wiley and Sons Inc., N.Y., N.Y., page 363.
24. Ibid. Page 229.
25. A patent has recently been filed covering some of the concepts described in this chapter, notably the removal of TNM from mixed acid systems, and destruction of TNM in weak nitric at elevated temperatures in the presence and absence of sodium nitrite or NOx.

RECEIVED January 5, 1996

Chapter 18

Modeling Nitronium Ion Concentrations in $HNO_3-H_2SO_4-H_2O$ Mixtures

Lyle F. Albright, M. K. Sood[1], and Roger E. Eckert

School of Chemical Engineering, Purdue University, West Lafayette, IN 47907

Nitronium ion (NO_2^+) data reported by Zaman (1972), for $HNO_3-H_2SO_4-H_2O$ mixtures have been modeled using four ionization reactions that produce NO_2^+, H_3O^+, HSO_4^-, and NO_3^-. Zaman had measured NO_2^+ concentrations for 62-66 acid mixtures at 20°, 30°, and 40°C. The equilibrium constants for the four reversible reactions were calculated using a non-linear least squares programming technique. The predicted NO_2^+ concentrations agree well with the experimental data. The application of the results to industrial nitrations and the extrapolation of the findings to mixed acids containing N_2O_5 and SO_3 are discussed.

For numerous industrial nitration processes, the first step in the nitration sequence is the reaction between the organic molecules to be nitrated and a nitronium ion, NO_2^+ (1,2). Organic molecules that are so nitrated include aromatics, alcohols, glycols, glycerine, cellulose, and amines.

NO_2^+'s are produced by the ionization of nitric acid. Strong acids promote such ionizations. Sulfuric acid is the strong acid most frequently employed. Perchloric acid, HF, selenic acid, and $BF_3 \cdot H_2O$ can also be used as the strong acid. Solid catalysts have also been suggested. NO_2^+'s are also produced by ionization of N_2O_5.

In a mixture of HNO_3, H_2SO_4, and water, the following reactions occur:

$$HNO_3 + H_2SO_2 \rightleftharpoons NO_2^+ + HSO_4^- + H_2O \quad (1)$$

$$2HNO_3 \rightleftharpoons NO_2^+ + NO_3^- + H_2O \quad (2)$$

$$H_2SO_4 + H_2O \rightleftharpoons H_3O^+ + HSO_4^- \quad (3)$$

$$HNO_3 + H_2O \rightleftharpoons H_3O^+ + NO_3^- \quad (4)$$

[1]Current address: Mobil Research and Development Company, Princeton, NJ 08540

0097-6156/96/0623-0201$15.00/0
© 1996 American Chemical Society

Both Chedin (3) and Zaman (4) measured NO_2^+ concentrations in mixtures of HNO_3, H_2SO_4, and water using Raman spectra readings. Chedin failed to report important details of his experimental techniques including the temperature investigated. His measurements are generally thought to have occurred at about 25°C. Zaman, however, reported more details on his experimental procedures including his method of determining the NO_2^+ concentrations as a function of the Raman spectra readings for 66 acid mixtures. On a non-ionized molar basis, he investigated the following ranges: 1 to 80% HNO_3, 1 to 90% H_2SO_4, and 5 to 50% water. Zaman reported NO_2^+ concentrations as gram mole/liter of the acid mixture at 20°, 30°, and 40°C. At 20° only, he reported for all acid mixtures the percent of the HNO_3 converted to NO_2^+. In order to report such a percent, he obviously had to know the densities of the acid mixtures at 20°C. Such density information was, however, not reported. Based on graphical plots of his data, Zaman's results are reproducible to within 5-10%. He reported for three acid mixtures HNO_3 conversions to NO_2^+ of over 100%, from 101.8 to 105.8%.

Zaman's results (4) are nevertheless considered the most reliable NO_2^+ information that are currently available. In this investigation, his data were mathematically modeled employing chemical equilibrium equations. This method also predicts the concentration of the other ions produced, and hence, provides information on the gegen ions present in the ionized mixtures of HNO_3, H_2SO_4, and water. The results reported here will aid in developing improved kinetic models for many nitration processes.

Development of NO_2^+ Model

Reactions 1, 2, 3, and 4 were assumed to be the only ionization reactions that occur in mixtures of nitric acid, sulfuric acid, and water. These reactions are further assumed to result in equilibrium concentrations of NO_2^+, H_3O^+, HSO_4^-, and NO_3^-. Some non-ionized H_2SO_4, HNO_3, and H_2O remain in the equilibrium mixtures. Only three of the four equations are thermodynamically independent, and Reactions 1, 3, and 4 are used in the modeling procedure employed in this investigation. K_1, K_3, and K_4 are the equilibrium constants for these three reactions. K_2 can be calculated since it equals $K_1 K_4 / K_3$.

An equilibrium equation was developed for each of the three reactions being considered. N, S, and W are the moles of HNO_3, H_2SO_4, and water that are present in each acid mixture assuming no ionization reactions had occurred. X_1, X_3, and X_4 are the moles of HNO_3 that reacted in Reaction 1, the moles of H_2SO_4 that ionized in Reaction 3, and the moles of HNO_3 reacted in Reaction 4 respectively. When N + S + W = 100, N, S, and W are also the mole % of HNO_3, H_2SO_4, and water respectively as reported by Zaman. Furthermore, the final moles in the ionized mixture equals $100 + X_1$. The equilibrium equations for Reactions 1, 3, and 4 are as follows:

$$K_1 = \frac{(X_1)(X_1 + X_3)(W + X_1 - X_3 - X_4)}{(N - X_1 - X_4)(S - X_1 - X_3)(100 + X_1)} \tag{5}$$

$$K_3 = \frac{(X_3 + X_4)(X_1 + X_3)}{(S - X_1 - X_3)(W + X_1 - X_3 - X_4)} \tag{6}$$

$$K_4 = \frac{(X_3 + X_4)(X_4)}{(N - X_1 - X_4)(W + X_1 - X_3 - X_4)} \tag{7}$$

The following steps were employed at at a given temperature using the non-linear computer program developed by Hartmann (5) to determine the best values of K_1, K_3, and K_4. Initial values of these three K's were assumed first at 20°C. Experimental information in the literature helped to make reasonable initial estimates. Next the three non-linear equations were solved to determine tentative values of X_1, X_3, and X_4 at 20°C for each acid mixture. X_1 equals the moles $NO_2^+/100$ moles of non-ionized acid mixture. The following were calculated for each acid mixture using the calculated X_1 values:

$$\% \text{ HNO}_3 \text{ converted to } NO_2^+ = 100 \, X_1/N \tag{8}$$

$$NO_2^+ \text{ mole } \% = 100(X_1)/(100 + X_1) \tag{9}$$

Experimental values of X_1 were calculated at 20°C for each acid mixture using Zaman's data as follows:

$$(X_1)_{\text{exp.}} = N(\% \text{ HNO}_3 \text{ ionized to } NO_2^+)/100 \tag{10}$$

Improved K values were then determined using the computational method based on the minimum sum of squares of the differences between experimental and predicted responses. In this investigation, three methods of reporting the NO_2^+ concentrations were tested; these methods are as follows: % HNO_3 converted to NO_2^+; moles NO_2^+ in ionized acid mixture per 100 moles of non-ionized acid mixture; and mole % NO_2^+ in ionized acid mixture. For each acid mixture for which Zaman had reported conversions of HNO_3 to NO_2^+ of over 100%, the conversion employed in the computer program was 100%.

At both 30° and 40°C, experimental values of the % HNO_3 converted to NO_2^+ were calculated using the following equation. Account was taken of the change of moles NO_2^+/liter and of the density of the acid mixture as the temperature increased from 20° to either 30° or 40°C:

$$(\% HNO_3 \text{ converted})_t = (\% \text{ HNO}_3 \text{ converted})_{20} (\text{conc.t/conc.20})/(\rho_t/\rho_{20}) \tag{11}$$

where

(% HNO_3 converted)$_t$ = % HNO_3 converted to NO_2^+ at 30° or 40°C

(% HNO_3 converted)$_{20}$ = % HNO_3 converted to HNO_3 at 20°, as reported by Zaman

conc. t = moles NO_2^+/liter at 30° or 40°C, as reported by Zaman

conc. 20 = moles NO_2^+/liter at 20°C, as reported by Zaman

ρ_t = density of acid mixture at 30° or 40°C

ρ_{20} = density of acid mixture at 20°C

Density values for H_2SO_4–H_2O mixtures and for HNO_3–H_2O mixtures are reported (6). The ρ_t/ρ_{20} ratios for H_2SO_4–H_2O mixtures are essentially 0.995 and 0.99 respectively at 30° and 40°C. The ratios for HNO_3–H_2O mixtures are 0.99 and 0.98 respectively at 30° and 40°C. The following equation was used to estimate the density ratio for the acid mixtures investigated by Zaman at 30° and 40°C.

$$\rho_t/\rho_{20} = [(H_2SO_4 \text{ ratio}) S + (HNO_3 \text{ ratio})N)]/(S+N) \qquad (12)$$

The ρ_t/ρ_{20} ratio for the acid mixtures at 40°C varied between 0.98 and 0.99.

Modeling Results

Table 1 indicates the values of K_1–K_4 determined using the NO_2^+ mole % as the objective function in the model. Table 2 indicates molar concentrations predicted for 66 acid mixtures investigated at 20°C. The difference between the experimental and predicted NO_2^+ concentrations are also shown. Figure 1 shows the predicted NO_2^+ concentrations as a function of acid composition (reported on a non-ionized basis).

Table 1. Chemical Equilibrium Constants for Reactions 1-4 at 20°, 30° and 40°C

Reaction	20°C	30°C	40°C
1	0.067	0.055	0.045
2	2.0×10^{-4}	1.6×10^{-4}	1.3×10^{-4}
3	67	65	59
4	0.20	0.19	0.16

The standard deviation of the individual NO_2^+ values is about 0.5 mole % in the ionized acid mixtures. For 18 of the 19 acid mixtures containing >5% NO_2^+, the concentrations are predicted within 10% on a relative basis. For all acid mixtures containing <0.5 mole % of NO_2^+, the predicted values are higher than the experimental values. Yet it is known that easily nitratable aromatics such as benzene and toluene are readily nitrated by many acid mixtures that show no detectable NO_2^+ as measured using Raman spectra. Apparently the Raman spectra readings often fail to detect NO_2^+ at concentrations less than 0.2 - 0.5 mole % NO_2^+. Possibly in such acids, the NO_2^+ may be hydrated as $NO_2^+ \cdot H_2O$ (or $H_2NO_3^+$). As shown in Figure 1, the contours of 0.1 and possibly even 0.5 mole % NO_2^+ occur for acid mixtures reported by both Chedin and Zaman as containing no NO_2^+.

Two other models were developed for Zaman's data in which the NO_2^+ concentrations were reported either as % HNO_3 ionized to NO_2^+ (as reported by Zaman at 20°C) or as moles NO_2^+/100 moles of non-ionized acid. The predicted NO_2^+ concentrations using these two models differ to a small extent as compared to predicted values using the K values of Table 1, which are considered to be the best ones. Reporting the NO_2^+ concentrations as mole % is considered most useful for kinetic equations to be used for nitration reactions.

Increasing the temperature resulted in slightly lower NO_2^+ concentrations. Figure 2 shows several NO_2^+ contours at 20° and 40°C as a function of acid composition on a non-ionized basis. For the acids containing >10 mole % NO_2^+, the predicted decreases on a relative basis were in the 2-5% range, as the temperature increased from 20° to 40°C. Decreases as the temperature was raised from 20° to 30°C were about half as large. The Arrhenius equation resulted in a nearly perfect fit for K_1; the energy of activation in this case was 3620 cal/gmole. K values for the

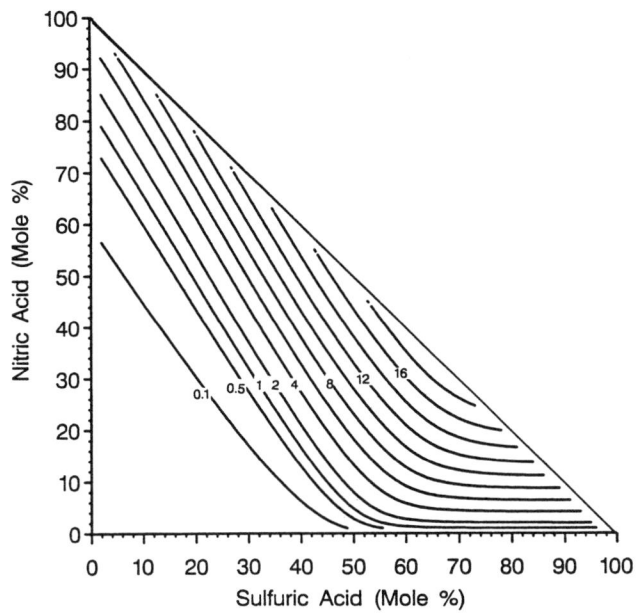

Figure 1. Nitronium Ion Contours Expressed as Mole % for HNO_3–H_2SO_4–H_2O Mixtures at 20°C

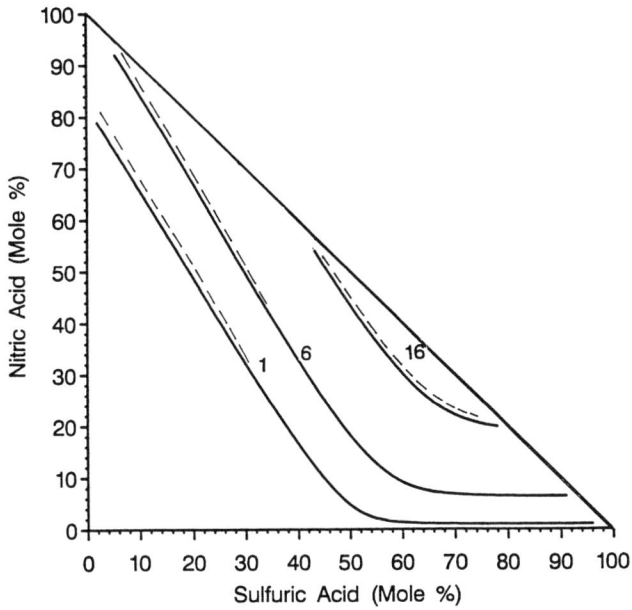

Figure 2. Nitronium Ion Contours at 20°C (—) and 40°C (----) Expressed as Mole % for HNO_3–H_2SO_4–H_2O Mixtures.

Table 2. Experimental and Predicted Concentrations of Ions and Non-Ionized Molecules at 20°C for 66 Acid Mixtures Investigated by Zaman (4)

Expt. Soln. No.	Composition (mole %)			% Dissoc. Nitric to NO2+ (Raman)	Ionized Composition (mole %)						Nitrate	Nitronium			No
	Nitric Acid	Sulfur Acid	Water		Water	Sulfuric Acid	Nitric Acid	Hydro-dronium	Bi-sulfate			Expt'l.	Pred.	Resid.	
1	5	90	5	96	0.030	75.038	0.004	9.966	14.962		0	4.580	4.758	-0.178	1
2	3	90	7	100	0.025	77.029	0.002	9.973	12.971		0	2.913	2.911	0.002	2
3	1	90	9	100	0.021	79.022	0	9.979	10.978		0	0.990	0.990	0	3
4	3	87	10	100	0.043	71.052	0.004	12.952	15.948		0	2.913	2.909	0.004	4
5	14	80	6	100	0.207	46.738	0.264	19.528	33.262		0	12.281	12.076	0.204	5
6	10	80	10	100	0.172	50.442	0.134	19.693	29.558		0	9.091	8.979	0.111	6
7	5	80	15	100	0.133	55.216	0.042	19.825	24.784		0	4.762	4.724	0.038	7
8	1	80	19	100	0.105	59.116	0.005	19.890	20.884		0	0.990	0.985	0.005	8
9	3	77	20	97	0.171	51.244	0.037	22.792	25.756		0	2.828	2.878	-0.050	9
10	24	70	6	83.3	0.737	23.936	3.569	25.674	46.064		0.020	16.661	16.951	-0.290	10
11	20	70	10	81.8	0.681	25.847	2.564	26.742	44.153		0.013	14.060	14.838	-0.778	11
12	15	70	15	90.5	0.599	28.665	1.523	27.871	41.335		0.006	11.952	11.871	0.081	12
13	10	70	20	96.4	0.508	32.029	0.756	28.733	37.971		0.003	8.792	8.459	0.333	13
14	5	70	25	100	0.414	35.944	0.264	29.321	34.056		0.001	4.762	4.522	0.240	14
15	2	70	28	100	0.360	38.525	0.082	29.557	31.475			1.961	1.882	0.079	15
16	3	67	30	100	0.526	31.989	0.230	32.243	35.011		0.001	2.913	2.694	0.218	16
17	5	65	30	100	0.708	26.960	0.622	33.667	38.040		0.003	4.762	4.192	0.570	17
18	30	60	10	57.5	1.383	13.721	11.005	27.502	46.279		0.109	14.712	15.886	-1.174	18
19	20	60	20	65	1.377	16.324	6.885	31.679	44.676		0.059	11.504	11.548	-0.044	19
20	15	60	25	69	1.349	16.286	4.910	33.702	43.714		0.039	9.379	9.133	-0.246	20
21	10	60	30	76.4	1.296	17.449	3.043	35.639	42.551		0.022	7.098	6.485	0.613	21
22	5	60	35	83.2	1.208	18.956	1.361	37.422	41.044		0.009	3.994	3.503	0.491	22
23	1	60	39	100	1.104	20.583	0.237	38.657	39.417		0.001	0.990	0.756	0.234	23
24	3	57	40	82	1.661	15.096	1.202	40.126	41.904		0.010	2.401	1.756	0.645	24
25	5	55	40	74	2.093	12.279	2.554	40.327	42.721		0.026	3.568	2.363	1.205	25
26	40	50	10	40.5	1.971	8.003	22.494	25.187	41.997		0.348	13.941	14.645	-0.704	26
27	30	50	20	37.2	2.219	8.354	17.678	29.844	41.646		0.260	10.040	10.764	-0.724	27
28	20	50	30	35.3	2.561	8.388	12.636	34.618	41.612		0.185	6.594	6.698	-0.104	28
29	15	50	35	34	2.821	8.193	9.961	37.068	41.807		0.150	4.853	4.661	0.191	29
30	10	50	40	30	3.220	7.745	7.091	39.575	42.255		0.114	2.913	2.719	0.194	30

31	5	50	45	24.8	3.938	6.879	3.863	42.127	43.121	0.071	1.225	1.054	0.171	31
32	10	45	45	5.4	5.912	4.134	8.729	42.104	43.866	0.254	0.537	1.006	-0.469	32
33	0	42	48	1.4	9.039	2.564	9.141	40.406	40.436	0.414	0.140	0.443	-0.303	33
34	10	40	50	0	11.806	1.853	9.187	39.449	38.147	0.558	0	0.255	-0.255	34
35	10	40	10	28	2.474	4.652	34.916	38.449	35.348	0.782	12.281	12.512	-0.232	35
36	50	40	10	24	3.020	4.631	29.799	26.510	35.369	0.671	8.759	8.701	0.058	36
37	40	40	20	16.5	3.938	4.230	24.240	31.219	35.770	0.604	4.717	4.903	-0.187	37
38	30	40	30	13.3	4.746	3.811	21.145	33.517	36.189	0.592	3.218	3.160	0.058	38
39	25	40	35	7	6.089	3.212	17.667	35.648	36.788	0.596	1.381	1.707	-0.326	39
40	20	40	40	19.5	2.948	2.509	47.479	18.039	27.491	1.534	10.474	9.899	0.575	40
41	60	30	10	14.6	3.946	2.358	42.028	22.574	27.642	1.452	6.803	6.121	0.682	41
42	50	30	20	11.5	4.738	2.171	39.005	24.783	27.829	1.474	4.920	4.325	0.595	42
43	45	30	25	8.3	5.941	1.895	35.640	26.861	28.105	1.558	3.213	2.725	0.488	43
44	40	30	30	1.2	10.717	1.202	27.346	30.005	28.798	1.931	0.359	0.718	-0.359	44
45	30	30	40	0.5	13.328	0.884	27.100	29.120	27.116	2.452	0.150	0.446	-0.296	45
46	30	28	42	0	16.193	0.656	26.673	28.095	25.344	3.039	0	0.287	-0.287	46
47	26	26	44	0	10.039	0.927	36.276	25.907	24.073	2.779	0.754	0.937	-0.182	47
48	40	25	35	1.9	3.593	1.141	58.962	14.570	18.859	2.874	7.233	6.685	0.548	48
49	60	20	11	11.3	0.987	0.682	53.313	18.380	19.013	3.027	4.306	3.531	0.775	49
50	60	20	20	7.5	9.259	0.419	45.065	21.912	19.318	3.764	0.892	1.157	-0.265	50
51	50	17	30	1.8	12.473	0.302	44.232	21.134	16.581	5.160	0.200	0.604	-0.404	51
52	50	15	33	0.4	14.889	0.217	43.373	20.515	14.698	6.222	0.150	0.403	-0.253	52
53	50	13	35	0.3	17.425	0.467	42.370	19.854	12.783	7.351	0	0.278	-0.278	53
54	60	15	37	0	8.367	1.039	53.671	18.043	14.533	4.920	1.127	1.390	-0.263	54
55	70	19	25	1.9	3.657	0.644	60.097	14.183	17.961	3.063	7.689	6.403	1.287	55
56	70	15	11	11.9	4.761	1.039	61.849	14.328	14.356	4.062	4.862	3.928	0.484	56
57	79	10	11	7.3	4.330	0.361	69.334	10.881	9.639	5.454	4.525	4.042	0.484	57
58	75	10	15	6	5.429	0.331	66.385	12.475	9.669	5.711	2.983	2.822	0.161	58
59	70	10	15	4.1	7.351	0.282	62.029	14.328	9.718	6.291	1.244	1.652	-0.408	59
60	65	10	20	1.8	9.998	0.232	57.004	15.923	9.768	7.075	0.324	0.912	-0.588	60
61	70	10	25	0.5	9.739	0.149	60.823	14.184	6.851	8.255	0.140	0.914	-0.774	61
62	80	7	23	0.2	4.418	0.309	70.222	10.544	8.691	5.816	3.920	3.811	0.109	62
63	80	9	11	5.1	6.206	0.127	70.037	10.808	4.873	7.949	1.029	1.974	-0.945	63
64	80	5	15	1.3	7.465	0.064	69.256	10.916	2.936	9.362	0.160	1.363	-1.203	64
65	80	3	17	0.2	8.953	0.018	68.093	10.995	0.982	10.960	0	0.938	-0.938	65
66	70	5	25	0	11.600	0.089	59.626	14.029	4.911	9.746	0	0.625	-0.625	66

other three reactions all decrease with increasing temperature, but the Arrhenius equation was not as good a fit suggesting that the values of K_2, K_3, and K_4 as shown in Table 1 are less accurate.

Various subsets of Zaman's data were also modeled. In one case, the data obtained for acid mixtures containing either the highest concentrations of sulfuric acid or of nitric acid were not used. The resulting acid mixtures are probably of more industrial interest for nitration. Only minor changes of the predicted NO_2^+ results were obtained in all cases, and the standard deviations decreased slightly, to about 0.4 mole % NO_2^+. There is hence no valid reason for adjusting the K values of Table 1.

For a given water concentration in acid mixtures, the maximum NO_2^+ concentrations occur when the molar ratio of H_2SO_4/HNO_3 is about 2, on a non-ionized basis. Zaman reported a NO_2^+ concentration of approximately 17 mole % when the water content was 6 mole %. The predicted results at 20°C indicate a NO_2^+ concentration of about 20% for an acid mixture containing no water.

The NO_2^+ concentrations were predicted for HNO_3-H_2O mixtures by using the equilibrium equations developed for Reactions 2 and 4. The predicted NO_2^+ concentrations are shown in Figure 3 at 20°C expressed on a HNO_3 mole % basis in the non-ionized acids. The NO_2^+ concentration increases from about 0.2 mole % at 60% HNO_3 to almost 6% at 100% HNO_3. No experimental information is known of NO_2^+ concentrations in strong HNO_3 solutions, but the values predicted here seem reasonable.

Concentration of Other Ions

The concentration of H_2O^+, HSO_4^-, and NO_3^- can be calculated at 20°, 30° and 40°C using the K values in Table 1. Figures 4, 5, and 6 show the predicted results at 20°C for the three ions respectively. The largest concentrations for both H_3O^+ and HSO_4^-, about 45 %, occur for $H_2SO_4-H_2O$ mixtures containing 50% H_2SO_4. Figure 6 indicates that the highest concentrations of NO_3^-, about 15%, occur for HNO_3-H_2O mixtures containing about 50% HNO_3. The predicted concentrations of the non-ionized HNO_3, H_2SO_4, and H_2O at 20°C are shown in Figures 7, 8, and 9 respectively. As expected, the highest concentrations of each increase to 100% as the non-ionized concentration of each approaches 100%. Mixed acids of most industrial importance for nitrations generally contain approximately the following reported on a molar basis: 20-35% H_3O^+, 20-40% HSO_4^-, 0.1-3% NO_3^-, 10-40% HNO_3, 1-15% H_2SO_4, and 2-10% H_2O.

Discussion of Results

Although the model developed in the current investigation predicts Zaman's data within experimental accuracy, more experimental data are needed for $HNO_3-H_2SO_4-H_2O$ mixtures. A wider range of temperatures needs to be investigated since many nitrations occur at >40°C. Hopefully information can be obtained to clarify if NO_2^+'s are present for acid mixtures that allow nitrations but for which neither Chedin (3) or Zaman (4) could detect NO_2^+'s. Measurements of the concentrations of H_3O^+, HSO_4^-, NO_3^-, or possible other ions would help clarify if any ionization reactions are occurring in addition to Reactions 1-4. Measurements of the densities of the mixed acids would aid in determination of the molar concentrations of NO_2^+.

Figures 1-6 are of importance in estimating the concentration of the ion complexes or gegen ions that occur in acid mixtures. These ion groups include $NO_2^+-HSO_4^-$, $NO_2^+-NO_3^-$, $H_3O^+-HSO_4^-$ and $H_3O^+-NO_3^-$. The reactivity of NO_2^+

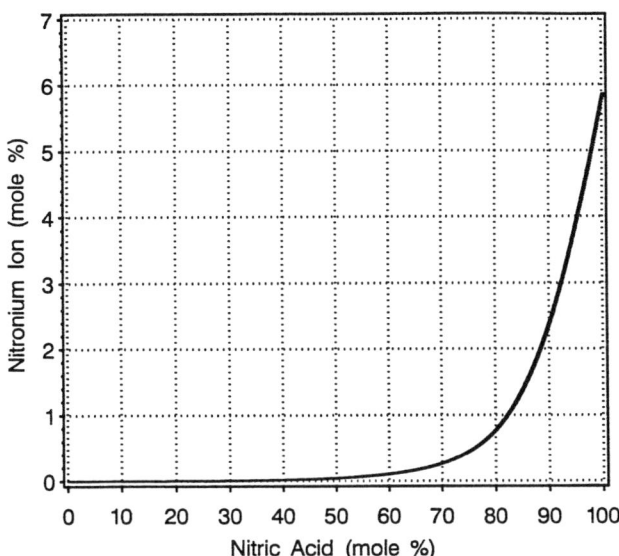

Figure 3. Nitronium Ion Contours Expressed as Mole % for HNO_3–H_2O Mixtures at 20°C

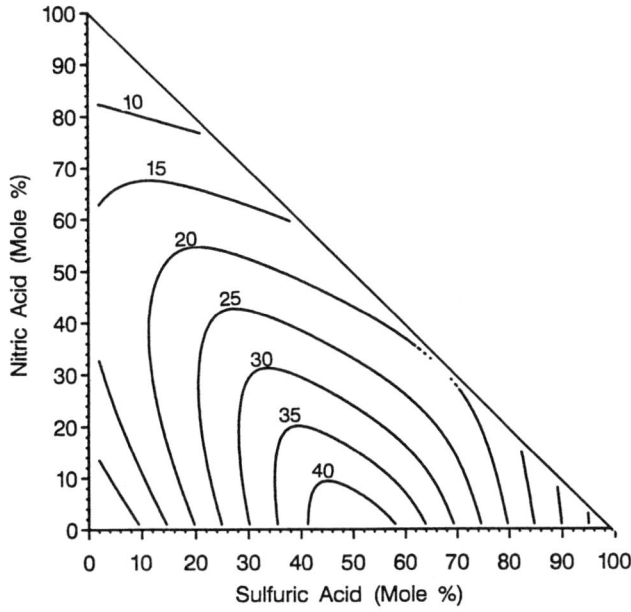

Figure 4. Hydronium Ion Contours Expressed as Mole % for HNO_3–H_2SO_4–H_2O Mixtures at 20°C

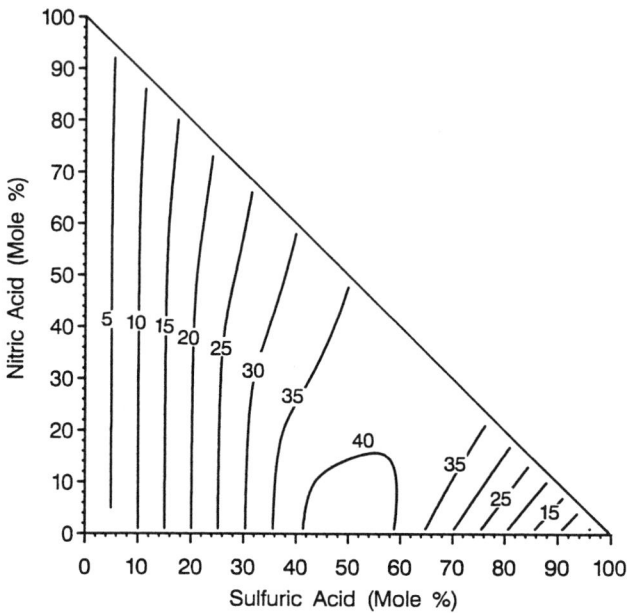

Figure 5. Bisulfate Ion Contours Expressed as Mole % for HNO_3–H_2SO_4–H_2O Mixtures at 20°C

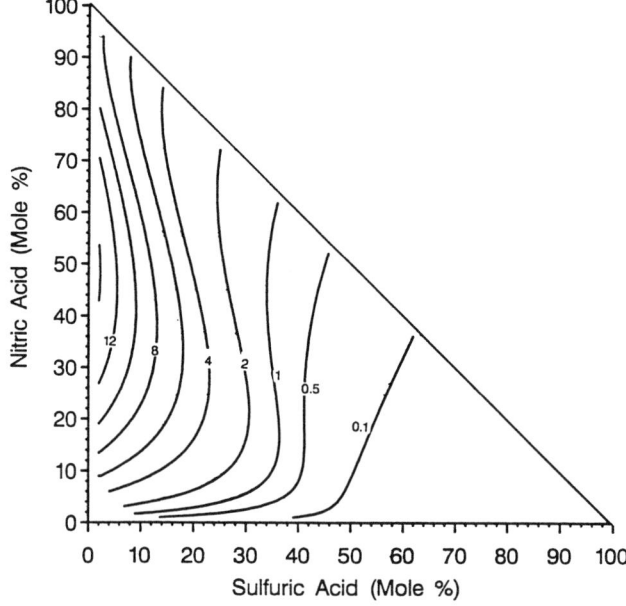

Figure 6. Nitrate Ion Contours Expressed as Mole % for HNO_3–H_2SO_4–H_2O Mixtures at 20°C

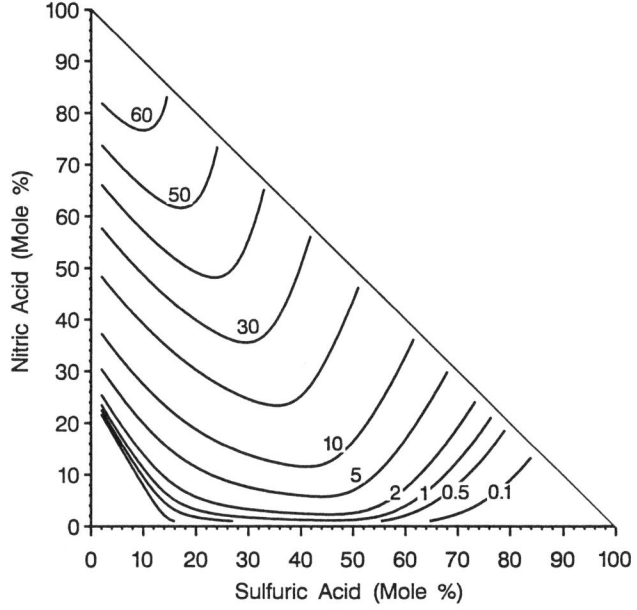

Figure 7. Non-Ionized HNO_3 Contours Expressed as Mole % for HNO_3–H_2SO_4–H_2O Mixtures at 20°C

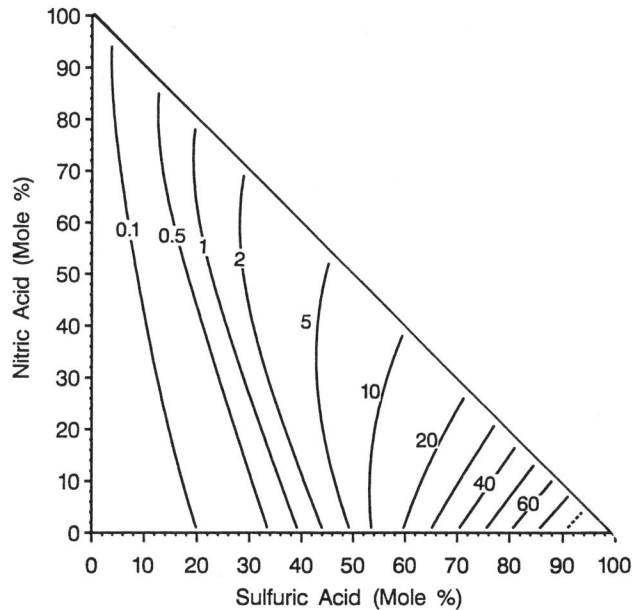

Figure 8. Non-Ionized H_2SO_4 Contours Expressed as Mole % for HNO_3–H_2SO_4–H_2O Mixtures at 20°C

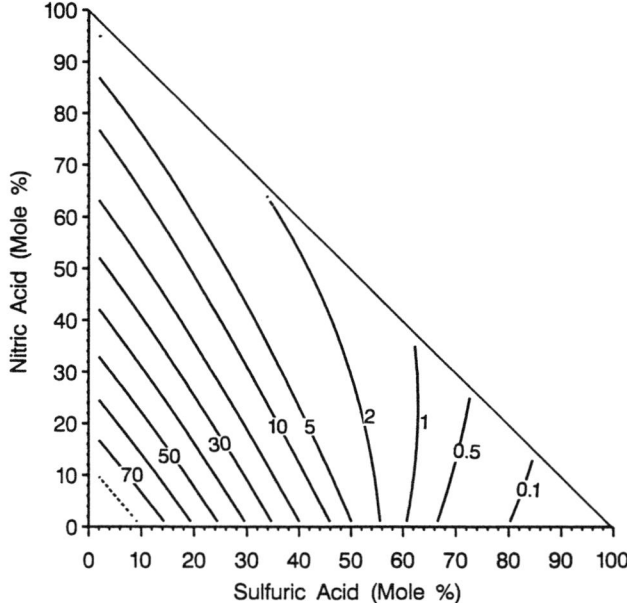

Figure 9. Non-Ionized Water Contours Expressed as Mole % for $HNO_3-H_2SO_4-H_2O$ Mixtures at 20°C

during nitration may differ substantially for NO_2^+–HSO_4^- as compared to NO_2^+–NO_3^-. The former gegen ion is obviously predominant in the mixed acids used in many nitration processes. The latter complex, however, is predominant, if not exclusive, when ionic nitrations are made using HNO_3–H_2O mixtures. Possibly the Raman spectra readings depend on the specific NO_2^+ complex being measured.

NO_2^+ contours in acid mixtures containing N_2O_5 and SO_3 can be predicted at 20°C using Figures 1-3. The concentration contours are extrapolated to < 0% water, i.e. to >100% HNO_3 (where N_2O_5 is present) or to >100% H_2SO_4 (where SO_3 is present). N_2O_5 is known to ionize often to a high degree producing NO_2^+ and NO_3^-. Mixtures of HNO_3, H_2SO_4, and SO_3 are highly effective for nitrations of toluene (7). There is obviously the need to obtain experimental information for NO_2^+ concentrations in mixtures containing N_2O_5 or SO_3.

The model developed in this investigation will be useful in developing improved kinetic equations. Based on the generally accepted chemical mechanism, the reaction between the hydrocarbon and NO_2^+ is the rate-controlling step in the nitration sequence. The model can probably be employed to calculate the NO_2^+ concentration at the reaction site (which is likely at or at least near the interface between the acid and the hydrocarbon phases). The assumption made is that the ionization reactions are extremely rapid so that equilibrium concentrations of NO_2^+ are always present.

The kinetics of nitration are of course often dependent on the interfacial area between the acid phase and the hydrocarbon phase. Increased interfacial areas have apparently been realized in many modern industrial reactors with the result that extremely short residence times are often needed to obtain essentially complete reactions. Increased levels of agitation result in larger interfacial areas and in improved transfer of reactants to and of products from the interfacial area. There is however the need to develop techniques to better predict interfacial areas.

Acknowledgments

The late Carl Hanson of the University of Bradford provided key suggestions in the early stages of this investigation. R. Goldstein of Picatinny Arsenal made preliminary calculations that demonstrated the modeling approach employed here gave promising results. Wolfgang M. Hartmann of SAS Institute, Inc. helped develop at our request the non-linear optimization program employed.

Literature Cited

1. Olah, G.S.; Malhotra, R.; Narang, S.C. *Nitration: Methods and Mechanism,* VCH Publishers, New York, N.Y., 1989.
2. Albright, L.F. in *Encyclopedia of Chemical Technology*, J.I. Kroschwitz, Ed., 4th Edition, John Wiley and Sons, Inc., New York, NY, Vol. 17, in press.
3. Chedin, J.: Ann. Chim. 1937, 302.
4. Zaman, M.B., *Nitronium Ions in Nitrating Acid Mixtures,* Ph.D. thesis, University of Bradford, Bradford, UK, 1972.
5. Hartmann, W.M., Applications of Nonlinear Optimization PROC NLP and SAS/IML Software, SAS Institute, Inc., Cary, NC 27513, pp. 98-105, (1995).
6. Perry's Chemical Engineering Handbook, Green, D.W., Ed., McGraw-Hill Book Co., New York, NY. 6th Ed., 1984, pp 3-83 to 3-86.
7. Hill, M.E.; Coon, C.L.; Blucher, W.G.; McDonald, G.J; Marynowski, C.W.; Tolberg, W.; Peters, H.M.; Simon, R.L.; Ross, D.L. in *Industrial and Laboratory Nitrations,* L.F. Albright and C.. Hanson, Eds., ACS Symposium Series 22, American Chemical Soc., Washington, D.C., 1976, pp 253-271.

RECEIVED January 5, 1996

Chapter 19

Commercial Dinitrotoluene Production Process

Allen B. Quakenbush and B. Timothy Pennington

Olin Corporation, P.O. Box 2896, Lake Charles, LA 70602

Since 1988, Olin has been developing a commercial dinitrotoluene (DNT) production process that does not use mixed acids for nitration. This new process avoids use of sulfuric acid by using high strength nitric acid to synthesize DNT, a precursor of toluene diisocyanate. The advantages of the new process include reduced raw material consumption and significantly lower capital investment for waste treatment facilities. These advantages are derived from careful control of the reaction conditions. Unwanted byproducts are minimized in the reactor and do not have to be removed in subsequent processing.

Olin has developed the chemistry and separation technology to produce dinitrotoluene (DNT) which is a precursor to toluene diisocyanate (TDI) manufacture. The new technology does not employ mixed acid for synthesis as used by current DNT manufacturers. The new process uses only nitric acid as the nitrating media. This offers several process advantages - especially in the reduction of unwanted toxic byproducts. The conventional process water effluent is classified as a hazardous waste. The dark-colored waste stream contains 2,4-DNT and 2,6-DNT, as well as soluble salts of the various nitro-aromatic byproducts such as polynitrated phenols and cresols. During the past several years, treatment costs for the DNT effluent have soared to meet Resource Conservation and Recovery Act (RCRA) and Clean Water regulations. The Olin DNT process technology avoids the problems of a conventional process effluent by producing fewer byproducts in the reactor. The new process has less effluent per unit of production. Even though the effluent is a distilled overhead product, operating and capital costs are favorable when compared to conventional manufacturing technology. In the discussion that follows, the conventional and new technologies will be described, and the development challenges related to using only nitric acid will be outlined. Finally, the advantages of the new process will be presented.

Conventional DNT Technology

DNT is conventionally produced by nitrating toluene in a series of reactors. The nitrating medium is a mixed acid that contains sulfuric and nitric acids. A conventional DNT process has 5 major operations: nitration, product separation, product washing, spent acid handling, and effluent purification. Feed stocks are nitric acid, sulfuric acid, and toluene. The conventional process can be designed to use dilute (65%) or concentrated (98%) nitric acid. Depending on the technology, the required sulfuric acid feed strength is from 93 to 98%. Sulfuric acid is not consumed by the nitration chemistry, but there is some sulfuric acid lost in the process effluent streams. Because of water formed during nitration and the water content of the nitric acid feed, spent sulfuric acid leaves the process at about 72%. The spent sulfuric acid leaving the process contains traces of organics and nitric acid. These must normally be removed before the spent acid can be re-concentrated or used by other processes. There are several methods used to handle the spent acid. Normally, the sulfuric acid is re-concentrated or used as a feed stock to an adjoining process. Nitration grade toluene (ASTM D841-85) is used instead of industrial grade toluene. This helps avoid emulsion formation during liquid-liquid phase separations and can reduce regulated substances in the process effluent.

The mononitration and dinitration take place in a series of reactors. As few as 2 reactors can be used, but there can be as many as 6. Two liquid phases make up the reaction mixture because the organics have a low solubility in the sulfuric acid. This property is desirable since product recovery from the nitration mixture can be accomplished by phase separation. The disadvantage of the heterogeneous system is that a high agitation level must be used to overcome mass transfer limitations during the reaction. Acid use is optimized by using counter current flow to the organic in the reaction system. After combining the strong sulfuric acid with nitric acid, the mixed acid is first used for the dinitration which requires a stronger nitrating media. The mixed acid is diluted by water formation during the reaction and is subsequently used for mononitration.

Most of the reaction byproducts are formed early during the mononitration reaction. The predominant byproducts are formed by the oxidation of toluene. Under these conditions, toluene oxidation produces phenol and cresol type compounds (*1*). Subsequently, these byproducts are nitrated to become nitrocresols and nitrophenols or higher nitrated analogs.

After the dinitration is complete, the crude product is separated by phase separation from the mixed acid. The crude DNT has several impurities that must be removed before the product can be used. Because there is some solubility of sulfuric and nitric acids in the organic phase, the crude product can contain up to 4% acid. Also, the crude product contains 600-2000 ppm nitrophenols and nitrocresols that can degrade catalysts in downstream processes. The acids and phenolic byproducts are removed from the DNT by a series of two or three aqueous washes. The first wash removes most of the acid. The second wash uses a weak carbonate solution to neutralize the remaining acid content and remove phenolic byproducts such as dinitrocresols and dinitrophenols. This wash is sometimes followed by a third wash with fresh water to remove excess carbonate. The aqueous extract from the third wash is typically recycled for use in the first wash to minimize water added for crude purification. The wash steps produce product DNT with less than 50 ppm acidity, but the wash water extract is contaminated with DNT isomers and the phenolic byproducts. This wash water extract generates most of the process effluent. Because of corrosivity, pH, and DNT concentration, the conventional effluent stream is classified as a hazardous waste (K111).

Olin DNT Technology

The new DNT technology can be divided into 5 major operations, as shown in Figure 1, that are similar to conventional technology nitration, product separation, product washing, nitric acid recycle, and effluent purification. Feed stocks are nitric acid and toluene. The process can be designed to use dilute (65%) or concentrated (98%) nitric acid. Nitration grade toluene is used instead of commercial grades to reduce phenolic byproducts, not in the effluent, but in the product.

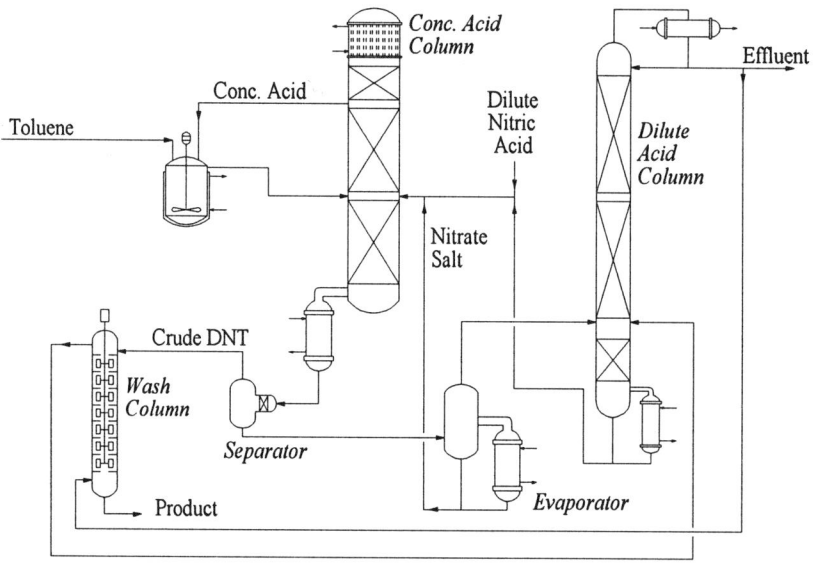

Figure 1. Olin DNT Process

Concentrated nitric acid (>95%) is fed to the reactor. Toluene is injected into the reaction solution which rapidly mononitrates the toluene. The mononitration of the toluene is so rapid that it can be considered to be almost instantaneous. The reactor is circulated or agitated to approach CSTR (continuous stirred tank reactor) behavior. Depending on reaction conditions, the dinitration is complete (99.9+%) in less than an hour. There is an excess of nitric acid in the reactor. DNT is soluble in nitric acid so the reaction mixture is homogenous. A key operating point, which will be discussed later, is that fewer byproducts are produced by toluene oxidation.

The next process step is to remove the DNT and recycle concentrated acid. The reactor effluent and a nitrate salt solution are fed to a distillation column that operates under partial vacuum. The concentrated nitric acid column uses nitrate salt to enhance the separation and recycle of the nitric acid from the water. The nitrate salt raises the volatility of the acid over the water. Concentrated acid is distilled overhead for recycle to the reactor. The addition of the nitrate salt alone to the reactor effluent will cause two phases develop - a lighter organic phase and a heavier aqueous phase. The combination of salt solution addition and acid evaporation improves the liquid-liquid separation. The salt is magnesium nitrate trihydrate, or a mixture of magnesium nitrate and zinc nitrate trihydrates. The two phase liquid mixture leaving the bottom of the column are separated into the lighter crude DNT and the heavier 'dilute salt'. Besides nitrate salt, the 'dilute salt' contains water and acid.

The crude product is water washed in an counter-current extraction to remove nitric acid. This water is recycled to the process to recover the nitric acid and trace amounts of DNT. The washed DNT is sent to storage.

The 'dilute salt', a dilute acid stream, and process feed acid are fed to an evaporator. During evaporation, the temperature is controlled so that the nitrate salt water content is reduced to a trihydrate, so the salt can then be recycled. The nitrate salt entering the evaporator contains less than 3% acid. Most of this residual acid goes with the evaporated water. The vapor stream leaving the evaporator feeds a column that recovers this acid by performing a nitric acid - water distillation. With low acid concentration and no nitrate salt, water is more volatile and so it becomes the overhead product of the rectification. This water stream is the process effluent. Water formed during the nitration and water entering with the acid feed stock leave the process in this stream. The acid leaving the bottom of the column is concentrated to near the azeotropic composition and then returned to the acid concentration column before use in the reactors. The dilute acid column is also used to partially concentrate and recycle acid from several minor weak acid streams in the process.

Single Acid Process Development Challenge

Single acid nitrations have always been of interest because of the inherent simplicity of using only one acid in the synthesis step. Crater (2) explored single acid use for DNT synthesis in the 1940's. There are several problems and concerns that have prevented commercialization of a single acid process:
- Suspected Increase in Oxidation Byproducts
- Slow Reaction Rates
- Separation of the Product
- Efficient Regeneration of Excess Nitric Acid

The easiest concern to address is that of greater oxidation using nitric acid alone. Work in the literature (3) and our own findings show that if the proper reaction conditions are selected there are fewer oxidation byproducts for single acid than for mixed acid nitrations. As discussed previously, the major oxidation byproducts of the mixed acid process are dinitrocresols in the range of 600-2000 ppm in the crude DNT. Such byproducts are unwanted in the DNT because of the detrimental effect they have upon the hydrogenation catalysts used to convert DNT to the intermediate toluene diamine (TDA). This level of byproducts must be removed by a sodium carbonate washing step for the product to be acceptable.

In the single acid process, the reactor conditions and feed ratios can be adjusted such that low cresol levels in the product can be attained without carbonate washing. In fact, cresol levels under 100 PPM can be met by adjusting the reactor conditions and reactant ratios. We have investigated the relationship between the water content of the reaction mixture during nitration and the cresol level in the resultant DNT as measured by UV-VIS absorbance of the caustic extract of the DNT. As seen in the figure, there is a good correlation between the water content during nitration and the cresol level. Of the independent variables considered, the CSTR steady state concentration of water gave the best correlation. Even though some HPLC data suggests that the wet UV-VIS absorbance method may overestimate the actual cresolic/phenolic content, the UV-VIS method seems to be a good predictor of the relative amount of cresolic impurities when different DNT samples are compared. The relationship is shown in Figure 2.

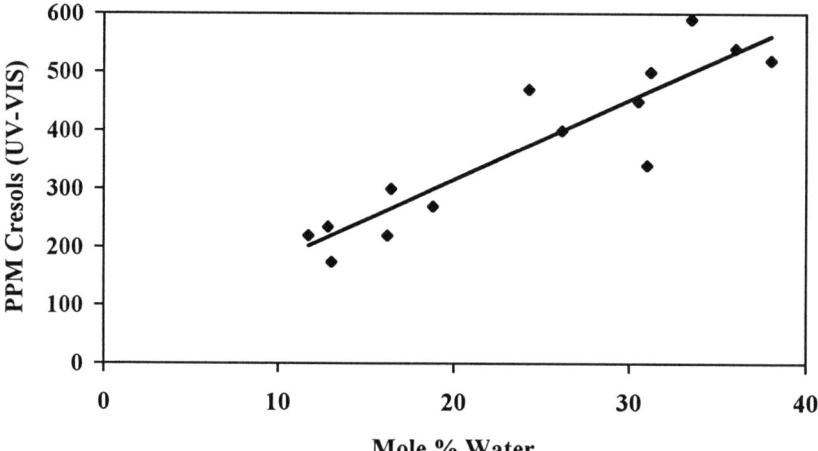

Figure 2. Cresol Formation vs. Reactor Water

The strong positive correlation of cresolic impurities with water mole fraction present in the reaction medium supports the mechanism of cresol formation proposed some time ago by Hanson, et. al (1). In this mechanism, water attacks the reaction intermediate formed after nitronium ion addition at the methyl group aromatic carbon (such nitronium addition itself is a rather rare event). Attack of water upon this intermediate leads to addition of the water on the aromatic ring and subsequently to elimination of nitrous acid leaving an OH group behind on the ring. This mechanism predicts an increase in NOx's in the reaction medium as a result of cresol formation. In actual practice, the amount of NOx's increase during nitration are much more than could be produced by cresol formation although the absolute amount of NOx's remains relatively small at 0.1 - 0.3%. Measured NOx formation while not nitrating indicates the acid decomposition rate is low, so this observed NOx level can only be explained by other oxidations that are also occurring and giving some of the byproducts discussed herein.

Implicit in the Hanson mechanism is that the rate of cresol formation will be dependent upon the rate of nitration at the methyl group aromatic carbon which will always be small and will also be dependent upon the water concentration. In the absence of attack by water, the intermediate rearranges to give normal nitrotoluene product. Cresol content of the DNT produced by the single acid process are seldom found to exceed 2000 ppm by the UV-VIS absorbance test and can be controlled to be as low as 100 - 150 ppm by reducing the water content in the nitration mixture.

Typical washed product purity levels are given for the two processes in Table I. As shown in this data, the general level of impurities is similar. Some other impurities have been found in the DNT from both processes, using capillary GC, GC-MS, GPC, and other analytical methods. These include carboxylic acids and higher molecular weight nitrated aromatics. The source of the carboxylic acids may be low levels of hydrocarbon impurities in the toluene feed or from oxidation of the toluene. The levels of these organic acid impurities are quite low and are about the same in DNT from either type of process. The impact on downstream performance of the DNT would be the same for each process. The higher molecular weight nitrated aromatics are present in DNT from either process at about 200 PPM. These higher molecular weight nitrated aromatics have not yet been fully characterized. The highest molecular weight present is about 300.

Table I. Typical Final Product Impurities in DNT From Mixed Acid and Single Acid Processes

Component	Mixed Acid Carbonate Washed ppm	Single Acid Water Washed ppm
Mononitrotoluene Isomers	600	~500[a]
Carboxylic Acids	25	35
Nitrocresols	75	130
Trinitrotoluene	150	150
High Molecular Weight Nitrated Aromatics	180	250
Water	5000	5000
Acidity as H_2SO_4	40	-
Acidity as HNO_3	-	50
Alkalinity as KOH	-	0

[a] Can vary greatly with reactor conditions

The second commercialization hurdle for a single acid process to overcome is a slow reaction rate. As nitration proceeds, water is formed in the reaction. The presence of free water in the nitration medium is known to inhibit the formation of the nitronium cation - the active nitrating agent. It is known that the nitric acid-water medium exhibits a complex set of equilibria involving self-dissociation and aqueous acid dissociation. Increased water levels reduce the concentration of the available nitronium ion, and therefore interfere with the rate of nitration. In a conventional process, water is bound, or tied up, by the sulfuric acid. For a single acid process in a CSTR (continuously stirred tank reactor), we have found that the water concentration during DNT synthesis should be kept below 9%. When the water concentration is low, nitration of toluene to DNT is quite rapid above 40°C. Reaction times to complete DNT synthesis are typically an hour or less. This is the same temperature range as used in the mixed acid process and less than half the required time. With excess nitric acid, reactor conditions can be selected to keep conversion time in the new technology to less than 10 minutes. However, the balance of optimum conditions in the overall process generally reduces the amount of excess recycled acid and therefore requires longer times to complete nitration. To use the excess acid, an efficient means of DNT recovery must be employed.

Efficient recovery of DNT from nitric acid phase is the third commercialization hurdle. Unlike in the mixed acid case, DNT is very soluble in strong nitric acid. Until now, no one had figured out a commercially viable way to recover the DNT from single acid nitration. Just adding water until the DNT phases out leads to a huge energy penalty when it comes time to re-concentrate and recycle the nitric acid. Direct distillation of the mixture also involves a unacceptable energy penalty and a potential safety hazard. A strong nitric acid solution containing DNT must not be overheated in order to prevent TNT formation and a thermal runaway situation. Key to the Olin DNT process is the innovative use of nitrate or mixed nitrate salts to phase separate or "salt out" the DNT from the nitric acid phase. Specifically, magnesium nitrate and mixtures of magnesium nitrate and zinc nitrate, in the form of their trihydrates, are added to the acid-DNT phase to cause the formation of an acid-salt phase and a DNT phase. Crude DNT can then be separated by simple liquid-liquid separation.

This leaves our final concern - how to efficiently regenerate the excess nitric acid. Resent methods (4) that manipulate nitrogen oxide chemistry to remove water have been studied, but have not been commercialized. In this process, regeneration of the recycle nitric acid is accomplished with the use of the nitrate salt. Nitric acid regeneration is accomplished in the acid concentration column. In this column, several different streams are fed into the middle of the column. These streams are diluted nitric acid-salt, wash water containing nitric acid and salt, and fresh trihydrate nitrate salt. Operation is similar to a magnesium nitrate based concentrated nitric acid process (5). The nitrate salt forms a higher hydrate with the water from the acid and moves down the column while the nitric acid is rectified overhead in the column. Contact with the nitrate salt dehydrates the acid to get it above the nitric acid - water azeotrope. Once above this azeotrope, the reflux ratio at the top of the column determines the nitric acid strength leaving the column.

Single Acid Process Advantages

In the conventional process, a large water stream originates from the need to remove oxidation byproducts from the crude DNT. This water stream contains both DNT and oxidation byproducts that are toxic to aquatic life. Limits on these compounds have been consistently reduced and their permitted discharge will become increasingly difficult. Regulations define conventional DNT effluent water to be a listed hazardous waste (K111). Rigorous treatment standards are required before discharge is allowed. Incineration or carbon treatment, which add cost to the final product, must be used to meet the discharge standards for the hazardous waste.

By contrast, the single acid process limits discharge cleanup costs by generating fewer byproducts in the nitration reaction. The byproduct levels are low enough that removal from the final product is not required which avoids the undesirable effects of the conventional product washing. The degree of byproduct formation is determined by reactor conditions. Key reactor parameters include the nitric acid to toluene molar feed ratio, nitric acid strength, and temperature. With the Olin process, conditions are selected to altogether avoid costly downstream cleanup. The chemistry of the conventional technology cannot use this optimization, so downstream cleanup and the associated water cannot be avoided.

Since water is generated by nitration, there will be a water effluent for any DNT process. The new technology includes a distillation step so only a distilled effluent leaves the process. Besides effluent quality, the new process also has a smaller effluent per unit of production. Based on dilute acid feed, conventional technology discharges roughly 1.5 pounds effluent per pound of product. The new technology does not require the addition of external water for product washing. This decreases effluent water to 0.6 pounds effluent per pound of product which reduces the requirement of any downstream treatment. Since the effluent is an overhead distilled product, the energy cost compared to conventional production is slightly higher. The operating and capital costs for the new process are still favorable because of reduced waste treatment and raw material costs.

Not handling the spent sulfuric acid is another advantage of the new technology. For every pound of DNT product, more than a pound of sulfuric acid must be handled and treated. Organics and nitric acid must be removed from the spent acid before it is recycled or used by other processes. Re-concentration of spent acid requires significant capital investment. Finding other uses for the spent acid can avoid some of the capital investment, but that will inevitably link the manufacture of two different product lines.

The new process also offers higher toluene yields. Because of less oxidation and effluent stream DNT losses, more toluene leaves as product DNT. Conventional yields are slightly over 96%. The new process yields have been measured at over 99% during pilot trials.

Conclusion

Olin holds several patents related to the technology - U.S. Patent 4,918,250, 5,001,272, and 5,099,078. Pilot studies are ongoing to optimize and determine the operational characteristics of the new process. Olin Corporation is moving forward with the design and construction of a Demonstration Unit at Lake Charles, LA.

The Olin DNT process, then, is a source reduction process with regard to unwanted and hazardous byproducts. Nitric acid and toluene are the feed stocks and DNT and distilled water are the exit streams. This helps to lower the overall environmental cost of producing TDI and therefore, lowers the environmental impact for the urethanes industry.

Literature Cited

1. Hanson, C.; Kaghazchi, T.; Pratt, M. W. T. *Industrial and Laboratory Nitrations*, Albright, L. F. and Hanson, C., ACS Symposium Series; Gould, R. F. , American Chemical Society, Washington, D. C., 1976, Vol. 22, pp 147-153.
2. Crater, W. de. C., U.S. Patent 2,362,743; (1944).
3. Olah, George A.; Malhotra, Ripudaman; Narang, Subhash C., *Organic Nitro Chemistry Series - Nitrations: Methods and Mechanisms*, Feuer, Henry, VCH Publishers, Inc., New York, New York, 1989, pp 11-12.
4. Carr, R. V. C.; Toseland, B. A.; Ross, D. S., U.S. Patent 4,642,396; (1987).
5. Mandelik, B. G., U. S. Patent 2,860,035; (1958).

RECEIVED January 5, 1996

Chapter 20

Recent Advances in the Technology of Mononitrobenzene Manufacture

A. A. Guenkel and T. W. Maloney

NORAM Engineering & Constructors Ltd., Granville Square, 200 Granville Street, Suite 400, Vancouver, British Columbia V6C 1S4, Canada

New process conditions for the production of MNB were identified and tested on a laboratory scale, with a view to elucidating the mechanism of this reaction and particularly the mechanism of by-products formation. Promising results led to further development in a commercial plant, using a novel reactor system with no moving parts. This reactor offers the potential for improved mass transfer in heterogeneous liquid/liquid systems, resulting in accelerated reaction rates with concomitant improvements in process efficiencies. The optimized reactor system resulted in low levels of by-products and high conversion efficiencies of benzene (99.8%) and nitric acid (99.1%). These results were measured in two world-scale MMB plants which also incorporate integral energy recovery and MNB purification. A process overview is presented with specific references to chemistry background, process safety, energy integration and environmental aspects.

Since 1990, two world-scale MNB plants have been commissioned and have achieved superior process efficiencies through a synergistic combination of a novel reactor, the Jet Impingement Nitrator (JINIT), and unique process conditions. Exceptionally high nitration rates have been demonstrated with rapid and essentially complete conversion of nitric acid. This has resulted in low by-products formation and high conversion efficiencies of benzene (99.8%) and nitric acid (99.1%). The small sulfuric acid process inventory, in conjunction with operation under atmospheric pressure, makes the NORAM plants inherently safe and a nitrator quench system and rupture disks on pressurized process vessels are not required. Total plant emissions of NO_x and volatile organics are well below 0.5 kg/hr. Aqueous streams can be segregated at their sources to allow economic treatment of effluents.

Process Overview

In this process, which is disclosed in a US Patent (1), sulfuric acid is circulated in a closed loop through a JINIT, a MNB-acid decanter, a Sulfuric Acid Flash

Evaporator (SAFE) and back to the suction of a sulfuric acid circulation pump (Figure 1). The JINIT incorporates jet impingement elements, which are described in a second US Patent (2) (Figure 2). These elements, which have no moving parts, create a fine dispersion of benzene and MNB in a continuous acid phase and they are spaced to achieve optimum conditions for mass transfer and process yield. Nitration reaction rates have been demonstrated that are at least an order of magnitude faster than those reported in prior art technologies (3, 4). Nitric acid enters the acid loop at the suction of the sulfuric acid circulation pump, while benzene is distributed evenly through a special Benzene Inlet Manifold (BIM) into the mixed acid stream at the entry of the JINIT.

The acid loop is hydraulically balanced such that most of the head developed by the sulfuric acid circulation pump is utilized to create the dispersion in the JINIT. Full acid circulation is maintained even when the production rate is cut back.

The heat of nitration released in the JINIT is absorbed by the large volume of circulating sulfuric acid and is subsequently utilized in the SAFE for acid concentration by flashing the spent acid under vacuum. The overall energy balance of the acid loop can be maintained by preheating the feed nitric acid, and with heat exchange between the crude MNB and the benzene feed. A start-up heater is required in the acid loop to bring the plant to normal operating temperature.

Vapors flashed in the SAFE are condensed and collected in a condensate barometric seal tank for transfer to the extraction and stripping areas of the plant. Inerts are evacuated through a vacuum pump or steam jet. Crude MNB overflows by gravity from the MNB-acid decanter directly into the first extraction stage, where mineral and organic acids are removed through contact with an aqueous alkaline solution. Extraction is promoted through the use of in-line mixing elements. The nitration process operates with a slight stoichiometric excess of benzene, which is recovered in a stripper for recycle to the JINIT. Aqueous condensate streams not contaminated by nitrophenols are steam stripped to recover dissolved MNB. Aqueous nitrophenols effluent is sent separately to battery limits for treatment. The combined process vents are sent to a NO_x scrubber.

The feature which distinguishes the NORAM process from earlier adiabatic processes (3, 4, 5, 6) is the use of a mixed acid where dissociation of nitric acid to nitronium ion is promoted. The ideal mixed acid is a nitronium ion solution. The process operates with a mixed acid of exceptionally high sulfuric acid concentration and low nitric acid concentration. This mixed acid, when employed under plug flow conditions in the JINIT, results in very low rates of by-products formation. Another interesting feature is that the process operates without the use of agitators in both nitration and extraction.

Chemistry

It is generally thought (7, 8) that, in aromatic nitration, the nitronium ion is the reactive species and that nitration proceeds in a four step sequence. The second reaction, shown below, is believed to be rate controlling. It is, therefore, evident that mixed acid compositions should have concentrations of nitronium ion as high as possible.

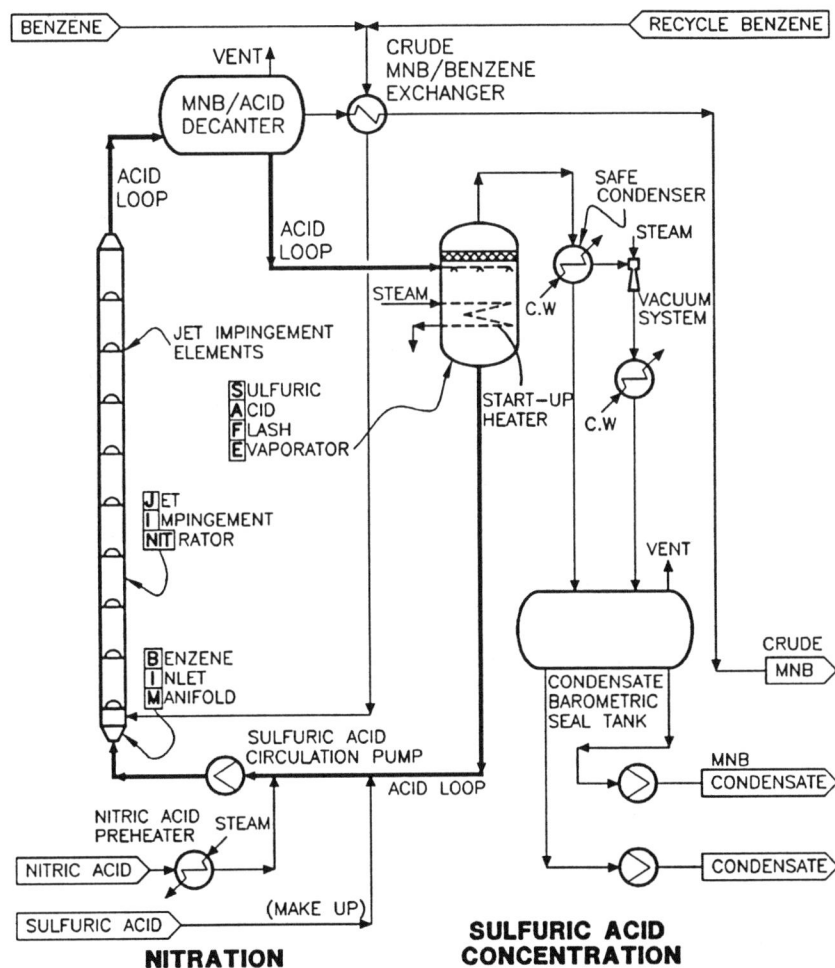

Figure 1. Flow Sheet for the NORAM MNB Process

$$HNO_3 + 2H_2SO_4 \rightleftharpoons NO_2^+ + H_3O^+ + 2HSO_4^- \qquad (1)$$

$$C_6H_6 + NO_2^+ \longrightarrow [C_6H_6(H)(NO_2)]^+ \qquad (2)$$

$$[C_6H_6(H)(NO_2)]^+ + HSO_4^- \longrightarrow C_6H_5NO_2 + H_2SO_4 \qquad (3)$$

$$H_3O^+ + HSO_4^- \rightleftharpoons H_2SO_4 + H_2O \qquad (4)$$

Figure 3 (9) shows contours of constant nitronium concentrating in the ternary system of nitric acid, sulfuric acid and water. A key feature of the NORAM process is its operation with a mixed acid containing less than 3% HNO_3 and a sulfuric acid strength in excess of 70% H_2SO_4. Under those conditions, dissociation of nitric acid to nitronium ion is favored. At the higher nitric acid concentrations used in conventional adiabatic technologies, a substantial fraction of nitric acid remains undissociated. This is of significance since oxidation side reactions involve undissociated nitric acid.

Dinitrobenzene (DNB) is formed when MNB is nitrated. This reaction is a homogeneous acid phase reaction because MNB has a significant solubility in sulfuric acid. The homogeneous MNB nitration competes for nitronium ion with the heterogeneous benzene nitration reaction. In the NORAM process, very low levels of DNB are obtained because of the very high heterogeneous nitration rates in the JINIT and of the high sulfuric acid strengths. In addition, nitric acid is totally consumed under the plug flow conditions of the JINIT. DNB formation in the MNB-acid decanter, the SAFE and process piping is, therefore, minimized.

A number of mechanisms have been proposed (10) to account for oxidation side reactions in MNB production. It has now been established in laboratory studies that nitrophenols are formed through a sequential reaction mechanism which takes place in both the organic and acid phases. In the first step nitric acid diffuses into the organic phase where oxidation of benzene to phenol takes place.

$$2 HNO_3 + 3 C_6H_6 \longrightarrow H_2O + 2NO + 3 C_6H_5OH \qquad (5)$$

Extraction of nitric acid from a mixed acid by nitroaromatics has been reported (11). It has now been shown in laboratory experiments, that mass transfer of phenol from an organic solvent into an acid phase is extremely rapid. In the MNB process, therefore, any phenol formed through oxidation rapidly diffuses back into the mixed acid phase where it is converted to

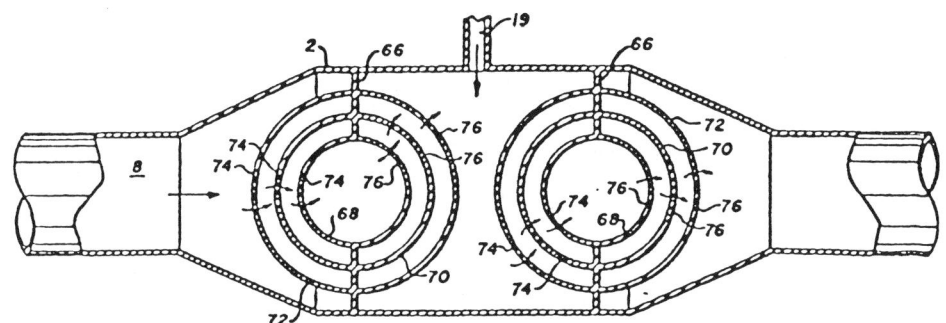

Figure 2: Example of a Jet Impingement Element
(Reproduced from U.S. Patent 4,994,242)

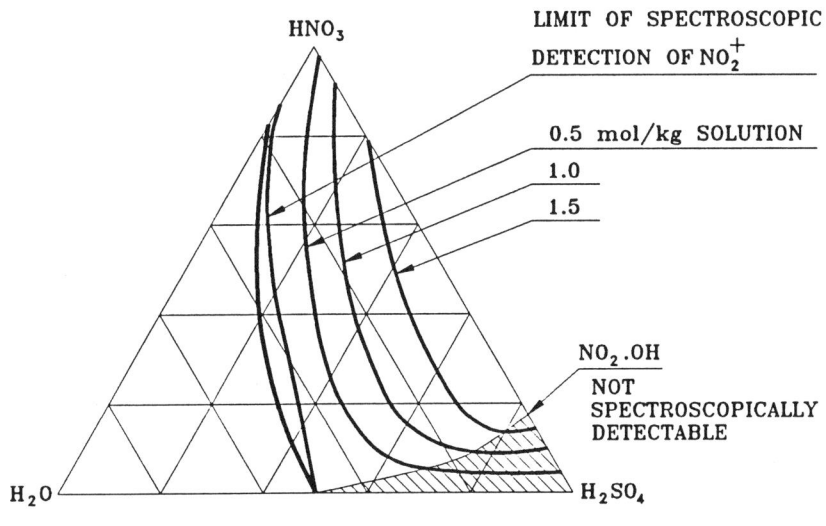

Figure 3: Nitric Acid / Sulfuric Acid / Water System Contours of Nitronium Ion Concentration
(Adapted from Ref.9)

mononitrophenol. This is the fourth step in the reaction sequence. Phenol has not been found in any sample taken from NORAM plants. Further nitrations to dinitrophenol and picric acid are much slower steps and traces of mononitrophenol have been found. The isomers formed under the conditions of this process are 2-nitrophenol (trace), 2,4-dinitrophenol (85%) and 2,6-dinitrophenol (15%). Picric acid, formed in trace quantities only, originates from dinitrophenol. Following formation in the acid phase, nitrophenols diffuse back into the MNB phase and the net nitrophenols production is purged from the nitration loop through the crude product. This is the last step in the reaction sequence.

In the NORAM process, only small amounts of nitrophenols are formed mostly because nitric acid is consumed rapidly under the plug flow conditions of the JINIT and because sulfuric acid strengths are high. Nitric acid oxidation of benzene is minimized, therefore, and it is this second reaction step which ultimately controls the rate of nitrophenol formation.

Nitration grade benzene also contains small amounts of aliphatic impurities, their level depending on the properties of the solvent used in BTX extraction. A typical analysis together with the respective boiling points are shown below (Table I). The boiling points are close to that of benzene and much lower than that of MNB. Since excess benzene is completely recovered for recycle in the benzene stripper, no escape route exists for aliphatic impurities. A significant build-up in the recycle benzene stream has not been observed in a plant environment and no purge has become necessary. It is deduced, therefore, that aliphatics are slowly oxidized in the organic phase as they will remain in this phase. Reaction products found are trace quantities of carbon dioxide and carboxylic acids.

Table I: Trace Impurities in Nitration Grade Benzene

	Typical Concentration (ppm)	B.P. (°C)
Cyclopentane	100	49.3
Methlcyclopentane	50	71.8
Cyclohexane	200	80.7
Methylcyclohexane	100	100.9
Toluene	100	110.6
Benzene	-	81.1
Nitrobenzene	-	210.6

Nitration Rate

The rate of aromatic nitrations has been the subject of numerous studies (12, 13, 14, 15). There is some evidence to support the hypothesis that mass transfer is important in heterogeneous industrial nitrations. For example, nitration rates can usually be increased by increasing agitator speed. However, there is also much evidence to support the hypothesis that nitrations are kinetically controlled. For example, different aromatic compounds react at significantly different rates under otherwise identical conditions of mixed acid composition, temperature and agitation (12). Substantial changes in nitration rates resulting from very small changes in sulfuric acid strength can only be rationalized in terms of a kinetic model, for example, by postulating a variation in the activity of nitronium ion (13). Physical properties simply do not change significantly over small ranges of mixed acid strength. Finally, the Arrhenius type temperature dependence of nitration rates also suggests kinetic control (13).

A model of aromatic nitration has been proposed (16) where the dispersed organic phase is surrounded by a "reaction zone" in which nitration occurs homogeneously. Reactants and products diffuse into and out of this reaction zone. This model makes allowances for both kinetic and mass transfer effects. It has also been suggested (17) that the reason for the substantial variation in nitration rates of aromatic compounds is the difference in their solubility in mixed acid. This theory can be merged with the above reaction zone model by taking solubility into account in this domain.

In industrial heterogeneous nitrations the situation becomes even more complex, because coalescence and redispersion of the organic phase occur simultaneously as process fluids circulate through a stirred nitrator. It has been shown (18) that dispersion of an organic phase in an aqueous phase in a stirred vessel is a surprisingly slow process, taking 10-20 minutes even on a bench scale. Evidently, fluids have to pass many times through the impeller domain before the final equilibrium drop size distribution is reached. On scale-up to an industrial size, this dispersion process will become much slower, which may have significant implications on rate and conversion efficiencies in industrial nitrations.

It is believed, therefore, that the fundamental knowledge required for scale-up of nitrators is inadequate. Degree of agitation, the reactants' feed method, mass transfer, kinetics and solubility all play a role in nitration processes and all parameters will affect yield and conversion efficiency. In addition, it has been shown through plant trials that the mechanism controlling rate and by-products formation changes as the reaction proceeds to completion.

The JINIT: A Novel Type of Heterogeneous Liquid/Liquid Reactor

The potential benefits of the unique process conditions of the NORAM system were initially identified in a bench scale stirred batch nitrator. Encouraging results, namely the observed low levels of by-products formation, provided the motivation for a further systematic experimental study of benzene nitration rates in mixed acid. In parallel, experiments were carried out aimed at verifying the postulated mechanism of by-products formation. It then became the objective to create an industrial scale reactor system which could achieve very high benzene nitration rates while minimizing by-products formation. This reactor system is the JINIT, otherwise known as the Electrophilic Reactor.

It was recognized early during the development that a realistic assessment of the rate of formation of by-products in any nitration process, reactions occurring in the sulfuric acid recovery process should also be taken into account. This fact is often overlooked in examples given in the patent literature, where nitrophenols dissolved in the spent acid are not accounted for (4, 6). In this process, the main nitration reaction as well as side reactions can take place anywhere in the sulfuric acid loop, including the JINIT, the acid-MNB decanter, the SAFE and all interconnecting piping, provided that nitric acid or nitronium ion and benzene or MNB are present. For a realistic assessment of the rate of formation of by-products it is absolutely necessary to examine the nitration process in conjunction with the sulfuric acid recovery system. If one were to undertake a program to pilot a new nitration process, nitric acid conversion rate and efficiency, residence time in various parts of the process, temperature and acid composition history would have to conform to those of the industrial scale plant in order to provide confidence in a projected rate of by-products formation. Since this is an almost impossible task, the NORAM process has been developed mostly in a full scale plant, with progress being made through stage-wise equipment replacement, dedicated plant trials, sample analysis and data interpretation followed by further equipment replacements. Similarly, the hydraulics of the JINIT were proven at full flow rates in a test facility. This learning process culminated in the subsequent construction of two world scale grass roots facilities, where operating performance exceeded expectation in every aspect. The following are some of the key features of the JINIT:

1. **Plug Flow.** The JINIT is a plug flow reactor, an important feature for a number of reasons. Under plug flow conditions, feed acid is exposed to feed benzene only resulting in rapid conversion of nitric acid. Also, plug flow reactors generally achieve better conversion efficiencies than back mix reactors. In the JINIT, nitric acid is consumed essentially to extinction so that dinitration in the acid-MNB decanter and SAFE is minimized.

2. **Benzene Inlet Manifold (BIM).** As a consequence of the very high nitration rates in the JINIT, it is important to obtain uniform distribution of benzene in the mixed acid stream at the entry point. The goal is to maintain close proximity between mixed acid entering the JINIT and a proportional amount of benzene so that high reaction rates are achieved. Injection under back mix conditions is not beneficial. The BIM achieves this objective with negligible pressure drop. To put the task into perspective, in a small commercial plant 200 litre per minute (50USGPM) of benzene must be distributed into 2500 litre per minute (650USGPM) of mixed acid.

3. **Jet Impingement Elements.** It has been demonstrated through plant trials under the conditions of this process that the nitration reaction is mostly mass transfer controlled, more specifically, by the rate of generation of interfacial area between the phases. However, kinetic control also dominates in parts of the nitrator. The Jet Impingement Elements create interfacial area by forming high velocity jets, by shearing the two phases and by allowing these high velocity liquid jets to impinge on the downstream segment of the elements. The requirement for interfacial area creation varies throughout the reactor, as the nitration proceeds to completion. This can be accommodated through appropriate spacing and design of the Jet Impingement Elements. Nitration rates have been demonstrated on a plant scale through the measurement of temperature profiles and volumetric throughput which are an order of magnitude higher than those reported in the patent literature for conventional adiabatic technology (2, 4). As a result of the high reaction rates it has become feasible to operate the process with a unique mixed acid containing unusually low concentrations of nitric acid and specifically of undissociated nitric acid, thereby reducing the rate of by-products formation.

The benefits of the system, therefore, are achieved through synergistic combination of unique process conditions operating in the JINIT. It has not been possible to segregate the relative importance of the process conditions and of the JINIT, the NORAM process requires both. In other applications the concepts of the JINIT alone may offer advantages to promote liquid/liquid reactions.

Process Safety

Because of the exothermic nature of nitration reactions, process safety must be considered carefully in the development of any new process. Therefore, to establish input data for extensive mathematical modelling of the NORAM process, differential scanning calorimeter and pressure bomb tests were carried out. It was found that, in confined systems, a reaction takes place at about 200°C between MNB and sulfuric acid, where one mole of MNB reacts with one or two moles of sulfuric acid forming a homogeneous dark solution together with an unidentified gaseous compound. The heat release in this one-molar reaction is about 1790 kJ/kg MNB. By comparison, the heat of nitration is 1170 kJ/kg MNB and the heat released during thermal decomposition of MNB at 360°C is about 3000 kJ/kg MNB (19).

The key safety feature of this process is its operation under atmospheric pressure. Boiling conditions would be reached in the MNB-acid decanter in the event of temperature excursions only a few degrees above the normal operating temperature. Also, following a process trip, the MNB inventory in the nitrators will quickly rise to the MNB-acid decanter. Additional heat input, for example from an external fire around the decanter, would safely be dissipated by vaporizing and venting benzene and MNB. Computer simulations have shown that all MNB will have been vented from process at temperatures well below those required to initiate secondary exothermic reactions. In addition, in the absence of MNB, these reactions cannot take place. There is, therefore, no justification for a nitrator quench system or a need for rupture disks on pressurized process equipment, as practiced with conventional adiabatic technology (20).

Another important safety feature of the process is its very small sulfuric acid hold-up, with the MNB-acid decanter accounting for more than half the process inventory. Even a minor spill, caused for example by a gasket failure, would quickly be detected through an alarm in the control room and would be followed by a process trip. This feature is important since MNB plants run essentially unattended. In a plant environment, a minor spill may be about 1 m^3 of sulfuric acid, which is easily contained in a sump.

A final notable safety feature is the very small process inventory of unreacted nitric acid, about 25 kg in a world-scale plant. This small inventory results from the very high nitration rates which are achieved in the JINIT. Following a process trip, even this small quantity would be consumed to extinction within a few seconds.

Environmental

The NORAM MNB process offers a number of environmentally attractive features which are expected to become the industry standard. Atmospheric operation of the nitration train minimizes permeation of benzene through gaskets. With the exception of a short pipe run at the discharge of the sulfuric acid circulation pump, the process operates at atmospheric pressure or vacuum conditions. Low rates of by-products formation result in low emission rates of nitrogen oxides. In a recently commissioned plant, total NO_x emission is consistently below 0.5 kg/hr. Finally, picric acid formation has been reduced to trace levels, a feature which greatly facilitates effluent treatment. The bulk of the condensate effluent from the plant is segregated at its source and can be treated in a conventional biological treatment facility. Nitrophenols produced in process are segregated in a very small "nitrophenols effluent" stream. A number of site specific economic methods are available to treat this stream.

Process Performance

The operating requirements for the NORAM MNB process are summarized in Table II. Benzene requirements correspond to 99.8% of stoichiometry, while nitric acid requirements correspond to 99.1%. Product MNB contains less than 100 ppm DNB and less than 10 ppm nitrophenols. The rate of total nitrophenols production corresponds to less than 2 kg per ton of MNB (2000 ppm) and includes only traces of picric acid.

Table II: Operating Requirements per ton of MNB Product

Raw Materials	Benzene	635 kg	
	Nitric Acid (100%)	515 kg	
	Sulfuric Acid (100%)	1.5 kg	
	Sodium Hydroxide (100%)	7 kg	a
Utilities	Cooling Water (30°C)	40 ton	
	Steam	0.2-0.4 ton	b
	Electricity	10 kWh	
	Process Water	not required	

a   Includes requirements to neutralize all aqueous streams at battery limits.
b   Includes requirements for sulfuric acid concentration, condensate and benzene stripping; consumption depends on nitric acid concentration.

Conclusions

A novel type of reactor, the Jet Impingement Nitrator, has been developed on a plant scale. This nitrator, operated at unique process conditions, has resulted in a new energy efficient MNB process. A substantial reduction in the rate of nitrophenols formation has been achieved with virtual elimination of picric acid. The process has a number of inherent safety features and environmental benefits, mostly resulting from atmospheric operation and from low process inventory. Exceptionally high nitration rates have been demonstrated with consequentially rapid conversion of nitric acid to extinction.

The Jet Impingement Reactor has potential applications in many other liquid/liquid and gas/liquid systems.

Literature Cited

(1) US Patent 5,313,009 (1994); Guenkel A.A; Hauptmann E.G.; Rae J.M.
(2) US Patent 4,994,242 (1991); Hauptmann E.G.; Rae J.M.
(3) US Patent 4,973,770 (1990); Evans C.M.
(4) US Patent 4,091,042 (1978); Alexanderson V.; Trecek J.B.; Vanderwaart C.M.
(5) US Patent 2,256,999 (1941); Castner J.B.
(6) US Patent 4,021,498 (1977);Alexanderson V.; Trecek J.B.; Vanderwaart C.M.
(7) Hanson C.; Marsland, J.G.; Wilson, G. *Chemy. Ind.* **1966**, pp. 675
(8) Hughes, E.D.; Ingold, C.K.; Reed, R.I. *J. Chem. Soc.* **1950**, pp. 2400
(9) Urbanski, T.; *Chemistry and Technology of Explosives*; The MacMillan Company: New York, N.Y., 1964; Vol. 1, pp 24

(10) Hanson, C.; Kaghazchi, T.; Pratt, M.W.T. In *Industrial and Laboratory Nitrations*; Albright, L.F.; Hanson, C., Ed.; ACS Symposium Series; American Chemical Society: Washington, D.C., 1976, pp. 132-155
(11) Schiefferle, D.F.; *Organic-Phase Aromatic Nitration*; M.S. Thesis; Purdue University, West Lafayette, IN., 1970
(12) Hanson, C.; Marsland, J.G. *Chem. Eng. Sci.* **1974**, 29, pp. 297
(13) Strachan, A.N.; In *Industrial and Laboratory Nitrations*; Albright, L.F.; Hanson, C., Ed.; ACS Symposium Series; American Chemical Society: Washington D.C., 1976, pp. 210-218
(14) Schiefferle, D.F.; Hanson, C.; Albright, L.F. In *Industrial and Laboratory Nitrations*; Albright, L.F.; Hanson, C., Ed.; ACS Symposium Series; American Chemical Society: Washington D.C., 1976, pp. 176-189
(15) Hanson, C.; Ismail, H.A.M. *Chem. Eng. Sci.* **1977**, 32, pp. 775
(16) Giles, J.; Hanson, H.; Ismail, H.A.M. In *Industrial and Laboratory Nitrations*; Albright, L.F.; Hanson, C., Ed.; ACS Symposium Series; American Chemical Society: Washington, D.C., 1976, pp. 190-209
(17) Hanson, C.; Ismail, H.A.M. *J. appl. Chem. Biotechnol*, **1975**, 25, pp. 319
(18) Narsimhan, G.; Nejfelt, G.; Ramkrishna, D. *AICHE J.* **1984**, 30, pp. 457
(19) Bretherick, L.; *Handbook of Reactive Chemical Hazards*; Butterworths, 1985; pp 579
(20) Silverstein, J.L.; Wood, B.H.; Leshaw, S.A. *Loss Prevention*, **61981**, 14, pp. 78

RECEIVED January 5, 1996

Chapter 21

Industrial Nitration of Toluene to Dinitrotoluene

Requirements of a Modern Facility for the Production of Dinitrotoluene

H. Hermann, J. Gebauer, and P. Konieczny

Josef Meissner GmbH and Company, Bayenthalgürtel 16–20, 50968 Cologne, Germany

Technical and chemical aspects of the industrial nitration of toluene to mono- and dinitrotoluene in mixed acid are discussed.
In modern mixed acid nitration plants for DNT the spent acid from the MNT-stage is purified, reconcentrated, and recycled back into the nitration process. Thus the consumption of sulfuric acid per one tonne of DNT is reduced to almost zero. Moreover, also the sulfuric-, nitric-, nitrous acid and MNT/DNT from the washing of the crude DNT and from the purification and reconcentration of the MNT spent acid are recovered and recycled back into nitration. By doing so not only the nitrate load of the waste water from a DNT nitration plant is reduced by 95 % but also the consumption figures for nitric acid are considerably improved. More than 98 % of the nitric acid needed for nitration can thus be converted to DNT.

Dinitrotoluene (DNT), a starting material for the manufacture of toluene diisocyanate (TDI) is one of today's basic chemical commodities. The production output of about 1 million metric tonnes per year and an estimated growth rate of 4 - 8 % of TDI per year (*1*) turn it into the most important product which results from the nitration of toluene. Thirty years ago, things were different. Main products from the nitration of toluene were the mononitrotoluenes (MNT), as intermediates for the manufacture of dyestuff, drugs, optical brighteners, etc. (*2*) and trinitrotoluene (TNT).
  Research work in the field of toluene nitration was consequently dealing with questions as to how to nitrate toluene to yield MNT or TNT, how to optimize the manufacturing process for these products, how to minimize the meta-MNT-content and how to shift the o/p-MNT ratio in favor of the p-MNT (*3*). This has changed dramatically since that time. TNT has been replaced by high-performance explosives. Compared to the amount of DNT manufactured, the quantity of various

industrially produced MNT-isomers and TNT is relatively small, about 80,000 tonnes of MNT-isomers were produced in 1985 (2).

In terms of economy, two-step nitration of toluene to DNT in mixed acid is still the most favorable procedure for manufacturing DNT.

Chemical Aspects of Nitration of Toluene to DNT in Mixed Acid

Industrial nitration of toluene to DNT in mixed acid is always performed in a heterogeneous two-phase system. Three chemical parameters are important: The rate of nitration, the isomer distribution, and the amount of by-products formed (HNO_2, nitrocresols, products of oxidation and decomposition).

Rate of Nitration. The conversion of toluene to MNT and MNT to DNT only takes place in the aqueous phase of the aqueous heterogeneous system. Two totally different types of kinetics are observed when nitrating toluene to DNT in mixed acid:

The nitration of toluene at a constant acid concentration in the constant flow stirred tank reactor (CFSTR) was thoroughly examined (4-12). Within the technically interesting range (70 - 72 % b.w. H_2SO_4 and 30 mole % respectively) the rate of nitration of toluene to MNT is governed by the transfer rate of toluene from the organic phase to the aqueous phase, and thus by the interfacial area between the two phases (8,9,13-15). Hence, the technical parameters like agitating speed, design of the reactor, and type of emulsion have a deciding influence on the nitration of toluene.

The nitration of MNT to DNT in the aqueous heterogeneous system with mixed acid has not been examined in such detail as has been the nitration of toluene to MNT. The nitration of the MNT to DNT is performed in mixed acid at acid concentrations in the range of about 78 - 82 % b.w. sulfuric acid. Its rate is no longer controlled by mass transfer but kinetically (14-18) and strongly depends on the acid strength (19-21).

In industrial nitration the excess of nitric acid in the DNT spent acid is selected in such a way that, at a given residence time and space velocity, the amount of MNT still unconverted and the amount of TNT in DNT are within product specification (*Table I*).

Isomer Distribution. Besides the product purity according to specification the isomer ratio of 2.4-DNT / 2.6-DNT is of major importance for TDI manufacture. It should be 4 : 1 (80/20) and is principally adjusted during nitration of toluene to MNT by the o/p-MNT-ratio in the MNT-isomer-mixture.

The isomer ratio on the MNT-stage depends on the nitration temperature and the acid concentration employed. In particular the amount of meta-isomer in MNT-isomer-mixtures and thus the portion of the so-called "ortho isomers" in the DNT-isomer-mixture will be reduced when nitrated at low temperatures in the MNT-stage (13,22). The o/p-MNT ratio varies little with temperature and is only slightly shifted in favor of the p-MNT (13).

In mixed acid nitration the o/p-MNT ratio will be shifted in favor of the p-MNT with increasing acidity in the aqueous phase (*13,16,21*), or in the presence of phosphoric acid in the mixed acid (*23*).

A significant shifting of the isomer ratio is not possible during the nitration of the MNT isomers to DNT within the technically accessible range of parameters (temperature, composition of the acid, phase ratio, type of emulsion, and agitating speed) (*13*).

Formation of By-Products. Besides being nitrated, the toluene will be partially oxidized to nitrocresols due to the oxidative power of the mixed acid (*24, 25*). During the nitration of toluene to MNT from 0.3 to 1.75 % (average 0,7 %) (*Table II*) nitrocresols are formed, which are mainly dinitro-p- and dinitro-o-cresol (80 % 2.6-dinitro-p-cresol). Per each mole of nitrocresol, one mole of nitric acid is consumed and converted to HNO_2. With increasing acidity in the mixed acid, for instance from 70 % H_2SO_4 to 75% H_2SO_4 in the spent acid, the amount of cresols formed decreases by about 25 % (*26*).

At a higher acid concentration in the DNT step (80 - 82 % H_2SO_4 b.w.) the nitrocresols formed in the MNT step (70 - 72 % H_2SO_4 b.w.) are further oxidatively degraded (*14,15,27-29*). The amount of nitrocresols will be reduced from about 0.7 % in the crude MNT to 400-1300 ppm (average 850 ppm) in the crude DNT (*Table II*). The nitrocresols which remain in the crude DNT are, besides trinitro-ortho- and trinitro-para-cresols, mainly trinitro-meta-cresol (*44*).

It is surprising that dinitrocresols are almost completely oxidized in the DNT nitration mixture which has a concentration of 82 % sulfuric acid, although nitrocresols are supposed to be stable under this condition (*30*). The high concentration of NO_2 in the crude DNT (*Table II*) might be responsible for the rapid oxidative degradation of the nitrocresols (*25,31*). The oxidation does not only yield CO_2 and H_2O, moreover small organic molecules like formic acid, acetic acid, hydrocyanic acid (*29*), oxalic acid, etc. are obtained. Assuming that about 11-12 mole of nitric acid are consumed during oxidation of nitrocresols on the DNT stage (as in case of the oxidative degradation of 3.5-DNT (*28*)), then almost all of the nitrous acid formed at the DNT stage originates from the oxidative degradation of the nitrocresols formed at the MNT-stage.

Industrial Manufacture of DNT

Each industrial nitration process of toluene to DNT consists of three parts:
- Nitration of toluene to DNT
- Purification of the crude DNT
- Treatment of the effluents.

Nitration. When toluene is nitrated to DNT in mixed acid, besides kinetic aspects, the following parameters determine the selection, design, and operation of a nitration plant for the manufacture of DNT:
1. The quality of the raw materials like sulfuric acid and nitric acid and their concentrations. The lower the respective concentrations, the more water will be fed into the mixed acid and the larger the amount of spent acid which has to be disposed.

2. The materials employed for all parts which come into contact with the mixed and spent acids must be corrosion resistant to sulfuric acid. Sulfuric acid of about 70 % b.w. is a limit value for most types of stainless steel.
3. The actual heat of reaction to be removed and hence the heat exchange capacities of the nitrators.
4. General safety aspects and legal safety requirements.
5. The following requirements should be met:
   - The quantity of spent acid obtained per each tonne of DNT should be as low as possible
   - Nitration should be performed in a two-phase liquid state
   - Nitration of toluene to MNT should be performed with fortified spent acid from the nitration of MNT to DNT.
   - Toluene should be completely converted to the desired product, and the residual content of MNT and TNT in the product should not exceed the values specified for the final product (*Table I*), so that additional purification processes are not required for removal of said substances.

Table I: DNT-Isomer-Mixtures (Technical Grade)

		Standard Specific.	Supplier I	Supplier II	US-Patent 4 224 249	Germ. Appl.[e] DE 3 705 091
Purity	%	min. 95.8[a]	96.01	96.06	95.33	96.65
Σ 2.4-,2.5-,2.6-DNT Isomer Ratio (2.4-/2.6-DNT)		80/20 ± 1	79.9/20.1	79.8/20.2	79.7/20.3	80.60/19.40
Other Isomers (tot)[c]		-	-	4.04	4.11	-
(2.3- +3.4-DNT)	%	max. 4.2	3.97	3.96	4.08	3.35
(2.5- +3.5-DNT)	%	max. 1.0	0.63	0.56	0.78	0.55[b]
MNT	%	max. 0.1	0.003	<0.01	0.05	traces
TNT	ppm	max. 500	20.0	<200	730±40	traces
H_2O -Content	%	max. 1.0	0.07[d]	0.54	0.44	-
Solidif. point	°C	55-57	56.62[a]	56.2-56.4[a]	56.2-56.3	-
Sulfur	ppm	max. 5	0.64	n.d.	n.d.	-
Nitrocresols	ppm	max. 100	8.9	5.4-10.2	8.0	-
Ashes	%	max. 0.1	-	-	nil	-
Alkalinity[f]	ppm	max. 40	-	-	nil	-
Acidity[g]	ppm	max. 40	-	-	1.0	-

a) sample dried; b) only 2.5-DNT; c) 2.3-, 3.4-, 3.5-DNT; d) product after drying; e) *C.A.* **1988**, *109*, 233 186z; f) as NaOH; g) as H_2SO_4

It was proposed to perform nitration of toluene to DNT with mixed acid in one step directly (*32-35*).

In all known industrial processes for the continuous manufacture of DNT in mixed acid, DNT is manufactured from toluene in a two-step, countercurrent process (*Figure 1*) (*36-39*).

In both steps nitration always takes place in the spent acid with an excess of nitric acid (about 1.05 - 1.08 mole %) and a given residence time in order to achieve a complete conversion of MNT to DNT in the DNT-stage and in order to avoid the formation of black acid in the MNT-stage. If nitrating with an excess of nitric acid in the spent acid, considerable amounts of nitric acid and NO_2 are obtained in the "crude DNT" and the "crude MNT". The amount of dissolved nitric acid and NO_2 in the organic phase increases with acidity (*Table II*) and parallels the increase of vapor pressure of nitric acid and NO_2 in the mixed acid with dissolved nitrosylsulfuric acid (*19,40*).

Table II: Industrial Nitration of Toluene to DNT - Classical Process, Equilibrium Stage at the MNT-and DNT-stage

		Extraction	MNT Stage [a]		DNT Stage [b]	
		Waste Acid (black acid)	Spent acid (MNT)	Crude MNT[c]	Spent Acid (DNT)	Crude DNT
MNT/DNT	kg/t	-	-	796.8[c]	-	1006[c]
H_2SO_4-100	kg/t	602.30	602.30	-	602.3	-
Spent acid	kg/t	854.80	858.80	-	774.6	-
Toluene	%	n.d.[d]	n.d.[d]	0.15	none	none
MNT	%	0.06	0.15	94.89	none	0.02
DNT	%	0.08	0,300	4.96	5.1	99.90
TNT	%	-	-	-	-	<0,01
Nitrocresols	%	0.1-0.25	0.04-0.15	0.3-0.75[i]	n.d.	0.04-0.13[k]
Oth. org.mat.	%	n.d.	abt .0.28[f]	n.d.	n.d.	abt. 0.1460[e]
H_2SO_4	%	70.46	70.5	-	77.96	0.6-0.7
HNO_3	%	0.01	1.10(0.82)[h]	0.91[g]	1.48(1.26)[h]	1.75[g]
HNO_2	%	1.98	2.09(0.32)[h]	0.69[g]	1.94(0.55)[h]	1.12[g]
COD[m]	mg/l	19,940	13,630	-	n.d.	-
TOC	mg/kg	n.d.	ab. 3200	-	n.d.	-

a) at 45°C; **b)** at 70°C; **c)** only Toluene, MNT, DNT, TNT; **d)** n. d. = not determined; **e)** from TOC in acidic waste water without DNT/MNT, as DNT equivalent; **f)** from TOC, without DNT/MNT, as DNT equivalent; **g)** calculated taking into account the equation $2NO_2 + H_2O \rightarrow HNO_3 + HNO_2$; **h)** apparent distr. coeff. org. phase/acid phase (weight %); **i)** = av. 0.72; **k)** = av. 0.0870; **m)** without COD for HNO_2 (only for carbon)

Nitration is performed by intensive mixing of both phases which form a homogeneous emulsion of the organic in the acid phase or of the acid in the organic phase. The type and stability of emulsion formed depend on the phase ratio, the mode of agitation, and the toluene quality. MNT and especially DNT/acid mixtures

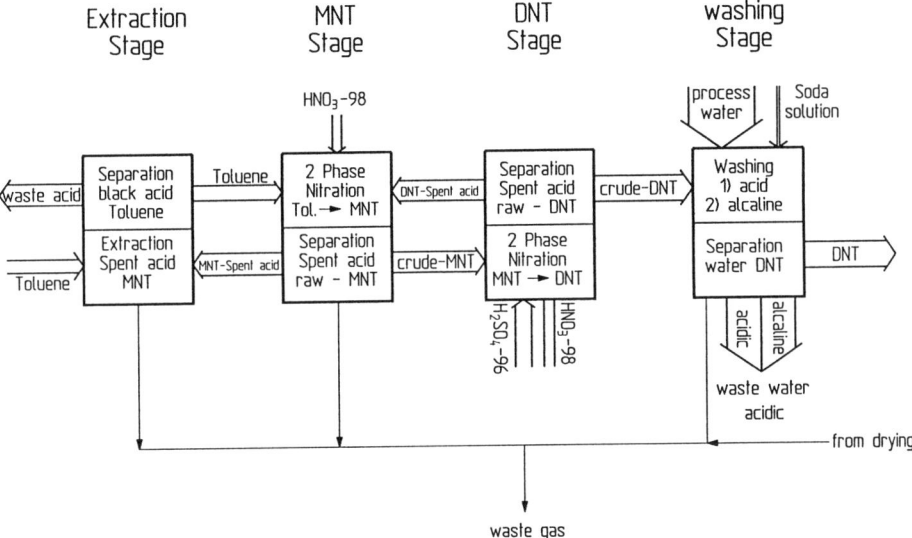

Figure 1. Classical 2-Stage Nitration of Toluene to DNT with Spent Acid Extraction

tend to form quite stable and difficultly separable emulsions with the acid phase dispersed in the organic phase (*41*).

In the first nitration step **(MNT- stage)**, at temperatures between 35 and 45 °C, selective conversion of toluene to MNT takes place in the MNT spent acid (70-72 % sulfuric acid with about 0.3 - 0.7 % b.w. nitric acid (*Table II*), isomer ratio o:m: p-MNT in a current sample is 58.6 : 4.25 : 37.15 on average). After phase separation the MNT spent acid is transferred to a final treatment for MNT spent acid in order to remove the excess of nitric acid, the dissolved MNT/DNT, and the nitrous acid.

The "crude MNT" (*Table II*) which still contains all nitrocresols, nitric acid, NO_2, and degradation products, is directly transferred to the second nitration step **(DNT-stage)** without further purification. There it is converted to DNT in the DNT spent acid (80 - 82 % sulfuric acid with about 0.7 - 1.2 % b.w. nitric acid) at 60 - 70 °C. The 2.6-dinitro-p-cresol dissolved in the crude MNT is oxidatively destroyed and nitrous acid (about 11 - 12 mole/mole dinitrocresol) is formed. After the crude DNT has been separated from DNT spent acid, the DNT spent acid is fortified with nitric acid (65 or 98 % b.w.) and is then used for the nitration of toluene at the MNT stage.

The "crude DNT" which, amongst nitrocresols, also contains nitric acid, sulfuric acid, NO_2, and dissolved degradation products from the oxidative degradation of nitrocresols from the MNT stage (*Table II*) is transferred to the washing stage.

This close link-up of both nitration steps without removal of the by-products dissolved in the "crude MNT" (e.g. nitrocresols) is responsible for a number of problems typical of toluene nitration. These problems are the formation of large amounts of nitrous acid and thus a loss of about 30 kg nitric acid per tonne DNT, the formation of hydrocyanic acid (HCN), reduced nitration rates and the formation of black acid in case the MNT spent acid is extracted with toluene. All these problems do not arise, if nitration of MNT to DNT is performed with purified MNT (*29*).

Acceptable nitration rates and selectivity are achieved, if the nitration of MNT to DNT with mixed acid is performed within the range of 38 to 51 mole % sulfuric acid (about 77 - 85 % b.w.) in the DNT spent acid. Industrial practice, however, has shown that nitration of MNT to DNT in the range of 39 to 42.5 mole % sulfuric acid in the mixed acid (80-82 % b.w. in the DNT spent acid) are optimal for selectivity, nitration rate, and consumption of sulfuric and nitric acid per tonne DNT.

With sulfuric acid 96 - 98 % and nitric acid 98 % (or 63 - 67 %) a volume-based phase ratio (acid phase/organic phase) at the DNT stage from 0.8 to 1.6 is achieved. Characteristic consumption figures for one tonne of DNT in a **classical plant** with extraction of spent acid (*Figure 1*) are: 518 kg toluene per tonne, 630 kg sulfuric acid 96 % per tonne, and 760 kg nitric acid 98 % per tonne. There are a number of technical solutions for performing industrial nitration.

If nitration is performed in loop reactors (*42*), as in the MEISSNER process (*39,43*), an optimal emulsification of the two phases and a very effective heat exchange are achieved. Dynamic separators (centrifuges) are an effective means for separating even difficultly separable emulsions (acid/organic type emulsion)(*41*).

Additionally, the use of dynamic separators for phase separation helps to reduce the amount of organic material in the nitration building to a minimum. This is nowadays an important safety asset.

Purification of the Crude DNT. The crude DNT still contains contaminants like nitrocresols (mainly trinitrocresols), degradation products from oxidation of dinitrocresols, sulfuric acid, nitric acid, and NO_2 (*Table II*). All these products can be removed by washing with water in three steps at temperatures of about 60-70 °C.

Step 1: Acidic Washing. This acidic washing removes all highly acidic components such as sulfuric acid, nitric acid, and NO_2 and acidic degradation products from the nitrocresols like formic acid, oxalic acid, etc. Due to the high content of NO_2 in the crude DNT (about 1 %) large amounts of NO are formed according to equation 1 which leads to intense bubbling during washing.

$$3 NO_2 + H_2O \Rightarrow 2 HNO_3 + NO \tag{1}$$

Step 2: Alkaline Washing. In step 2 the DNT is washed in one or two stages with a solution which contains an excess of sodium carbonate to remove all nitrocresols (mainly trinitrocresols) (*44*), last traces of NO_2, and all other less acidic components. The solution resulting from this alkaline washing has a deeply red/orange color. Washing at pH 6 for selective removal of the more acidic trinitrocresols has been proposed (*44*).

Step 3: Neutral Washing. All salts entrained in the DNT are washed out in the last washing step. The washing water from this step is transferred to the alkaline washing step.

By means of countercurrent washing and other process measures the total amount of water needed for washing can be reduced from the originally 2.5 to 3 m^3/tonne DNT to less than 1 tonne/tonne DNT, without affecting the product quality.

Drying. After it has left washing step 3, the liquid DNT (at about 60 - 70 °C) meets the specification (*Table I*) but is still saturated with water (0.6 - 0.8 %). For storage and transport purposes the water content of the product should be as low as possible, since the storage and transport containers are mostly made of standard mild steel and in case of longer storage times this could lead to considerable corrosion.

Treatment of the Effluents. Apart from the product three main effluent streams result from nitration which require treatment: spent acid, waste waters, and waste gas.

Spent Acid. After separation of the MNT, the yellowish/green up to light orange colored MNT spent acid (the color depends on the nitric acid content) still contains a residual amount of nitric acid (0.4 - 0.8 % b.w.), HNO_2 (0.4 - 2 % b.w.), dissolved MNT/DNT (0.3 - 0.5 % b.w.), and other degradation products from oxidation of the nitrocresols. The total organic carbon (TOC) of the MNT spent acid mentioned above is about 3200 to 3400 mg/kg (the portion of dissolved MNT/DNT corresponds to a TOC about 2400 mg / kg).

Before being further used, e.g. in the fertilizer industry, or before being further processed to sulfuric acid 96 % this acid needs to be purified. If this spent acid is extracted with the toluene in order to recover the residual nitric acid and the dissolved nitro compounds, as it is common practice in industrial nitration, a deeply black colored acid "black acid" is formed (*16*).

Compared to the original MNT spent acid, the amount of MNT/DNT dissolved in the black acid is reduced, however undefined highly polar and deeply colored decomposition products are formed. This is indicated by the COD value of the black acid which is higher than the COD value of the original MNT spent acid before extraction (*Table II*).

When black acids with high concentrations of nitrous acid are stored for a longer time they tend to spontaneously decompose with sudden degassing. Black acids are difficult to purify.

Waste Waters. All waste waters are saturated with DNT at 60 - 65 °C. The acidic waste waters contain considerably more dissolved DNT than the alkaline or neutral wash waters (2.7 - 5.9 DNT g/l at 65°C with total acid 4.5 - 23.0 % b.w. as HNO_3 compared to 1.2 - 1.5 g/l at 65 °C in the alkaline and neutral washing solution).

Apart from organic impurities (DNT, nitrocresols, etc.) the combined waste waters from the first and second washing step contain large amounts of nitrate (about 1 - 3 %, 25 - 30 kg/tonne DNT), traces of HNO_2 (about 0.1 - 0.2 %), and sulfuric acid (0.2 - 0.6 % , 6.0 - 7.5 kg/tonne DNT). These combined waste waters are always acidic. During cooling down from washing temperature (60-70 °C) the dissolved DNT at first separates as an emulsion which extracts the partly dissolved nitrocresols and other deeply colored impurities and begins to crystallize when the temperature reaches about 40 °C.

Not only DNT, but also the trinitrocresols and other deeply colored traces of organic compounds of unknown structure and degradation products from oxidation of dinitrocresols are difficult to remove or to degrade biologically. Moreover, large amounts of nitrate have to be removed from the waste waters before they can be discharged into an outfall ditch. Since a separate self-supporting waste water treatment for the DNT waste water is always very expensive, it is recommended to have the waste water from a DNT plant treated in an already existing waste water treatment system.

Waste Gas. Large amounts of waste gases containing nitric acid, NO_x, DNT, MNT, and toluene will be obtained as fume exhaust of nitrators and washers and from the drying of DNT.

In older plants the combined waste gases are scrubbed with water to remove the contaminants. The water from the scrubbing is transferred to the washing section.

Since DNT, like all other nitroaromatics, is volatile in steam and since the 2.6-DNT is classified as being carcinogenic, great efforts are required for the treatment of waste gas, waste water and spent acid.

Modern Plant Design

The classical industrial nitration process of toluene with mixed acid has specific disadvantages like the accumulation of large amounts of contaminated diluted sulfuric acids, the formation of unwanted by-products like nitrocresols, nitrous acid, and the formation of black acid, in case the MNT spent acid has been extracted with toluene.

Due to the generally growing awareness of environmental problems and strict pollution control regulations of the last 10 to 15 years, it has become the main objective of research and process design for DNT manufacture to overcome these disadvantages connected with nitration in mixed acid. Attempts were made to avoid the use of sulfuric acid in nitration by nitrating in pure nitric acid (*24,45-47*).

To make further use of the unsurpassed properties of sulfuric acid as catalyst, water-binding agent, and solvent with low vapor pressure for the nitration process, however, development was geared to reducing the effluent streams resulting from mixed acid nitration to zero.

Spent Acid. The main problem in all industrial mixed acid processes is how to deal with the spent acid and especially the black acid. Before the MNT spent acid can be used further for other industrial processes or before it can be recycled into the nitration process after reconcentration, the spent acid has to be purified. Various processes for treating MNT spent acid with and without extraction with toluene were proposed.

Complete Purification/Reconcentration. The most effective method to purify and reconcentrate spent acid for re-use in nitration of toluene and for other uses is the traditional treatment of spent acids in the Pauling vessel (*48,49*). All organic impurities are completely degraded oxidatively. A modern version of this complete purification is the feeding of spent acid into a cracking plant for waste sulfuric acid. The sulfuric acid is split up into SO_2 and H_2O, and virgin sulfuric acid or oleum can be recovered (*48*). All these purification measures are very costly.

Partial Purification/Reconcentration. Complete purification of spent acid is not required for re-use of a reconcentrated spent acid in the nitration of toluene, provided that the quality of the DNT does not change.

A direct reconcentration of MNT spent acid without additional purification prior to reconcentration, e.g. extraction with toluene, and its re-use in nitration is described in EU Pat. 0 155 586 (*50*). In the German Patent Application DE Offenl. 4 238 390 und 4 309 149 (*32*) partial reconcentration of spent acid is described using the heat of nitration for the evaporation of water.

To recover the excess of nitric acid and the MNT and DNT still dissolved in the MNT spent acid by extraction with toluene and to avoid formation of black acid, it was proposed to perform the extraction with a stoichiometric amount of toluene (*51*) or to remove the nitrous acid prior to extraction by treating the MNT spent acid with oxidizing or reducing chemicals like hydrogen peroxide, urea or sulfamic acid (*52, 53*).

Another solution is described in US Pat. 4 496 782. After fortification of the spent acid to 2 % b.w. nitric acid this fortified acid will be reacted adiabatically with MNT at temperatures of up to 114 °C. All the nitric acid will be consumed and the MNT will be converted to DNT (*27*).

To avoid extraction of MNT spent acid with toluene or MNT it has been proposed to recover the nitric acid by flash evaporation at about 200 °C followed by steam stripping of the remaining sulfuric acid for recovery of the MNT/DNT still dissolved. The dilute nitric acid obtained can be recycled back into the nitration after it has been reconcentrated to 65 % b.w. (*54*).

By all these measures only the excess of nitric acid and MNT/DNT from the MNT spent acid are recovered. Until today, however, no process has been described for a complete recovery and re-use of the nitrous acid in the MNT spent acid (the nitrous acid in the MNT spent acid corresponds to about 20 - 25 kg nitric acid/tonne DNT).

The purified MNT spent acid can be reconcentrated in a one- or two-stage reconcentration unit as to yield a sulfuric acid with a concentration of 86 - 93 % b.w. (*27, 50*). This acid still contains 0.1 - 0.4 % nitrous acid and traces of organic contaminants (characterized as TOC = 150 - 250 mg/kg), mainly DNT.

If this reconcentrated MNT spent acid is recycled into nitration of toluene to DNT, the amount of fresh H_2SO_4 96 % required per tonne of DNT is reduced from about 630 kg per tonne of DNT to 20 - 30 kg per tonne of DNT. This quantity compensates the amount of sulfuric acid which is entrained into the waste water from the washing of the DNT and lost during reconcentration of spent acid.

Due to the reconcentration of spent acid and its re-use, the quantity of sulfuric acid needed as catalyst is no longer decisive for the economy of DNT production. Merely the costs for evaporation of the water generated per tonne of DNT and of the water introduced into the spent acid by means of the nitric acid have to be considered.

Waste Water. The quantity of the waste water discharged per tonne DNT can be reduced to less than one tonne / DNT by countercurrent washing and other measures. Before discharging the water into an outfall ditch, DNT and nitrocresols have to be removed. Also nitrite, nitrate and sulfate contents have to be reduced to the set limit values (e.g. 50 ppm for nitrate, 600 ppm for sulfate and 175 mg for COD).

It has been proposed to extract the dissolved DNT from the combined or from each individual waste water stream with toluene (*55*) which can be recycled into nitration (*25*). Thus not only 50 % of the organic load in the waste water are removed but also valuable products can be recovered.

Biological degradation of nitrocresols is difficult. Hence methods of physical treatment are required to remove them from the waste water. Activated carbon or resins are employed to absorb the nitrocresols (*56*), but also the oxidative degradation by means of simple thermolysis at elevated temperature and pressure (*57, 58*) as well as the treatment with hydrogen peroxide (*59*) or ozone followed by biological treatment have been proposed.

After DNT, nitrocresols and other deeply colored organic impurities have been removed, it is still necessary to reduce the amount of nitrate and nitrite in the waste water from the washing of the crude DNT and the steam stripping of the MNT spent acid. In general this is done by biological degradation.

An alternative to biological degradation of nitric acid in the waste water (about 40 - 50 kg HNO_3/t DNT) is the recovery of the nitric and nitrous acid dissolved in the crude DNT and in the MNT spent acid and their recycle back into nitration.

As described in EP Pat. 279 312, up to 10 % of water (preferably about 5 %) are mixed with the crude DNT. After phase separation the acid extract with an acid concentration of about 45-65 % is directly returned to the MNT stage. Thus about 96 % of the entrained and dissolved sulfuric acid and 55 - 72 % of the nitric acid can be recovered (*60*).

Another route is described in US Pat. 4 257 986 in which the crude DNT is extracted with the purified spent acid. A recovery of about 50 % of nitric acid seems possible (*52*).

MEISSNER developed a solution for recovering nitric acid, nitrous acid, and MNT/DNT from all waste streams which result from DNT manufacture, like the waste water from the washing of the crude DNT, from the stripping of the MNT spent acid, and the waste gas from the nitration plant and from the acid cleaning operations (*Figure 2*) (*61*):

More than 98 % of the nitric acid, NO_2, and sulfuric acid dissolved or suspended in the crude DNT are recovered by countercurrent extraction of the crude DNT with a dilute nitric/sulfuric acid mixture in the first washing stage. An acid stream with 35 - 40 % total acidity (expressed as nitric acid % b.w.) is obtained. This acid stream can be refed into nitration either directly or after having been reconcentrated to about 60 % total acidity.

To recover the nitric and nitrous acid from MNT spent acid the latter is stripped with steam or steam/air mixtures. This yields a dilute nitric acid (20 - 30 % b.w.). The excess of nitrous oxides (NO_x) from the stripping of the spent acid and the nitrous oxides in the waste gas of the nitration plant (mainly NO from the washing stage) are recovered in a NO_x-treatment unit.

All these nitric acid streams which result from the washing of crude DNT, the stripping of MNT spent acid and NO_x-treatment unit can be fed back into nitration either directly or after reconcentration. Simultaneously, the MNT/DNT dissolved in the MNT spent acid and in the waste water from the first washing step are also recovered and returned to nitration.

This concept for recovery of nitric, sulfuric, and nitrous acid from crude DNT, MNT spent acid and waste gas treatment was realized in an existing plant. With this concept not only the nitrate load in the waste water from a DNT plant can be reduced dramatically but also more favorable consumption figures are achieved. The reconcentration and recycle of MNT spent acid were integrated in the concept mentioned above, so that the consumption of sulfuric acid is reduced to almost zero. Discharge of sulfuric acid and its replacement by virgin sulfuric acid within this closed loop is only required, if the limit of solubility for salts in the sulfuric acid is exceeded or, if the losses of sulfuric acid during reconcentration have to be compensated. Consumption figures for HNO_3 can be improved considerably. More than 98 % of nitric acid fed into the nitration are recovered as DNT.

Waste Gas. In a modern nitration plant all units and apparatuses are contained and inerted by nitrogen, and the carrier gas applied for drying is conveyed in a circuit. Hence, the amount of spent gas to be treated can be reduced to some cubic meters

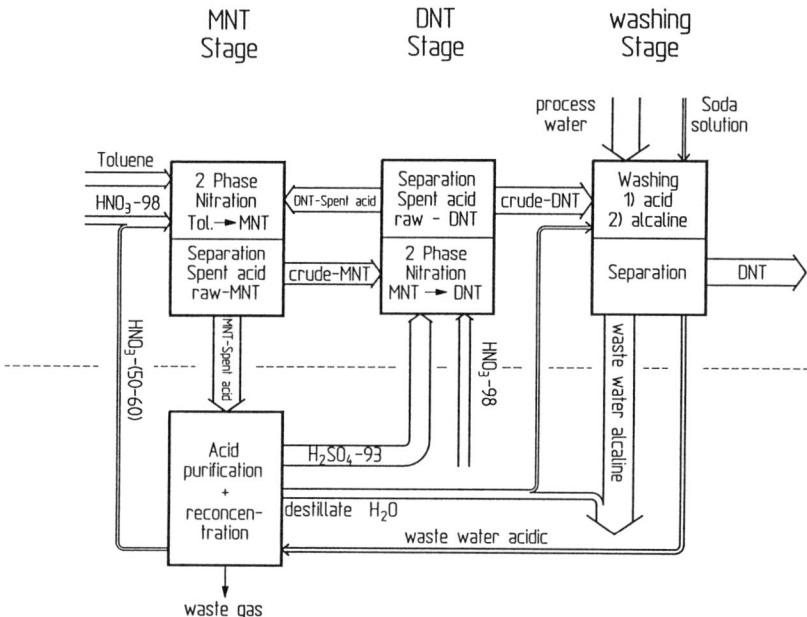

Figure 2. Modern 2-Stage Nitration of Toluene to DNT with Spent Acid Extraction

per hour (62). Before discharging the waste gas into the open air, the concentration of nitrous oxides (mainly NO) (e.g. max. 0.5 g/m^3 NO_2) as well as the organic contaminants like toluene, MNT, and the steam volatile DNT have to be reduced to the legally permitted limit values, e.g. in a NO_X-treatment unit.

Safety Measures.

An integral part of a modern plant for DNT manufacture is the complete safety engineering. Extensive safety measures are required to prevent thermal explosions caused by uncontrolled nitrations and other kinds of non-routine operating conditions and the uncontrolled leakage of reactants into the open air. To ensure the required safety conditions the following prerequisites must be fulfilled:
The reactants - toluene, nitric acid and sulfuric acid - have to be admixed in defined order and specified proportions
- The heat of reaction generated during nitration, mixing and dilution of acid must be removed efficiently
- A homogeneous emulsion of the type o/w or w/o must be maintained permanently.
- The concentrations of the product to be nitrated (toluene or MNT) in the nitration mixture should always be kept at such a low level that, in case of circulation failure, the nitration mixture does not reach decomposition temperature or spontaneous degassing and evaporation are avoided.
- Amounts of critical substances (acids, DNT) inside the plant must be kept as small as possible.

This is achieved by:
- Exact control of mass and energy flow in the plant
- Exact supervision of the raw material feed with redundant flow measurement
- Redundant temperature control
- Use of conditioned cooling water and redundant cooling water control
- Monitoring of proper emulsion conditions in the nitrators by true watts measurement of the circulation pump or stirrer and, in case loop reactors are concerned, by means of flow control
- Redundant temperature and speed control of dynamic separators
- Blanketing with nitrogen to minimize hazard of ignition
- Contained plants to minimize critical escape of spent gases or product vapors (2.6-DNT is a carcinogenic substance)
- Application of a gas pendulum system, e.g. for dynamic separators (centrifuges)
- Spent gas scrubbing with an acid suitable for nitration to prevent toluene, MNT, DNT and nitrous fumes from escaping
- Numerous actions to control level and overflow

All safety measures described above are interlocked according to a hierarchical order of safety. Exceeding the set limit values always effects an immediate disconnection of the raw material dosage. All these measures help to avoid uncontrolled nitration and hence prevent personal injury and harm to the environment.

Literature Cited

1. *Chemische Rundschau* Nr. 50, 16. Dezember 1994, Verlag Vogt-Schild, 4501 Solothurn/Switzerland; *European Chem. News,* January, 23.-29. **1995**, p 10
2. Guenkel, A.A., In *Encyclopedia of Chemical Process and Design,* McKetta, J.J.; Cunningham, W.A.; Eds.: Marcel Dekker Inc.: N.Y./Basel. 1990, Vol. 31, 183
3. *Industrial and Laboratory Nitration*; Albright; L.F.; Hansen, C., Eds.; ACS Symposium Series, American Chemical Society: Washington 1976; 22
4. McKinley, C.; White, R.R., *Trans. A. I. Ch. E.* **1944**, *40*, 143
5. Barduhn, A.J.; Kobe K.A., *Ind. Eng. Chem,* **1956**, *48*, 1305
6. Brennecke, H.M.; Kobe, K.A., *Ind. Eng. Chem.* **1956**, *48*, 1298
7. Kobe, K.A.; Lakemeyer, *J. Ind. Eng. Chem.,* **1958**, *50*, 1691
8. Giles, J.W.; Hanson, C., Marsland, J.G.; *Proc. Ind. Solv. Extr. Conf.* **1971**, *1*, 94 (Soc. Chem. Ind., London)
9. Giles, J.W.; Hanson, C.; Ismail, H.A.M, *ACS-Symp. Series*, **1976**, *22*, 190
10. Hanson, C.; Marsland, J.; Wilson, G.; *Chem. Ind.,* **1976**, 675
11. Strachan, A.N.; *ACS-Symp. Series*, **1976**, *22*, 210 , (s.ref.3)
12. Field, J.P.; Strachan, A.N.; *Ind. Eng. Prod. Res. Dev.,* **1982**, *21*, 352
13. Milligan, B.; *Ind. Eng. Chem. Fundam.,* **1986**, *25*. 782 - 789
14. Kobe, K.A.; Fortman, J.T.; *Ind. Eng. Chem.,* **1982**, *53*, 269
15. Kobe, K.A.; Skinner, C.G.; Prindle, H.B.; *Ind. Eng. Chem.,* **1955**, *47*, 785
16. Milligan, B., *J. Org. Chem.,* **1983**, *48*, 1495
17. Tillett, J.G.; *J. Chem. Soc.* **1962**, 5142
18. Vinnik, M.L.; Grabovskaya, Z.E.; Azamovskova, L.N., *Russ. J. Phys. Chem.,* **1967**, *41*, 580
19. Murthy, A.S.; *Comput.Appl. Chem. Eng. Process, Des. Simul.,* ACS-Symp. Series, American Chemical Society, Washington, 1980, *124*, 403
20. Ross, D.S.; Blucher, W.; *Report ARO-13831.3-CX, AD-A O85 324,* **1981**, *p 23*, C.A. **1981**, *94*, 120 451k
21. Schofield, K. *Aromatic Nitration,* Cambridge University Press, Cambridge, 1980, *Chapter 3* and p. 236
22. Roberts, R.M.; Heiberger, P; Watkins, J.D.; Browder, P.R.; Kobe, K.A.; *J. Amer. Chem. Soc.*; **1958**, *80*, 4285
23. Harris, G.F.P.; *ACS-Symp. Series*, **1976**, *22*, 313, (s. ref. 3)
24. Olah, G.A.; Malhotra, R.; Narang, S.C.; *Nitration, Methods and Mechanism*, Organic Nitrochemistry Series; VCH Publ., New York, 1989; pp 11 ff
25. Albright, L.F.; *Kirk-Othmer, Encycl. Chem. Technol.*, 3rd Edit.; Grayson, M.; Eckroth, D., Eds.; John Wiley & Sons, N.Y. 1980, Vol.15, 845
26. Hansen, C.; Kaghazchi, T.; Pratt, M.W.T.; *ACS-Symp. Series,* **1976**, *22*, 132
27. Carr, R.V.C., *US Pat.* 4 496 782
28. Ross, D.S.; Kirshen, N.A., *ACS Symp. Series,* **1976**, *22*, 114, 1976 (s. ref. 3)
29. Herman, F.L.; Sawicki, J.E.; *US Pat.* 4 361 712
30. Combs, R.G.; Golding, J.P.; Hadjigeorgios, P., *J. Chem. Soc. Perkin 2,* **1979**, 1451
31. Albright, L.F.; *Chem. Eng.* **1966**, *194*

32. Schieb, Th.; Wiechers, G.; Sundermann, R.; Zarnack, U., *German Patent Appl.* DE 4 238 390, A1, 43 09 140 A1
33. Hautze, R.C.; *US PAT.* 3 178 481
34. Brokden, M.E.; Milnes, G.; Pinkerton, H.; *US Pat.* 3 243 466
35. Pol. PL 151 340, C.A. **1991**, *114*, 209 564d
36. Booth, G. In *Ullmann's Encycl. Ind. Chem.*; Elvers, B.; Hawkins, S.; Schulz, G.; Eds.; VCH-Verlagsges., Weinheim, Germany 1991, Vol. *17*, 422
37. Urbanski, T.; *Chemistry Technol. of Explosives*, Pergamon Press, New York, 1984, Vol. 4, p 154
38. Brzezcki, A.; Ploszaj, A.; Stajsczyk, M.; *Pol. Tech. Rev.* (1982) 10- 11, C.A. **1982**, 97 146 532z
39. Toischer, E.L.; Bergmann, H.S.; *US Pat.* 3 434 802
40. Zinkov, Z.W.; Budrina, E.A.; Zakharova, N.C., *Zhur Priklad.Chim.*, **1962**, *35*, 139
41. Somer, T.G.; *J. appl. Chem.* **1963**, *334*
42. Humphry, S.B.; et al, *US Pat.* 3 092 671
43. Meissner, F.; Wandschaff, G.; Othmer, D.F.; *Ind. Eng. Chem.,* **1954**, *46*, 718
44. Toseland, B.A.; Carr, R.V.C.; *US Pat.* 4 482 769
45. Mason, R. W.; et al. *US Pat.* 5 001 272, 4 918 250
46. Quakenbush, A..B., *US Pat.* 5 099 080, 5 099 078
47. *Europ.Chem. News* **1995**, *15 - 21 May*, p. 24
48. Müller, H. In *Ullmann's Encyclopedia of Industrial Chemistry*, 5th Ed.; Elvers, B.; Hawkins, S.; Russey, W.; Eds.; VCH-Verlagsges., Weinheim, Germany, 1991, Vol. 25, 679 ff
49. Geisler, K.; Lailach, G.; Brändle, K.; Pütz, G., *German Pat. Appl.* DE 4 230 099
50. Gerken, R.; Lailach, G.; Becher, D.; Witt, H.; *EP Pat.* 155 586
51. Pohl, M.C.; Carr, R.V.C.; Sawicki, J.E.; *US Pat* 4 650 912
52. Milligan, B.; Huang, D.S.; *US Pat.* 4 257 986
53. De la Mare, G.B.; Milligan, B.; *US Pat.* 3 856 673
54. Parks, D.L.; Martin, J.C.; *E.P. Pat.* 0 415, 354
55. *Pol. PL* 128 198, C.A. **1986**, *105,* 193 325c
56. Cornel, P., *Chem. Ing. Tech.* **1991**, *63*, 969
57. Larbig, W.; *EP Pat.* 005 203
58. Baur, K.C.; Dockner, T.; Kanne, U.; Papkalla, Th.; *US Pat.* 5 232 605
59. Carr, R.V.C.; Martin, C. J.; Gonzalez, R.; Albanese, T.A., *US Pat.* 4 604 214
60. Witt, H.; Beckhaus, H.; *EP Pat.* 279 312
61. Hermann, H.; Gebauer, J.; *German Patent Appl.* DE 195 12 114.7
62. Langecker, G.; Gebauer, J.; Hermann, H., *US Pat.* 4 310 500

RECEIVED January 5, 1996

Chapter 22

Practical Considerations in Concentration and Recovery of Nitration-Spent Acids

C. M. Evans

Chemetics International Company Ltd., 1818 Cornwall Avenue, Vancouver, British Columbia V6J 1C7, Canada

The majority of organic nitrations require the use of sulphuric acid or oleum in the nitration acid. Even in relatively rare nitric acid only nitrations, sulphuric acid is often used as the dehydrating agent to produce 99% nitric acid. The sulphuric acid is never consumed, but is discharged in a diluted form contaminated with organic components and nitric/nitrous species. Pressures are increasing to reconcentrate and reprocess such spent acids.

Acid recovery and concentration is an expensive operation. This paper discusses some of the aspects which must be considered when contemplating acid recovery. In the current climate, acid recovery and recycle should be regarded as an integral part of a nitration process development rather than an afterthought. Case histories will be given in which such considerations influenced the course of the development of the nitration process itself. Emphasis will be placed on the importance of well planned bench and pilot scale test programmes.

Many authors have made reference to the fact that although sulphuric acid is the largest volume chemical used in developed countries, the only process in which it is consumed in significant quantities is that of phosphate fertilizer manufacture which yields a gypsum byproduct. (1-4)
In most other applications sulphuric acid is present as a catalyst, solvent or processing aid and a diluted, contaminated sulphuric acid byproduct stream is produced. Examples of this in the inorganic chemical industry include the manufacture of titanium dioxide pigment by the sulphate process, which produces a 20% strength sulphuric acid stream contaminated with ferrous sulphate,(5) and the pickling of steel with sulphuric acid. In the production of organic chemicals there are numerous examples of the use of sulphuric acid

including alkylation in the petroleum industry, methyl methacrylate production, olefin hydration and sulphonation. However, the purpose of this paper is to consider the area of organic nitrations, in which there are many processes which utilize sulphuric acid or oleum in the nitration step and which produce a contaminated, diluted spent acid stream.

Nitration spent acid traditionally resulted from the manufacture of explosives such as TNT and nitroglycerine. However, significant quantities of such acids now arise from the manufacture of dinitrotoluene and mononitrobenzene which are precursors in the production of TDI and MDI respectively for the polyurethane industry. Nitration is also a common feature in the manufacture of intermediates in herbicide and dyestuffs production.

Increasingly, as environmental legislation becomes more stringent, nitration spent acids must be treated and recovered, either for recycle to the process of origin or for use in another process. The purpose of this paper is to provide an overview of technologies which are available for treating such spent acid streams and to present examples of cases in which, during the development of a nitration process, an integrated approach to the combined nitration/acid recovery process from the outset resulted in a less expensive process.

Nitration Processes

There are three main types of nitration systems; mixed acid nitrations which require the use of oleum, mixed acid nitrations which use concentrated sulphuric acid and nitric acid only nitrations using strong nitric acid.

Oleum based nitrations require oleum for the nitration mix and produce a spent acid which is typically between 70% and 80% H_2SO_4 and contaminated with nitric and nitrous species and with organics. Mixed acid nitrations typically use sulphuric acid of between 86% and 96% H_2SO_4 in the nitration mix and produce a similar spent acid, (with the exception of the adiabatic mononitro- benzene process which utilizes large volumes of sulphuric acid over the 65% to 68% concentration range).

Nitric acid only nitrations use a large excess of 99% HNO_3 resulting in a dilute spent nitric acid stream contaminated with organics.

Disposal Options for Spent Sulphuric Acid

In general, there have traditionally been four principle techniques for handling spent sulphuric acid: (6)
- Discharge without Treatment
- Neutralization
- Direct Use
- Treatment and Recovery

In the context of nitration spent acids, direct discharge into water sources is not an acceptable solution. Neutralization with lime to form gypsum is still practised, particularly in cases in which the quantity of acid is relatively small; for example from multi-product dyestuff production. However, in many cases the gypsum is not of saleable quality and must be stockpiled leading to concerns both about storage space and concerns that contaminants in the gypsum may be leached into the ground water.

Direct use of nitration spent acids in other industries is rarely possible due to the presence of residual nitric/nitrous species and organic contaminants, particularly where there are concerns about the stability of the spent acid on storage and transportation. Therefore, the majority of spent nitration acids must be processed for recycle or reuse in other applications. In such cases a partial treatment is almost always the first step. The objective of the partial treatment is either to render the spent acid suitable for reuse in other applications or as a conditioning stage prior to recovery and recycle.

Partial Treatment

Partial treatment processes need to address several issues, the most important of which are usually:
- Spent acid stability/safety
- Removal of nitric/nitrous species
- Recovery of nitric acid values
- Removal and/or destruction or organics

The most common method for pretreatment of spent nitration acids is denitration via steam stripping. This recovers nitric acid values from the spent acid both by evaporation of residual nitric acid and breakdown of nitrous species to release NOx which can be recovered as nitric acid by absorption. Denitration is usually carried out in glass or glass lined steel packed columns with the heat required for the process being supplied through a combination of indirect heat transfer (via tantalum heat transfer surfaces) and through the injection of live steam. In order to ensure a low residual nitric/nitrous species content the sulphuric acid product from the nitration column should not be more than about 70% strength. In addition to recovery of nitric acid values, denitration in this manner will frequently recover volatile organic compounds which can be recycled (such as MNT in the denitration of spent acids from MNT, DNT or TNT production) and/or decompose organic contaminants as in the case of denitration of nitroglycerine spent acid.

The combined effect of these actions makes the acid more stable for subsequent storage and transportation by minimizing the chances of further reactions occurring and makes it more suitable for further processing. The economic benefits of recovering nitric acid values and potentially recyclable organics can also be appreciable. For processes where 99% HNO_3 is required the denitration step can be combined with the production of the strong acid.

Other techniques which may be applied separately or together with the denitration process described above include solvent extraction, organic oxidation and chemical denitration.

Solvent extraction (for example the contacting of nitrotoluene spent acid with toluene) is a useful technique which must be applied with caution as reactions can occur between the solvent and residual nitrous compounds in the spent acid to produce black nitroso-organic compounds which can cause severe frothing during subsequent processing steps. Organic oxidation can be applied as a distinct separate process or as an after treatment to the denitration step. It typically involves addition of an oxidizing agent (usually nitric acid or hydrogen peroxide) to the spent acid under defined residence time/temperature conditions. This has been applied as an after treatment for DNT spent acid prior to concentration.(7)

A final option is the possibility of a chemical denitration. This is relatively rarely applied but typically involves the addition of a compound from the urea/sulphamic acid family to the nitration spent acid to react with nitrous species to form nitrogen.

Treatment and Recovery

The most common existing methods of spent acid recovery can be categorized into atmospheric pressure concentration, utilizing direct heating by combustion gases (as in the Chemico submerged combustion process) or indirect heating (Pauling process); vacuum concentration (for example the Chemetics process); and total regeneration (thermal decomposition, for example the Stauffer (Rhône-Poulenc) process. No single process can be considered a universal solution for all types of nitration spent acid and each process has its own drawbacks be they economic, environmental or process related.

Regeneration. The major example of an acid which is treated by regeneration is the spent acid which is produced during alkylation in the petroleum industry. Dilution or neutralization or concentration of this acid leads to formation of a tarry organic mass. Following the installation of the first sulphuric acid alkylators in the early 1940's, sulphuric acid regeneration was first practised commercially. In this process, the waste acid is sprayed into a brick-lined furnace and burned in conjunction with a second fuel (either H_2S, hydrocarbon or sulphur). Through operating the regeneration plant at 1000°C, the organics are completely destroyed, and the resulting SO_2 gas stream can be used to produce fresh sulphuric acid.(8)

Other acids which are typically processed via regeneration technology include those from methyl methacrylate manufacture, sulphonation processes and olefin hydration processes.

The principal advantage of this technology is that a relatively pure acid can be produced of any desired strength (including oleum). However, the plants are relatively expensive and spent acids containing substantial amounts of inorganic impurities (except ammonium salts) are undesirable due to the fouling which occurs in the furnace and boiler.

Nitration spent acids are only rarely treated by regeneration. The reasons are twofold. Firstly, nitration spent acids are relatively weak compared to (say) alkylation spent acids which adds to the fuel requirements for the furnace and leads to increased costs, both of operation and for transportation if shipped to a toll regenerator. Secondly, even if already denitrated, a spent nitration acid will contain nitrated organic compounds which will release NOx on decomposition and may pose problems in the regeneration acid plant.

Nonetheless, regeneration may be the only option for certain acids and examples exist for dedicated, on-site regeneration plants for nitration spent acid.(2) A block diagram of an acid regeneration plant is attached as Figure 1.

Concentration. In many nitration processes, the spent sulphuric acid needs only to be reconcentrated, usually after denitration, in order to be reused or recycled. The traditional methods of achieving this have been drum concentration (Chemico process) and pot concentration (Pauling process) both of which operate at atmospheric pressure.

In a drum concentrator, hot combustion gases are used to heat the spent acid to the desired boiling point and remove the evaporated moisture. Although many such concentrators still exist, the brick lined drums can be high maintenance items and removal of volatile organics and acidic components from the exhaust gases can be difficult and expensive.

In pot concentrators, the sulphuric acid is boiled in a silicon iron pot, suspended over a heat source. The heat transfer in such units is poor due to the thermal properties of the pot resulting in severe limitations on capacity. Of greater concern, the pots are subject to catastrophic failure resulting in the boiling acid entering the furnace.

Therefore, for the above reasons atmospheric concentration processes based on the above technologies are not now commonly installed, although many older plants are still operating. Almost all modern sulphuric acid concentrators (SAC's) operate under vacuum, which gives significant advantages. The major benefit is that the acid boiling point is considerably reduced (170°C/10 mms Hg. abs. for 96% H_2SO_4) which extends the range of suitable materials of construction and allows steam to be used as a heat transfer medium resulting in improved efficiency.

The first vacuum concentration unit was the Simonson-Mantius concentrator developed in 1921.(9) Modern SAC's, when operating over the 70% - 96% H_2SO_4 range normal for spent nitration acids, generally comprise an external tantalum reboiler with glass or glass lined steel piping and vessels.

SAC's can operate with either natural thermosyphon or forced circulation reboilers. Chemetics preference, for units operating at high temperatures and strengths, is for natural thermosyphon reboilers, which avoid the risks of leakage which can occur with pumped loops. A schematic flowsheet of a typical natural thermosyphon type single stage concentration unit is shown in Figure 2. Figure 3 is a photograph of a 2-stage SAC for the concentration of nitration spent acid.

Acid concentration is generally used for acids contaminated with a relatively small amount of organics, a typical duty being the concentration of nitration acids to 93% or 96% strength for recycle. It has limited value for processing highly contaminated acids, such as alkylation spent acid, and acids from processes such as TNT manufacture, in which oleum, rather than strong sulphuric acid, is required for the nitration.

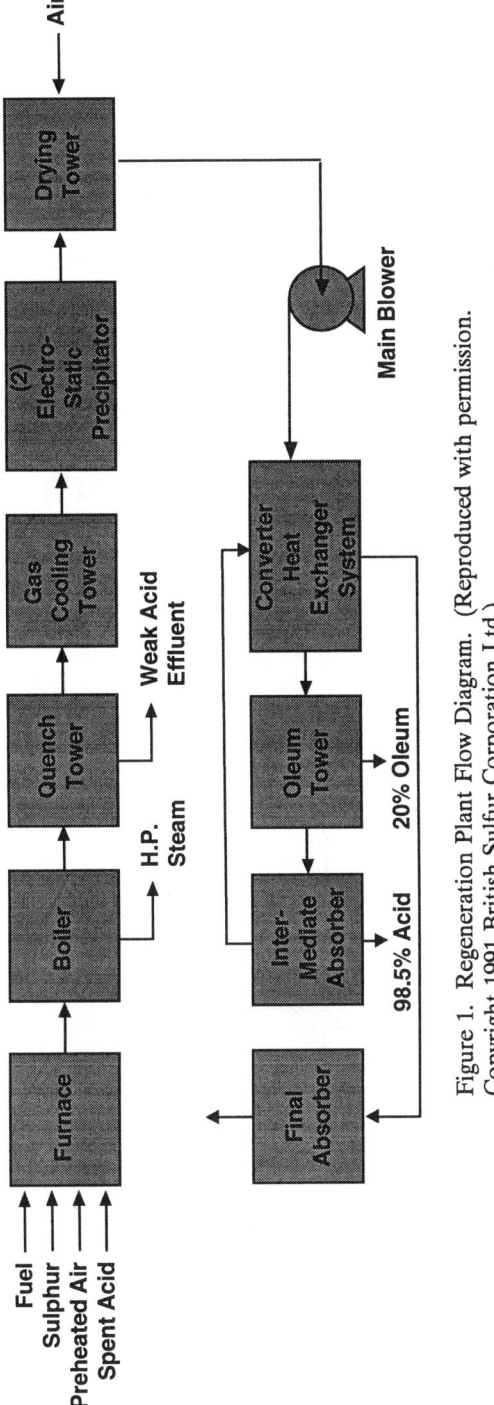

Figure 1. Regeneration Plant Flow Diagram. (Reproduced with permission. Copyright 1991 British Sulfur Corporation Ltd.)

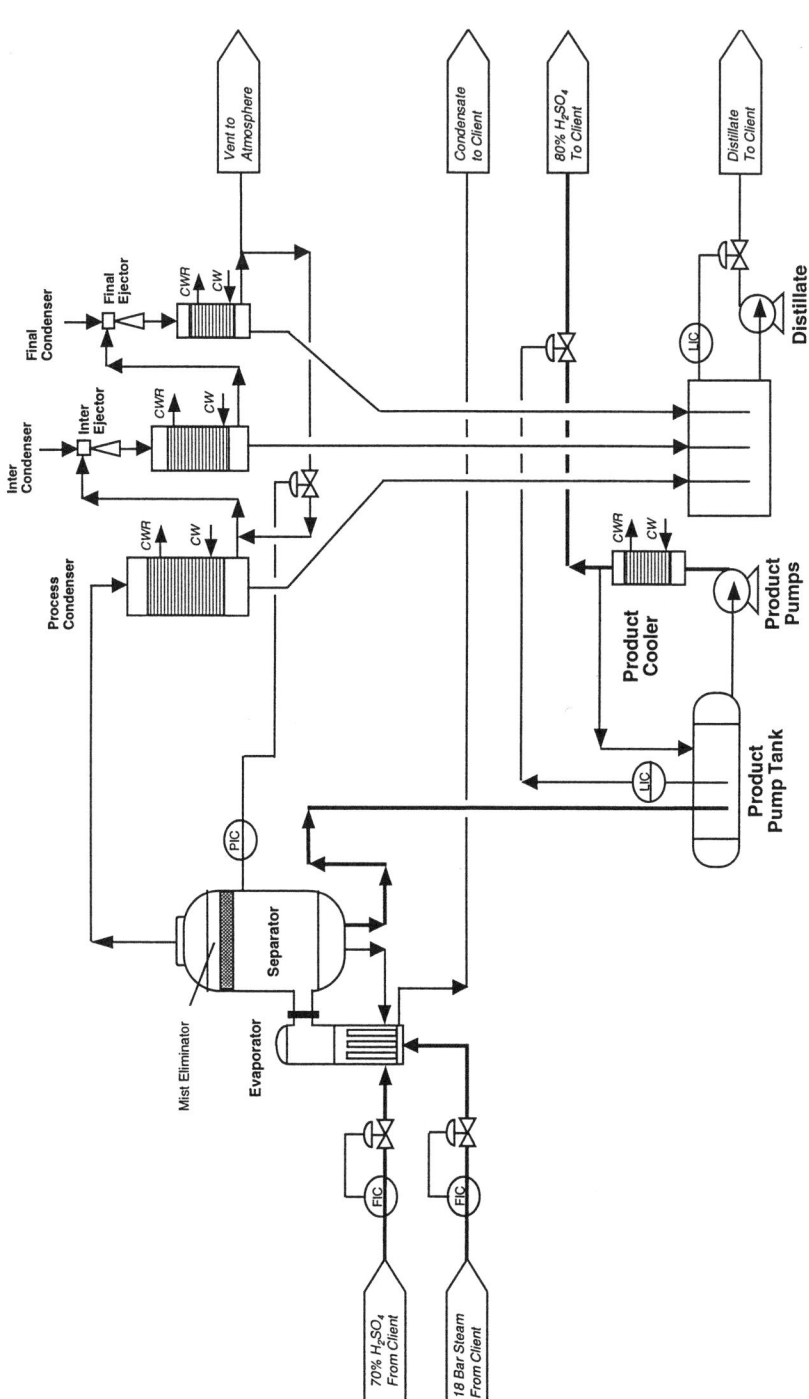

Figure 2. Schematic Flowsheet of Typical Single Stage SAC. (Reproduced with permission. Copyright 1991, Minerals, Metals, and Mining Materials Society.)

Figure 3. 2-Stage Nitration Spent Acid Concentrator. (Reproduced with permission. Copyright 1993 Canadian Institute of Mining, Metallurgy, and Petroleum.)

Application to Nitration Spent Acids

The three major types of nitration systems and resultant spent acid characteristics have been described previously.

In the case of oleum based nitrations, then the only option for spent acid treatment is regeneration, either in-situ or via a toll regeneration, if oleum is to be recovered for recycle. However, in practice most operations prefer to partially treat the spent acid to render it suitable for reuse elsewhere (usually by a combination of denitration, organic decomposition/oxidation and partial concentration) and purchase commercial grade oleum for the nitration.

When the nitration acid mix can be prepared using concentrated sulphuric acid (usually in the 86% to 96% strength range) rather than oleum, then treatment and recovery of the spent acid for recycle is extensively practised, for example for the recovery of spent acid from large scale nitrations such as MNB, DNT, etc. In these cases the acids will usually be denitrated and concentrated with the occasional incorporation of an organic oxidation stage. It is not often necessary to remove or destroy all of the organics when the acid is to be recycled to the process of origin. However, a certain amount of organics will normally need to be removed to enable indefinite recycling of the acid with no impact on operation of either the nitration or the acid recovery stages.

For the case of nitric acid only nitrations utilizing strong nitric acid, usually 99%, then concentration of the nitric acid is via extractive distillation using either sulphuric acid or magnesium nitrate as the dehydrating agent. In those cases where sulphuric acid is used, the weak nitric acid byproduct from nitration is combined with fresh weak nitric acid and contacted with strong sulphuric acid, usually 86%, in a packed column. The water and, frequently, the residual organic components in the spent nitric acid will pass into the sulphuric acid which then must be denitrated and concentrated for recycle to the nitric acid concentration column.

Thus, similar operations are required in nitric only nitrations as for spent acid recovery from a mixed acid nitration. Indeed, when a mixed acid nitration requires the use of both strong nitric and sulphuric acids to prepare the nitration mix then the recovery of spent sulphuric acid is often combined with the production of 99% nitric acid. As a further potential process integration strong sulphuric acid may be used as an effective medium for tail gas scrubbing to remove NOx. This eliminates the requirement for caustic tail gas scrubbing and formation of an aqueous effluent stream and also enables additional recovery of nitric acid values from the NOx which is absorbed.

Therefore, it is evident that technologies exist which are able to treat spent acid streams from the majority of nitrations. These range from simple denitration only to denitration combined with acid concentration, or denitration combined with regeneration in ascending order of cost. None of these options are inexpensive and none provide economic benefit to the operator. However, as has been said,(2) the price of spent acid treatment "is by no means a cheap option compared to the international going price for sulphuric acid, but perhaps a price worth paying to stay in business".

Bearing this in mind it is certain that future nitration developments will have to take into account the treatment of spent acid at an early stage in the development and not merely as an afterthought. Advance consideration of the options for spent acid recovery can lead to an integrated nitration/acid recovery concept which may produce savings compared to a more piecemeal approach.

Therefore, the remainder of this paper will discuss aspects of spent acid recovery which should be taken into account at an early stage of the nitration development together with three case histories which illustrate the benefits which can be obtained from an integrated approach. Some of the considerations which must be taken into account during a development programme are highlighted and the need for cooperation between chemists and chemical engineers is stressed.

Nitration Development Case Histories

The development of a nitration process in many current instances begins with the identification of a nitrated compound as a useful intermediate in the manufacture of a herbicide or an engineering plastic. Studies then begin on how to carry out the nitration. Such studies will mainly concentrate initially on the yield and quality of the desired product. Little thought will be given at this stage to practical means of recovering the product or treating and recovering the spent acid on a large scale. However, bearing in mind that product recovery may involve significant expense requiring several unit operations and that acid treatment and concentration costs can be up to 50% of those for the combined nitration and product recovery stages; the sooner consideration is given to such issues the better. This is illustrated by the following case histories:

Herbicide Intermediate. One example of cost savings resulting from an integrated approach is that of a nitration intermediate for a herbicide. The bench scale process originally developed by the client required use of oleum in the nitration acid. This would have meant that expensive regeneration was the only option available for treatment of the spent acid to produce oleum.

However, after significant bench and pilot scale development effort at the laboratories of CIL Explosives (formerly Chemetics owner and now ICI Explosives, Canada) the nitration process was redeveloped such that the nitration could be satisfactorily carried out using 96% sulphuric acid. This meant that the spent acid could be denitrated and concentrated for recycle, still an expensive process, but a much more economical option than regeneration. In addition, the acid recovery could be integrated with the production of 99% nitric acid required for the nitration. During pilot testing of the intermediate process a subsequent discovery was made that the nitration process was tolerant of certain impurities in the recycled acid but not of others. Therefore, additional residence time was incorporated into the acid recovery process in the form of an organic destruction unit (ODU) to ensure destruction of sufficient of the "bad actors" to allow indefinite recycling of the acid. The pilot plant nitration vessels used for the development are shown in Figure 4 and the subsequent full scale nitration and acid recovery units are shown in Figure 5. This

Figure 4. Pilot Plant Nitrators.

Figure 5. Nitration and Acid Recovery Plant.

illustrates visually the relative sizes of the nitration and acid recovery sections of the plant.

Engineering Plastic Intermediate. In this case the client had developed a mixed acid nitration to produce an intermediate in the manufacture of an engineering plastic. Unfortunately, recovery of the product required such a substantial dilution of the spent acid that the cost of recovery of the acid was prohibitive.

After testwork at the CIL laboratory it was discovered that the nitration could be successfully carried out using 99% nitric acid only. Product recovery from the nitric acid was relatively straightforward with a significant proportion of the acid directly recoverable as strong acid. The remaining contaminated dilute nitric acid was then combined with an incoming supply of make up clean dilute nitric acid and used to produce 99% nitric acid by extractive distillation with sulphuric acid. The organic components in the spent nitric acid passed into the sulphuric acid and were largely destroyed in the subsequent sulphuric acid denitration and concentration process. As a further example of integration of sulphuric acid into the process the concentrated sulphuric acid was used to scrub NOx from the nitric absorber tail gas prior to return to the nitric acid concentration column, thus recovering nitric acid values.

Adiabatic Mononitrobenzene. A final example is that of the development of the adiabatic mononitrobenzene process. Conventional benzene nitration required removal of the nitration exotherm with cooling water followed by concentration of the spent sulphuric acid requiring energy input. American Cyanamid had developed a conceptual adiabatic process in which the nitration was completed with no heat removal, resulting in a spent acid which could be concentrated by flashing, requiring only a small proportion of the energy needed in the conventional process.

The process was further developed in the aforementioned laboratory [10] and a significant portion of the developed world's nitrobenzene is now produced by the adiabatic process.

General Guidelines

The above examples of successful technical developments all illustrate how careful thought about the nitration, product recovery and spent acid recovery stages can result in a well integrated process. Further integration can be achieved by utilizing the sulphuric acid both for production of 99% HNO_3 when required and as part of the NOx absorption system, thus producing a fully integrated plant as illustrated in Figure 6.

When developing a nitration process the following guidelines will usually apply:
> If a nitration requires the use of oleum, then the spent acid can only be recovered by regeneration, the most expensive option, unless further use can be made of the partially treated spent acid in another process and commercial grade oleum purchased for the nitration.

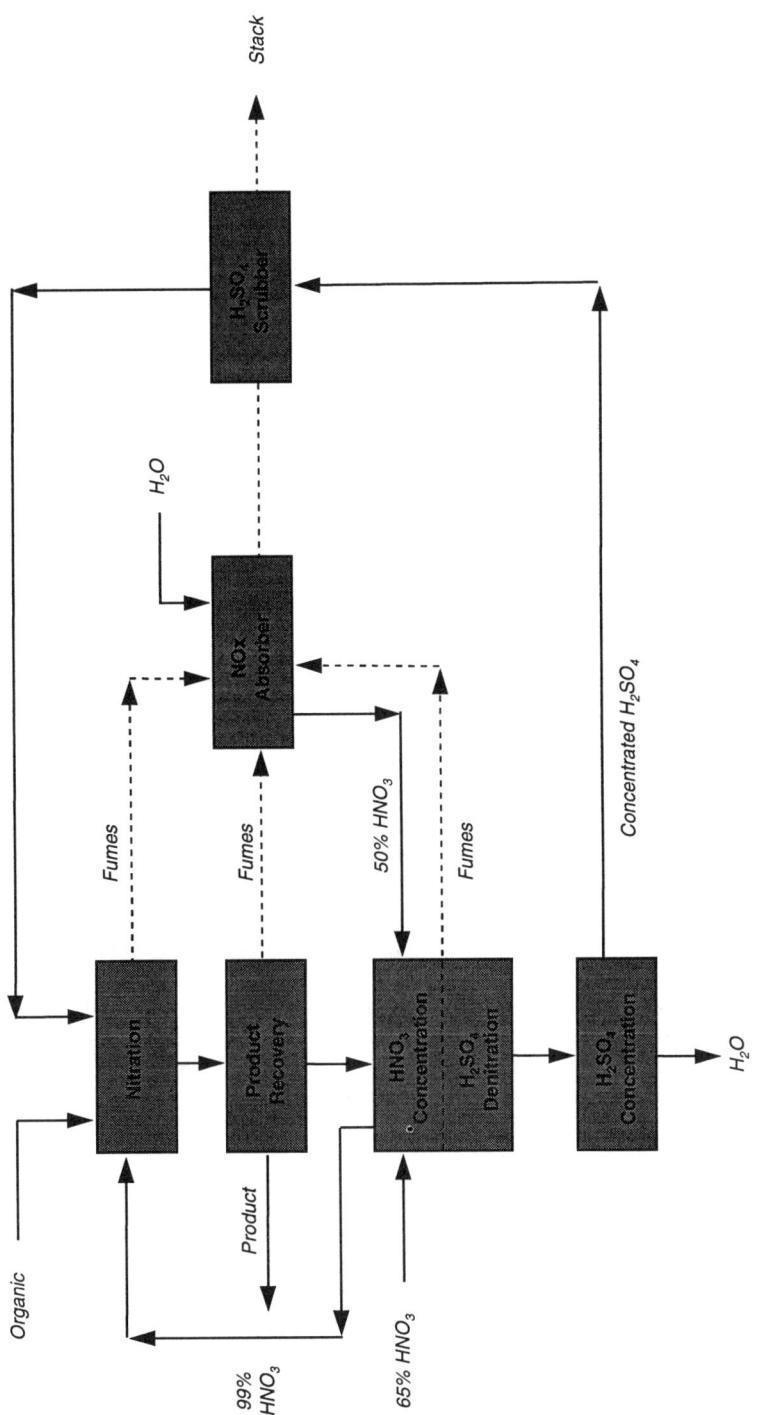

Figure 6. Flowsheet for Integrated Plant.

- Therefore, consideration should always be given as to whether the nitration can be carried out with strong sulphuric acid rather than oleum, thus allowing for concentration and recycle of the above acid.
- Acid concentration, although less expensive than regeneration is, nonetheless a costly exercise. Costs, both capital and operating, are reduced the lower the product acid strength required. Therefore, consideration should always be given as to whether the nitration can be performed using (say) 93% H_2SO_4 rather than 96% H_2SO_4 and so on.
- If a nitration process requires the use of concentrated sulphuric acid and acid recovery, consideration should always be given to the integration of other steps for which sulphuric acid can be utilized such as the production of 99% HNO_3, if required, and NOx absorption.

The effort required to develop the processes described above was considerable and costly. However, there are few short cuts available in the development of a nitration process with integrated acid recovery and recycle. Some aspects of such process development activities are summarized in the following section with emphasis on the cooperation required between chemist and chemical engineers.

Process Development Considerations

As has been described in the previous sections, acid treatment and recovery processes are always expensive and can account for a significant portion of the overall cost of a nitration plant. However, as illustrated by the case histories which have been discussed, careful, well planned process development can result in an optimized, integrated process which minimizes the total cost.

Such developments are painstaking, lengthy and expensive exercises which should not be carried out in isolation. The processes selected for the nitration and product recovery stages can have a significant impact on the feasibility and cost of acid recovery and recycle. Consideration of these issues at an early stage will lead to a well planned development programme. It is never too early for a chemical engineer to become involved with a chemist in a nitration development. Even at the very early stages of bench scale nitration development discussions can begin of those items which will have a significant impact on the viability of the process. The following is a partial, but by no means comprehensive, list of some of those issues:

- Scale up of the nitration
    - Batch vs continuous
- Selection of product recovery and purification process
    - Phase separation
    - Solvent extraction
    - Crystallization
    - Evaporation
    - Filtration
    - Product washing et al

- Spent acid treatment/recovery/recycle
  - Denitration
  - Concentration
  - Regeneration
- NOx absorption
  - HNO_3 recovery
  - Tail gas scrubbing
- Safety issues
  - Runaway reactions
  - Oxygen balanced mixtures/detonability
  - By-product formation (e.g. tetranitromethane (TNM))
  - Spent acid stability
- Materials of construction
- Effluent treatment

These are not isolated issues and information in one area will have an impact on others: for example; the tendency of a system to undergo runaway decomposition reactions will usually result in a continuous rather than a batch nitration and will lead to considerable analysis of relief system design.

As the nitration development proceeds from bench scale to pilot scale, possibly to semi-works scale, and finally to design and construction of a full scale plant, continuous evaluation of these issues should take place. Even on the bench scale potential product and acid recovery processes can be simulated and acid recycled to the nitration. At each stage, the key objective should be to obtain as full a component mass balance as possible. This will be difficult initially as many by-products will not be able to be identified. However, Total Carbon Content (TOC) analysis can be used to at least produce an overall carbon balance.

It is also never too early to commence hazards evaluation and testing. Once the conceptual bench scale process has been established samples can be obtained for testing for runaway decompositions, detonability, etc. Such testing will, in any case, have to be carried out before design of a pilot unit and the earlier hazards evaluation is commenced the better.

Therefore, as stated previously, nitration development is a complex, lengthy and expensive process. It is of paramount importance that chemists and chemical engineers cooperate from an early stage and that, by constant evaluation and consideration of all the relevant issues, an optimized, safe and environmentally acceptable process will be developed.

Acid Concentration Development

In addition to process development of simultaneous nitration/acid recovery processes, acid recovery, most especially concentration, is frequently applied retrospectively to existing nitration plants. In both cases bench and pilot scale testing is vital and a number of specific guidelines apply, (11) including the following:

- The acid should be analyzed as fully possible. As a minimum analysis should be performed for % H_2SO_4, % HNO_3, % HNO_2, TOC and major inorganic contaminants.

- Each stream should be analyzed for potential known hazardous components, e.g. TNM.

- The quality of product acid which is required must be defined, particularly in terms of concentration and TOC content, to avoid an unnecessarily expensive process.

- As many recycles as possible of the concentrated acid to the nitration process should be performed. Any build up of the acid organic content during recycling should be monitored to evaluate if an organic destruction stage or acid purge will be required.

- Any tendency for the acid to foam or cause fouling of heat transfer surfaces during concentration must be closely observed.

- A mass balance of organics over the concentration stage must be obtained to establish how many are destroyed, how many report to the overhead condensate and how many to the product.

- The overheads condensate must be fully analyzed to understand effluent treatment requirements and to evaluate if partial recycle of this stream within the nitration complex is feasible.

- The off gases from the concentration stage must be analyzed at least for NO, NO_2, N_2O, CO and CO_2 to understand NOx absorption and tail gas scrubbing issues.

Even in cases when the acid concentration development is to retrofit an existing nitration unit, an integrated approach is required. This is particularly so where the recycle and potential build up of impurities is concerned. As an example of the importance of pilot plant testing in acid concentration Figure 7 illustrates the behaviour of the same acid under slightly different conditions. Figure 7a shows the separator vessel of a pilot plant concentrator at design operating conditions. Figure 7b shows the separator at a slightly different set of conditions at which severe and unacceptable foaming occurred.

Conclusions

As environmental standards become more stringent it is inevitable that nitration spent acids will have to be treated for recycle or reuse. Proven methods exist to successfully treat and recover such acids but application of these techniques can result in a significant increase in the capital and operating costs of the nitration complex.

Figure 7. Pilot Plant Acid Concentration. (Reproduced with permission. Copyright 1990 British Sulfur Corporation Ltd.)

Therefore, in any new nitration development consideration must be given at an early stage to the issue of spent acid treatment and recovery. This can result in an optimized process in which the spent acid treatment is integrated with the nitration to minimize the increase in costs due to acid recovery.

In order to arrive at an optimized, integrated nitration and acid recovery process extensive bench and pilot scale testing is required. Close cooperation between chemists and chemical engineers is vital and must commence at early stage of the initial bench scale nitration development.

Literature Cited

1. Sander, U.H.F., Fischer, H., Rothe, U., and Kola, R., in "Sulfur, Sulfur Dioxide and Sulfuric Acid", British Sulfur Corporation, London, U.K., 1984.
2. Anonymous, Recovering Used Sulphuric Acid, Fertilizer International 326 34-41 (1993).
3. Rodger, I., Recovery and Reconcentration of Spent Sulfuric Acid, Proceedings of the British Sulfur Corporation's Third International Conference on Fertilizers, London, U.K., 1979.
4. Anonymous, Making Cents of Acid Recovery, Chemical Engineering, April, 1993.
5. Smith, I.M., Peterson, H.C. and Cameron, G.W., Sulfur Value Recovery from TiO_2 Waste Acid, Preprints of the International Sulfur Conference, London, U.K., 1985.
6. Douren, L., The Basic Sulfuric Acid Recovery System, Proceedings of The British Sulfur Corporation's Fifth International Conference, London, U.K., 1979.
7. De La Mater, M., and Milligan, B., U.S. Patent 3,856,673 (1974).
8. Gloster, A.J., Recovery of Sulfuric Acid from Organic Wastes, Duecker, W.W. and West, J.R., "The Manufacture of Sulfuric Acid", ACS Monograph #144, Reinhold Publishing Co., 1959.
9. Mantius, E., and Rinckhoff, J.B., Concentrating Sulfuric Acid, in Duecker, W.W., and West, J.R., "The Manufacture of Sulfuric Acid", ACS Monograph #144, Reinhold Publishing Co., 1959.
10. Guenkel, A.A., Prime, H.C., Rae, J.M., Nitrobenzene Via An Adiabatic Reaction, Chemical Engineering, August 10, 1981.
11. Küng, H.R., Reiman, P., Reconcentration and Purification of Sulfuric Acid Contaminated with Organic Impurities, Proceedings of The British Sulfur Corporation's Fifth International Conference, London, U.K., 1979.
12. Al Samadi, R., Evans, C.M., and Smith, I.M. Reprocessing Spent Sulfuric Acid, Sulfur 207 43-50 (1990).
13. Sander, U., and Daradimos, G., Regenerating Spent Acid, Chemical Engineering Progress, p. 57-67 September, 1978.

RECEIVED February 29, 1996

INDEXES

Author Index

Albright, Lyle F., 1,201
Bashir-Hashemi, A., 51
Burrichter, Arwed, 10
Carr, Richard V. C., 1,174
Case, J. L., 174
Chapman, Robert D., 78
Colclough, M. E., 97,104
Coombes, Robert G., 19,134
Coon, Clifford L., 151
Dave, Paritosh R., 51
Desai, H., 104
Devendorf, T. E., 68
Eckert, Roger E., 201
Evans, C. M., 187,250
Fukuyama, S. M., 187
Gebauer, J., 234
Gelber, Nathan, 51
Golding, P., 104
Graindorge, H. R., 43
Guenkel, A. A., 223
Guggenheim, T. L., 187
Hadjigeorgiou, Panicos, 19
Hermann, H., 234
Honey, P. J., 104,134
Jensen, Doron G. J., 19
Jessop, Edward S., 151
Konieczny, P., 234
Lescop, P. A., 43
Li, Jianchang, 51
Malhotra, Ripudaman, 31
Maloney, T. W., 223
Millar, R. W., 104,122,134
Mitchell, Alexander R., 151
Morris, David L., 19
Odle, R. R., 187
Olah, George A., 10
Pagoria, Philip F., 151
Paul, N. C., 97,104,165
Pennington, B. Timothy, 214
Pouet, M. J., 43
Prakash, G. K. Surya, 10
Quakenbush, Allen B., 214
Rasul, Golam, 10
Rodgers, M. J., 58
Ross, David S., 31
Sanderson, A. J., 104
Schmidt, Robert D., 151
Schmitt, Robert J., 1
Simpson, M. S., 174
Smith, Glen D., 78
Sood, M. K., 201
Stacy, J. R., 68
Stewart, M. J., 104
Swinton, P. F., 58
Terrier, F., 43
Warner, G. L., 187

Affiliation Index

Air Products and Chemicals, Inc., 1
Brunel University, 19,134
Chemetics International Company Ltd., 187,250
City University, (England), 19
Defence Research Agency, (England), 97,104,122,134,165
Ecole Nationale Supérieure de Chimie de Paris, 43
GE Corporate Development and Research Center, 187
GE Plastics, 187
Geo-Centers, Inc., 51
ICI Explosives, 58
Josef Meissner GmbH and Company, 234
Lawrence Livermore National Laboratory, 151

NORAM Engineering & Constructors Ltd., 223
Naval Surface Warfare Center, 68
Olin Corporation, 214
Purdue University, 1,201
SRI International, 1,31,174
Société Nationale de Poudres et Explosifs, 43
TPL, Inc., 78
U.S. Army Research, Development, and Engineering Center, 51
University of Southern California, 10

Subject Index

A

Acetonitrile, role in N_2O_5 separation from solutions in nitric acid, 82–87
Acetophenone, synthesis, 142–143
Acid washing, purification of crude dinitrotoluene, 241
Adamantane, selective functionalization, 51–56
Adiabatic calorimeters, advances, 178–179
Adiabatic calorimetry, standard techniques, 177–178
Adiabatic mononitrobenzene, nitration development, 262
Aliphatic nitro compounds, synthetic methods, 134–136
Alkaline washing, purification of crude dinitrotoluene, 241
Anhydrous N_2O_5
  applications, 2
  synthesis, 2
Aniline
  reaction with 2,3,5,6-tetrabromo-4-methyl-4-nitrocyclohexa-2,5-dien-1-one, 24,27–28
  synthesis from nitration, 1
Arenes, nitration using nitronium salts, 10
Aromatic nitration
  in liquid N_2O_4 promoted by metal acetylacetonates
    apparatus, 32,35f
    experimental description, 32
    fast mode, 32,36–37
    nitration of benzene and toluene, 32,34f
    nitrato complex reactions, 37–39
    one-electron oxidation mechanism, 32,33
    promotion by β-dicarbonyls, 39–41
    slow mode, 32,36–37

Aromatic nitration—*Continued*
  role of nitrocyclohexadienones, 19–29
  solvent effect, 10
  use of electrochemically generated N_2O_5, 58–66
Atmospheric pressure concentration, description, 254,256–257f

B

Benzene, nitration in liquid N_2O_4 promoted by metal acetylacetonates, 32,34f
Benzoylnitromethane, synthesis, 142–143
Binders
  examples, 97–98
  function, 97
N-tert-Butoxycarbonyl derivatives, use for nitramine synthesis, 159–162

C

C–nitro functionality
  building block for energetic materials, 104
  structure, 104
Calorimetrically determined kinetic constants
  adiabatic data, 177–178
  isothermal data, 175–177
Carboxycubane, synthesis, 53
Cerium(IV) acetylacetonate, promotion of aromatic nitration in liquid N_2O_4, 31–41
Chemistry of nitration
  free radical reactions, 2
  ionic mechanism using acid mixtures, 1
  nitronium ion salts, 2
Chlorobenzene, isomer synthesis, 2

INDEX

Chlorocarbonylation, photochemical, *See* Photochemical chlorocarbonylation

Clean Air Act Amendment of 1990, guidelines for $(NO)_x$ emissions, 7

Cobalt(III) acetylacetonate, promotion of aromatic nitration in liquid N_2O_4, 31–41

Concentrated sulfuric acid based nitrations, concentration and recovery of spent acids, 250–267

Concentration
  disposal of spent sulfuric acid, 254,256–257*f*
  nitration spent acids, 250–267

Constant flow stirred tank reactors, description, 165

Conventional dinitrotoluene technology
  byproducts, 215
  impurity removal, 215–216
  operations, 215
  reactors, 215
  sulfuric acid content, 215

Copper(II) acetylacetonate, promotion of aromatic nitration in liquid N_2O_4, 31–41

Cubane, selective functionalization, 51–56

D

Decomposition of dinitrotoluene, thermal, modeling, 174–186

Density functional theory, function, 12

Destruction, tetranitromethane from nitric acid, 187–198

β-Dicarbonyls, promotion of aromatic nitration in liquid N_2O_4, 39–41

1,4-Dideoxy-1,4-dinitro-*neo*-inositol-2,5-dinitrate, synthesis, 157–158

1,3-Dihydroxybenzene, nitration using N_2O_5, 61–65

N,N-Dimethylaniline, reaction with 2,3,5,6-tetrabromo-4-methyl-4-nitro-cyclohexa-2,5-dien-1-one, 28,29*f*

Dinitrate esters, synthesis, 113–115

gem-Dinitroarylmethoxymethanes, nitration using N_2O_5, 65

gem-Dinitro compounds, synthesis using Ponzio reaction, 134–148

1,1-Dinitro-1-(4-nitrophenyl)ethane, synthesis, 142–143

1,1-Dinitro-1-phenylethane, synthesis, 142–143

1,3-Dinitro-1,3-diazacyclopentane, synthesis, 161

2,5-Dinitro-7,9-dihydro-2,5,7,9-tetraaza-bicyclo[4.3.0]nonane
  experimental procedure, 45
  synthesis, 48

4,6-Dinitro-2-(dinitromethyl)phenol, synthesis, 144–145

9,9-Dinitrofluorene, synthesis, 138–141

Dinitroparaffins, synthesis, 3

Dinitrotoluene
  modeling of thermal decomposition, 174–186
  synthesis, 1,4,234–247

Dinitrotoluene process, Olin, *See* Olin dinitrotoluene process

Dinitrotoluene technology, conventional, *See* Conventional dinitrotoluene technology

Direct use, disposal of spent sulfuric acid, 252

Discharge without treatment, disposal of spent sulfuric acid, 251

Drying, purification of crude dinitrotoluene, 241

E

Electrochemically generated N_2O_5
  experimental description, 58–59
  manufacturing process, 59–61
  nitration
    1,3-dihydroxybenzene, 61–65
    gem-dinitroarylmethoxymethanes, 65
    organic solvent effect, 65–66
    pyridine and methylpyridines, 65

Electrolysis of nitric acid with N_2O_4, N_2O_5 synthesis, 107,108*f*

Electrolytic oxidation of N_2O_4 solutions in nitric acid, N_2O_5 synthesis, 79

Energetic compounds, role of nitration in synthesis, 97

Energetic material(s)
  building blocks, 104
  synthesis using nitration and nitrolysis procedures, 151–162
Energetic material synthesis using N_2O_5
  N_2O_5–nitric acid nitrations, 109–112
  N_2O_5–organic solvent nitrations, 112–117
  N_2O_5 synthesis
    electrolysis of nitric acid with N_2O_4, 107,108f
    ozonation of N_2O_4, 107,109
  nitration potential of N_2O_5, 105–107
  selective nitration for energetic binder macromolecules, 117–118
Energetic polymers, synthetic methods, 98
Energetics, protosolvation in nitrations with superacidic systems, 15–16
Engineering model
  function, 174–175
  strength, 174–175
Engineering plastic intermediate, nitration development, 262
Environmental concerns, industrial nitration plants, 7
Ethylenedinitramine, synthesis, 161

F

Flow reactor(s)
  advantages, 165
  ideal system, 166
  ideal types, 165
Flow reactor nitrations using N_2O_5
  advantages, 173
  applications, 173
  batch procedure, 169–170
  continuous quench and neutralization, 168
  flow nitration system, 167
  flow reactor design, 166
  glycidol nitration, 170
  glycidyl nitrate synthesis, 171
  N_2O_5 purity effect on yield and purity of products, 172
  N_2O_5 synthesis, 168
  3-(nitratomethyl)-3-methyloxetane synthesis, 170–171
  operating parameters and procedure, 168
  workup procedure and analysis, 168–169

Free radical reactions, paraffin nitration, 2–3
Fundamental reaction model, function, 174–175

G

Geometries, protosolvation in nitrations with superacidic systems, 15–16
GIAO–MP2 method, function, 12
Glycidol, energetic binder macromolecule synthesis, 117–118
Glycidyl nitrate
  energetic binder macromolecule synthesis, 118
  synthesis, 171

H

H_2SO_4–HNO_3–water mixtures, NO_2^+ concentration model, 201–213
Hazards, nitration, 5–7
Herbicide intermediate, nitration development, 262
Hexahydro-1,3,5-trinitro-1,3,5-triazine
  description, 43
  structure, 43
  synthesis, 162
2,4,5,7,9,9-Hexanitrofluorene, synthesis, 139–141
HNO_3–H_2SO_4–water mixtures, NO_2^+ concentration model, 201–213
3-(Hydroxymethyl)-3-methyloxetane, energetic binder macromolecule synthesis, 117–118
3-(Hydroxymethyl)-3-methyloxetane nitrate, energetic binder macromolecule synthesis, 118

I

Industrial nitration of toluene to dinitrotoluene
  chemistry of nitration in mixed acid
    byproduct formation, 236,238t
    isomer distribution, 235–236
    rate, 235,237t
  history, 234–235

INDEX

Industrial nitration of toluene to
  dinitrotoluene—*Continued*
  industrial manufacture
    effluent treatment
      spent acid, 241–242
      waste gas, 242
      waste waters, 242
    nitration, 236–241
    purification of crude dinitrotoluene
      acid washing, 241
      alkaline washing, 241
      drying, 241
      neutral washing, 241
  modern plant design
    spent acid
      complete purification/
        reconstruction, 243
      partial purification/reconstruction,
        243–244
    waste gas, 245,247
    waste water, 244–245,246*f*
    safety measures, 247
Industrial nitration plants, environmental
  concerns, 7
Industrial processes, nitration, 4
Ionic nitrations, physical steps, 3–4
Iron(III) acetylacetonate, promotion of
  aromatic nitration in liquid N_2O_4, 31–41
Isothermal calorimetry, standard
  techniques, 175–177

J

Jet impingement nitrator
  benzene inlet manifold, 230
  development, 229–230
  elements, 230
  example, 224,227*f*
  plug flow, 230

K

Kinetic constants
  adiabatic data, 177–178
  isothermal data, 175–177

L

Liquid N_2O_4 promoted by metal
  acetylacetonates, aromatic nitration,
  31–41

M

Manganese(III) acetylacetonate, promotion
  of aromatic nitration in liquid N_2O_4,
  31–41
Metal acetylacetonates, promotion of
  aromatic nitration in liquid N_2O_4, 31–41
Methylpyridines, nitration using N_2O_5, 65
Mixed acid(s), advantages and
  disadvantages as nitrating agents, 4
Mixed acid nitration, tetranitromethane
  removal and destruction from nitric
  acid, 187–198
Model
  NO_2^+ concentrations in HNO_3–H_2SO_4–
    water mixtures, 201–213
  thermal decomposition of dinitrotoluene
    actual and representative chemistry
      intermediate isolation, 182
      isolated explosive coke analysis,
        182–183
      reaction regimes, 181–182
      thermochemistry, 183,185
    multiple decomposition reaction model,
      prediction vs. experimental data,
      184*f*,185–186
    single reaction decomposition model
      advances in adiabatic calorimeters,
        178–179
      engineering model of PHI–TEC II
        device, 180
      prediction vs. experimental data,
        180–181,184*f*
      reaction, 178
    standard techniques for adiabatic and
      isothermal calorimetry, 175–178
Mononitrobenzene, adiabatic, nitration
  development, 262
Mononitrobenzene manufacture technology
  chemistry, 224,226–228

Mononitrobenzene manufacture technology—*Continued*
  environmental features, 231
  jet impingement nitrator, 224,227f,229–230
  nitration rate, 228–229
  performance, 232
  process, 223–224,225f
  process efficiencies, 223
  safety, 230–231
Mononitropyridines, synthesis, 112

N

N_2O_4, activation by ozone, 31–32
N_2O_5
  advantages as nitrating agent, 4
  applications to energetic material synthesis, 107
  electrochemically generated, *See* Electrochemically generated N_2O_5
  N_2O_5–nitric acid nitrations, 109–112
  N_2O_5–organic solvent nitrations, 112–117
  nitration potential, 105–107
  pilot plant scale continuous manufacturing, 68–77
  role in Ponzio reaction, 134–148
  selective nitration for energetic binder macromolecules, 117–118
  synthesis, 168
  use in nitramine synthesis in nonacidic media, 122–132
  See also Solid N_2O_5
N_2O_5 nitrations, classes, 104
N_2O_5 separation from solutions in nitric acid
  acetonitrile effect, 82–87
  advantages, 93,95
  cost, 95
  example of absorptive capacity of NaF for HNO_3, 79–80
  example of extraction technology, 79
  experimental procedure, 80–81
  N_2O_5 synthetic routes, 78
  NMR chemical shift analyses, 92–93,94f
  optimization, 95

N_2O_5 separation from solutions in nitric acid—*Continued*
  relative N_2O_5 content vs. relative nitric acid content, 86,88,89f
  residual acid vs. contact time, 82
  solution transfer effect, 88–92
N_2O_5 synthesis
  electrolysis of nitric acid with N_2O_4, 107,108f
  electrolytic oxidation of N_2O_4 solutions in nitric acid, 79
  ozonation of N_2O_4, 107,109
  ozonolysis of N_2O_4, 79
National Emissions Standards for Hazardous Air Pollutants, regulation of nitro compounds, 7
Neutral washing, purification of crude dinitrotoluene, 241
Neutralization, disposal of spent sulfuric acid, 251–252
Nitramides, synthesis in nonacidic media, 122–132
Nitramine(s)
  applications, 122
  of azaheterocycles, 43
  problems with synthetic methods, 122–124
  synthetic methods, 122
  use of *N-tert*-butoxycarbonyl derivatives for synthesis, 159–162
Nitramine functionality
  building block for energetic materials, 104
  structure, 104
Nitramine–nitrate products, synthesis, 116–117
Nitramine synthesis in nonacidic media
  advantages, 127–128
  apparatus, 129
  experimental procedure, 129–130
  nitration procedures, 130–132
  previous studies, 124
  silylamides, 127–128
  silylamines, 124–127
Nitrate ester functionality
  building block for energetic materials, 104
  structure, 104

INDEX

Nitrate–nitramine products, synthesis, 116–117
Nitrated hydroxy-terminated polybutadiene
  advantages, 103
  C=C bond conversion to epoxide groups, 98–99
  cure chemistry, 101
  differential scanning calorimetry, 101
  epoxide analysis, 100
  experimental procedure, 99–100
  Fourier-transform IR spectroscopy, 100–101
  miscibility with energetic plasticizer, 102
  nitration using N_2O_5
    concentration effect, 98
    reaction, 98
  NMR spectroscopy, 100
  preliminary hazard assessment, 101–102
  size exclusion chromatography, 101
  synthesis via nitromercuration/demercuration route, 98
  vacuum stability test, 101
Nitration
  aromatic
    role in nitrocyclohexadienones, 19–29
    use of electrochemically generated N_2O_5, 58–66
  chemistry, 1–3
  classes, 104–105
  commercial products, 1
  definition, 1
  hazards, 5–7
  in liquid N_2O_4 promoted by metal acetylacetonates, aromatic, *See* Aromatic nitration in liquid N_2O_5 promoted by metal acetylacetonates
  ionic, *See* Ionic nitrations
  new and potential industrial processes, 4
  of organic molecules, steps, 201
  of toluene to dinitrotoluene, industrial, *See* Industrial nitration of toluene to dinitrotoluene, 234–247
  using nitronium salts, previous studies, 58
  with superacidic systems, protosolvation, 10–17

Nitration and nitrolysis procedures in energetic material synthesis
  1,4-dideoxy-1,4-dinitro-*neo*-inositol-2,5-dinitrate, 157–158
  experimental description, 151
  ^{18}O-labeled 2,4,6-trinitrotoluene, 158–159
  2-oxo-1,3,5-trinitro-1,3,4-triazacyclohexane, 155–156
  structures, 151–152
  2,5,7,9-tetranitro-2,5,7,9-tetraazabicyclo[4.3.0]nonan-8-one, 153–154
  2,4,6,8-tetranitro-2,4,6,8-tetraazabicyclo[3.3.0]octane, 156–157
  2,4,6,8-tetranitro-2,4,6,8-tetraazabicyclo[3.3.0]octan-3-one, 152–153
  2,5,7-trinitro-2,5,7,9-tetraazabicyclo[4.3.0]nonan-8-one, 153–154
  2,6,8-trinitro-2,4,6,8-tetraazabicyclo[3.3.0]octan-3-one, 152–153
  use of *N-tert*-butoxycarbonyl derivatives, 159–162
Nitration development case histories
  adiabatic mononitrobenzene, 262
  engineering plastic intermediate, 262
  herbicide intermediate, 259–262
Nitration plants, environmental concerns, 7
Nitration spent acid concentration and recovery
  acid concentration development, 265–266,267*f*
  concentrated sulfuric acid based nitrations, 258
  disposal of spent sulfuric acid
    discharge without treatment, 251
    neutralization, 251–252
    partial treatment
      recovery of nitric acid values, 253
      removal/destruction of organics, 253
      removal of nitric/nitrous species, 252–253
    spent acid stability/safety, 252

Nitration spent acid concentration and recovery—*Continued*
  disposal of spent sulfuric acid—*Continued*
    treatment and recovery
      concentration, 254,256–257f
      regeneration, 253–254,255f
    nitration development case histories, 259–262
    nitration development guidelines, 262–264
    nitration processes, 251
    nitric acid based nitrations, 258
    occurrence, 251
    oleum-based nitrations, 258
    process development considerations, 264–265
Nitrato complexes, nitrations, 37–39
3-(Nitratomethyl)-3-methyloxetane, synthesis, 170–171
Nitric acid
  N_2O_5 separation from solutions, 78–95
  tetranitromethane removal and destruction, 187–198
Nitric acid based nitrations, concentration and recovery of spent acids, 250–267
Nitro compounds, regulation by National Emissions Standards for Hazardous Air Pollutants, 7
4-Nitroacetophenone, synthesis, 142–143
Nitroadamantanes
  applications, 51
  synthetic methods, 51–52
N-Nitroazetidine, synthesis, 115–117
Nitrobenzene, synthesis, 4
Nitrocyclohexadienones in aromatic nitration
  catalyzed rearrangements, 21–22
  reaction
    with aniline, 24,27–28
    with *N,N*-dimethylaniline, 28
    with phenols
      concurrent bromination, 24,25–26f
      nitration, 23–24
  structure, 19,20f
  uncatalyzed rearrangements, 19–21
  use as nitrating agents, 23,25

4-Nitro-2-(dinitromethyl)phenol, synthesis, 144–145
Nitrogen heterocycles, synthesis of energetic materials, 115–117
Nitroglycerine, synthesis, 112–113
Nitrolysis of secondary amide or *tert*-butylamines to nitramines, reagents, 151
Nitrolysis procedures in energetic material synthesis, *See* Nitration and nitrolysis procedures in energetic material synthesis
1-Nitropiperidine, synthesis, 160
Nitronium ion
  bent configuration, 11
  IR and Raman spectroscopy, 12
  ^{15}N-NMR chemical shifts, 11
  ^{17}O-NMR spectroscopy, 12
  role
    protonitronium dication in reactivity, 10–17
    reactive electrophile in acid-catalyzed nitration with nitric acid and derivatives, 10
  solvent effect, 11
Nitronium ion concentration model in HNO_3–H_2SO_4–water mixtures
  applications, 213
  development of model, 202–203
  equilibrium constants for ionization reactions, 204
  H_2O concentration, 208,212f
  H_3O^+ concentration, 208,209f
  H_2SO_4 concentration, 208,211f
  HNO_3 concentration, 208,211f
  HSO_4^- concentration, 208,210f,213
  nitration reactions, 201
  NO_2^+ concentration
    vs. acid composition, 204,205f
    vs. temperature, 204–205,208
    without acid, 208,209f
  NO_3^- concentration, 208,210f,213
  predicted and experimental molar concentrations, 204,206–207t
  previous studies, 202

INDEX

Nitronium salts
  arene nitration, 10
  chemistry of nitration, 2
  solvent effect on aromatic nitration, 10
Nitroparaffin synthesis
  nitration, 1
  Victor Meyer processes, 3
Nitrosonium ion, function, 2
NO_2^+, See Nitronium ion
NO_2BF_4, use as nitrating agent, 2
$NO_2H_2^+$, See Protonitronium dication
NO_2PF_6, use as nitrating agent, 2
$(NO)_x$ emissions, guidelines from Clean Air Act Amendment of 1990, 7
Nonacidic media, nitramine synthesis, 122–132

O

Octahydro-1,3,5,7-tetranitro-1,3,5,7-tetrazocine
  analogue studies, 43–44
  description, 43
  structure, 43
  synthesis, 111–112
Olin dinitrotoluene process
  advantages, 214,221–222
  conventional dinitrotoluene technology, 215–216
  description, 214
  materials, 216
  operations, 216–217
  single acid process
    advantages, 221
    development challenge
      efficient regeneration of excess nitric acid, 220
      increase in oxidation byproducts, 218–219
      product separation, 220
      slow reaction rates, 220
Organic molecules, nitration steps, 201
2-Oxo-1,3,5-trinitro-1,3,5-triazacyclohexane
  structure, 151–152
  synthesis, 155–156

Oxygen heterocycles, synthesis of energetic materials, 113–115
Ozonation of N_2O_4, N_2O_5 synthesis, 107,109
Ozone-mediated nitration with NO_2, advantages, 4
Ozonolysis of N_2O_4, N_2O_5 synthesis, 79

P

Paraffin nitration, free radical reactions, 2–3
Partial treatment, disposal of spent sulfuric acid, 252–253
2,4,7,9,9-Pentanitrofluorene, synthesis, 139–141
Phenol
  isomer synthesis, 2
  reactions with 2,3,4,6-tetrabromo-4-methyl-4-nitrocyclohexa-2,5-dien-1-one, 23–24,25–26f
Photochemical chlorocarbonylation
  carboxycubane synthesis, 53
  experimental description, 53
  1,3,5,7-tetranitroadamantane synthesis, 53–55
  1,3,5,7-tetranitrocubane synthesis, 55–56
Pilot plant scale continuous manufacturing of solid N_2O_5
  advantages, 77
  apparatus, 69
  bench top scale process development
    modified process, 70–72,
    original process, 69–70,71f
    plug flow reactor process, 72–73,74f
  reaction, 69
  pilot plant scale process development
    analytical determination of N_2O_5 solutions, 75–77
    N_2O_4 supply, 73,75
    ozone generation, 73
    process, 73,74f
    solid N_2O_5 trapping, 75
  program objectives, 68

Plug flow reactor(s)
 description, 165
 process, 72–73,74f
 safety, 165–166
Polybutadiene, nitrated hydroxy terminated, *See* Nitrated hydroxy-terminated polybutadiene
Polynitro aliphatic compounds, studies, 134
Polynitro cage molecules, applications, 51
Polynitro polycyclic compounds, synthetic studies, 51
Polynitroadamantanes, synthesis, 51–56
Polynitrocubanes, synthesis, 51–56
Ponzio reaction for *gem*-dinitro compound synthesis
 advantages, 134
 applications of products, 148
 polynitrofluorene derivatives, 138–141
 polynitrophenol derivatives, 143–145
 polynitrophenylethane derivatives, 142–143
 reaction mechanism, 145–148
 substituent vs. yield, 137
 synthetic potential, 137–138
Properties, nitrated hydroxy-terminated polybutadiene, 97–103
Protonitronium dication
 formation, 11
 identification, 12–17
 ^{15}N-NMR chemical shifts, 11
Protosolvation in nitrations with superacidic systems
 energetics, 15–16
 experimental calculation procedure, 12
 experimental description, 12,17
 geometries, 15–16
 IR frequencies, 16–17
 ^{17}O-NMR chemical shifts, 12–15
 Raman frequencies, 16–17
 total energies and relative energies, 12,13t
Pyridine, nitration using N_2O_5, 65

R

Reaction model, fundamental, function, 174–175
Recovery, nitration spent acids, 250–267
Regeneration, disposal of spent sulfuric acid, 253–254,255f
Removal, tetranitromethane from nitric acid, 187–198
Resorcinol, nitration using N_2O_5, 61–65

S

Safety, nitration, 5–7
Silylamides, synthesis in nonacidic media, 127–128
Silylamines, synthesis in nonacidic media, 124–127
Solid catalysts, advantages over nitrations, 4
Solid N_2O_5, pilot plant scale continuous manufacturing, 68–77
Solution transfer, role in N_2O_5 separation from solutions in nitric acid, 88–92
Solvent, role in aromatic nitration, 10
Spent acids, concentration and recovery after nitration, 250–267
Strongly acidic medium, role in nitronium ion reactivity, 11
Sulfuric acid
 applications, 250–251
 concentration and recovery after nitration, 250–267
Sulfuric acid based nitrations, concentration and recovery of spent acids, 250–267
Superacidic systems, protosolvation in nitrations, 10–17
Synthesis
 nitrated hydroxy-terminated polybutadiene, 97–103
 polynitroadamantanes, 51–56
 polynitrocubanes, 51–56
 2,5,7,9-tetraazabicyclo[4.3.0]nonanone, 44–45
 2,5,7,9-tetranitro-2,5,7,9-tetraazabicyclo[4.3.0]nonanone, 45–47

T

2,5,7,9-Tetraazabicyclo[4.3.0]nonanone
 experimental procedure, 48–49

INDEX

2,5,7,9-Tetraazabicyclo[4.3.0]nonanone—
Continued
  nitration, 45–47
  synthesis, 44–45
2,3,4,6-Tetrabromo-4-methyl-4-nitrocyclohexa-2,5-dien-1-one reactions
  with aniline, 24,27–28
  with N,N-dimethylaniline, 28,29f
  with phenols, 23–24,25–26f
2,5,7,9-Tetrahydro-2,5,7,9-tetraazabicyclo[4.3.0]nonanone
  experimental procedure, 48
  synthesis, 45
1,3,5,7-Tetranitroadamantane, synthesis, 52–55
2,7,9,9-Tetranitrofluorene, synthesis, 138–141
Tetranitromethane
  formation, 187
  safety, 187–188
  structure, 187
Tetranitromethane removal and destruction from nitric acid
  all nitric acid nitration of N-methylphthalimide, 189
  experimental procedure, 198
  known removal methods, 190–193
  mixed acid nitration
    benzene, 189
    toluene, 188
  removal using mixed acid approach, 196–198
  safety of phase separation, 189–190
  solubility, 193–195
  stability, 195–196
2,5,7,7-Tetranitro-2,5,7,9-tetraazabicyclo[4.3.0]nonanone
  experimental procedure, 48–49
  structure, 151–152
  synthesis, 45–47,153–154
2,4,6,8-Tetranitro-2,4,6,8-tetraazabicyclo[3.3.0]octane
  structure, 44,151–152
  synthesis, 44,156–157

2,4,6,8-Tetranitro-2,4,6,8-tetraazabicyclo[3.3.0]octan-3-one
  structure, 151–152
  synthesis, 152–153
Thermal decomposition of dinitrotoluene, modeling, 174–186
Toluene
  industrial nitration to dinitrotoluene, 234–247
  isomer synthesis, 2
  nitration in liquid N_2O_4 promoted by metal acetylacetonates, 32,35f
1,3,5-Triacetylhexahydro-s-triazine, synthesis, 161
Trimethylolethane trinitrate, synthesis, 112
1,2,2-Trinitroadamantane, synthetic methods, 52
1,1,2-Trinitro-1-phenylethane, synthesis, 142–143
2,5,7-Trinitro-2,5,7,9-tetraazabicyclo[4.3.0]nonan-8-one
  structure, 151–152
  synthesis, 153–154
2,6,8-Trinitro-2,4,6,8-tetraazabicyclo[3.3.0]octan-3-one
  structure, 151–152
  synthesis, 152–153

V

Vacuum concentration, description, 254,256–257f
Victor Meyer processes, synthesis of nitroparaffins, 3

W

Water–HNO_3–H_2SO_4 mixtures, NO_2^+ concentration model, 201–213

Highlights from ACS Books

Good Laboratory Practice Standards: Applications for Field and Laboratory Studies
Edited by Willa Y. Garner, Maureen S. Barge, and James P. Ussary
ACS Professional Reference Book; 572 pp; clothbound ISBN 0–8412–2192–8

Silent Spring Revisited
Edited by Gino J. Marco, Robert M. Hollingworth, and William Durham
214 pp; clothbound ISBN 0–8412–0980–4; paperback ISBN 0–8412–0981–2

The Microkinetics of Heterogeneous Catalysis
By James A. Dumesic, Dale F. Rudd, Luis M. Aparicio, James E. Rekoske, and Andrés A. Treviño
ACS Professional Reference Book; 316 pp; clothbound ISBN 0–8412–2214–2

Helping Your Child Learn Science
By Nancy Paulu with Margery Martin; Illustrated by Margaret Scott
58 pp; paperback ISBN 0–8412–2626–1

Handbook of Chemical Property Estimation Methods
By Warren J. Lyman, William F. Reehl, and David H. Rosenblatt
960 pp; clothbound ISBN 0–8412–1761–0

Understanding Chemical Patents: A Guide for the Inventor
By John T. Maynard and Howard M. Peters
184 pp; clothbound ISBN 0–8412–1997–4; paperback ISBN 0–8412–1998–2

Spectroscopy of Polymers
By Jack L. Koenig
ACS Professional Reference Book; 328 pp;
clothbound ISBN 0–8412–1904–4; paperback ISBN 0–8412–1924–9

Harnessing Biotechnology for the 21st Century
Edited by Michael R. Ladisch and Arindam Bose
Conference Proceedings Series; 612 pp;
clothbound ISBN 0–8412–2477–3

From Caveman to Chemist: Circumstances and Achievements
By Hugh W. Salzberg
300 pp; clothbound ISBN 0–8412–1786–6; paperback ISBN 0–8412–1787–4

The Green Flame: Surviving Government Secrecy
By Andrew Dequasie
300 pp; clothbound ISBN 0–8412–1857–9

For further information and a free catalog of ACS books, contact:
American Chemical Society
Customer Service & Sales
1155 16th Street, NW, Washington, DC 20036
Telephone 800–227–5558

Bestsellers from ACS Books

The ACS Style Guide: A Manual for Authors and Editors
Edited by Janet S. Dodd
264 pp; clothbound ISBN 0–8412–0917–0; paperback ISBN 0–8412–0943–X

Understanding Chemical Patents: A Guide for the Inventor
By John T. Maynard and Howard M. Peters
184 pp; clothbound ISBN 0–8412–1997–4; paperback ISBN 0–8412–1998–2

Chemical Activities (student and teacher editions)
By Christie L. Borgford and Lee R. Summerlin
330 pp; spiralbound ISBN 0–8412–1417–4; teacher ed. ISBN 0–8412–1416–6

*Chemical Demonstrations: A Sourcebook for Teachers,
Volumes 1 and 2,* Second Edition
Volume 1 by Lee R. Summerlin and James L. Ealy, Jr.;
Vol. 1, 198 pp; spiralbound ISBN 0–8412–1481–6;
Volume 2 by Lee R. Summerlin, Christie L. Borgford, and Julie B. Ealy
Vol. 2, 234 pp; spiralbound ISBN 0–8412–1535–9

Chemistry and Crime: From Sherlock Holmes to Today's Courtroom
Edited by Samuel M. Gerber
135 pp; clothbound ISBN 0–8412–0784–4; paperback ISBN 0–8412–0785–2

Writing the Laboratory Notebook
By Howard M. Kanare
145 pp; clothbound ISBN 0–8412–0906–5; paperback ISBN 0–8412–0933–2

Developing a Chemical Hygiene Plan
By Jay A. Young, Warren K. Kingsley, and George H. Wahl, Jr.
paperback ISBN 0–8412–1876–5

Introduction to Microwave Sample Preparation: Theory and Practice
Edited by H. M. Kingston and Lois B. Jassie
263 pp; clothbound ISBN 0–8412–1450–6

Principles of Environmental Sampling
Edited by Lawrence H. Keith
ACS Professional Reference Book; 458 pp;
clothbound ISBN 0–8412–1173–6; paperback ISBN 0–8412–1437–9

Biotechnology and Materials Science: Chemistry for the Future
Edited by Mary L. Good (Jacqueline K. Barton, Associate Editor)
135 pp; clothbound ISBN 0–8412–1472–7; paperback ISBN 0–8412–1473–5

For further information and a free catalog of ACS books, contact:
American Chemical Society
Customer Service & Sales
1155 16th Street, NW, Washington, DC 20036

RC
548
C37
1991

Date Due

DEC 15 '94	MAR 0 1 2001		
FEB 24 '96	OCT 23 2001		
MAR 18 '96			
MAR 27 '96			
MAY 02 '98			
MAR 10 1998			
JUL 10 1998			
AUG 10 1998			
NOV 09 1998			
NOV 29 1998			
NOV 28 1999			
DEC 10 1999			
NOV 14 2000			
DEC 08 2000			

BRODART, INC. Cat. No. 23 233 Printed in U.S.A.